T0282637

CAMBRIDGE LIBRARY COLLECTION

Books of enduring scholarly value

Earth Sciences

In the nineteenth century, geology emerged as a distinct academic discipline. It pointed the way towards the theory of evolution, as scientists including Gideon Mantell, Adam Sedgwick, Charles Lyell and Roderick Murchison began to use the evidence of minerals, rock formations and fossils to demonstrate that the earth was older by millions of years than the conventional, Bible-based wisdom had supposed. They argued convincingly that the climate, flora and fauna of the distant past could be deduced from geological evidence. Volcanic activity, the formation of mountains, and the action of glaciers and rivers, tides and ocean currents also became better understood. This series includes landmark publications by pioneers of the modern earth sciences, who advanced the scientific understanding of our planet and the processes by which it is constantly re-shaped.

The Climate of London

The 'student of clouds' Luke Howard (1772–1864) published this work of statistics on weather conditions in London in two volumes, in 1818 and 1820. Howard was by profession an industrial chemist, but his great interest in meteorology led to his studies on clouds (also reissued in this series), and his devising of the system of Latin cloud names which was adopted internationally and is still in use. Volume 1 begins with an introduction to the work, explaining his intention to make available in one place consistent records of weather events. He argues that for the benefit of 'agriculture and navigation', a systematic approach is required, and he outlines his methods and equipment in some detail. The tables of observations taken at Plaistow, near London, in the years 1806–9 then begin, and are interspersed with notes and a commentary which includes accounts of similar weather phenomena observed elsewhere.

The Climate of London

Deduced from Meteorological Observations

VOLUME 1

LUKE HOWARD

CAMBRIDGE UNIVERSITY PRESS

Cambridge, New York, Melbourne, Madrid, Cape Town,
Singapore, São Paolo, Delhi, Mexico City

Published in the United States of America by Cambridge University Press, New York

www.cambridge.org
Information on this title: www.cambridge.org/9781108049511

© in this compilation Cambridge University Press 2012

This edition first published 1818
This digitally printed version 2012

ISBN 978-1-108-04951-1 Paperback

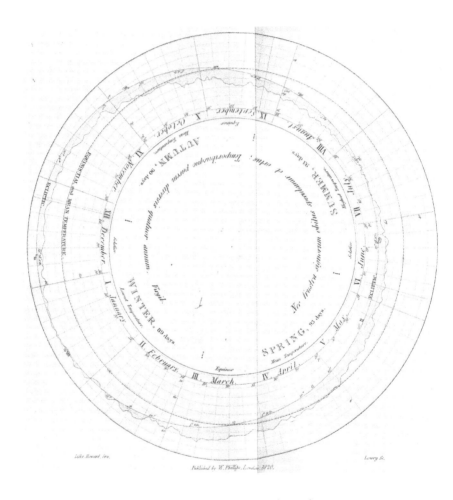

THE

CLIMATE OF LONDON,

DEDUCED FROM

𝕸𝖊𝖙𝖊𝖔𝖗𝖔𝖑𝖔𝖌𝖎𝖈𝖆𝖑 𝕺𝖇𝖘𝖊𝖗𝖛𝖆𝖙𝖎𝖔𝖓𝖘,

MADE AT DIFFERENT PLACES

IN THE

NEIGHBOURHOOD OF THE METROPOLIS.

By LUKE HOWARD.

IN TWO VOLUMES.

VOL. I.

Containing an Introduction relative to the Construction and Uses of several Meteorological Instruments; Tables of Observations for Ten Years, with Notes and Results; Accounts of collateral Phenomena in other Parts of the World; and occasional Dissertations.

LONDON:

PUBLISHED BY W. PHILLIPS, GEORGE YARD, LOMBARD STREET: SOLD ALSO BY J. AND A. ARCH, CORNHILL; BALDWIN, CRADOCK, AND JOY, AND W. BENT, PATERNOSTER ROW; AND J. HATCHARD, PICCADILLY.

1818.

C. Baldwin, Printer,
New Bridge-street, London.

INTRODUCTION.

—◆—

General Observations.

METEOROLOGY, though greatly advanced of late years, especially in what regards the perfection of its instruments, and the art of observing the changes of the atmosphere, is yet far from having acquired the regular and consistent form of a science. Its facts lie for the most part scattered, or rather buried, in volumes chiefly taken up with other more cultivated branches of natural philosophy : and it is only where detached publications have been ventured on, by individuals engaged in the study of particular classes of phenomena, that its principles have been developed with the clearness and method of which they are susceptible. A pretty large number of observers have been long engaged in doing for this science the office which the Chaldean shepherds are thought to have performed for astronomy. We may now probably venture, with safety, to anticipate some of those conclusions which posterity will otherwise have to draw from our data—to lay the ground-work of the edifice, if not to proceed to build, with the present materials. Should it be inquired, for what end—the answer (without travelling to more remote consequences) may be—For the benefit of *agriculture* and *navigation :* two objects of that magnitude that the most distant prospect of the smallest permanent addition to our store of knowledge and experience concerning them, will be slighted by none but those, who have not duly considered the influence of science on the arts, and of these on the well-being of society.

An extensive co-operation of observers in different countries has been justly deemed essential to the perfection of meteorological research.* But if we except the single instance of the Society of *Manheim*, patronized by the Elector Palatine, the voluminous Transactions of which, compiled from an extensive correspondence, include the years from 1781 to 1785, there seems to have been nothing done on a great scale to attain this object.

In the mean time observations continue to be made and published throughout Europe : and it is probable that many individuals have acquired, at least, a knowledge of the peculiar features of their own climate, and of the facts which, properly arranged, would form its *history*. The production of such a work for each of those districts, in which the requisite observations have been made, would greatly abridge the labour, if it did not remove the principal difficulty, of a general view of the phenomena of our atmosphere, in their various localities and relations through the year : which being obtained, we might proceed to constitute, on sound principles, the theory of the science.†

The volume, which on such considerations is now offered to the public, is composed chiefly of the observations of *ten years*, from 1807 to 1816 inclusive, made in the neighbourhood of *London*. They have appeared, for the

* Kirwan on the Variations of the Atmosphere, Dublin, 1801.

† In the spring of last year, I attempted to give a coup d'œil of the facts and principles of this interesting department of knowledge, in the way of lectures to a circle of friends. The best sketch which I was able to get ready for the occasion, aided by the globe, some graphic representations, and a few experiments with instruments, proved so far satisfactory, that I have been induced to give expectations of enlarging and publishing it. But I cannot promise, as my friend William Phillips (whose " Lectures on Astronomy " instructed and gratified the same audience) has ventured to do for me, that this shall be done " at no distant period."

most part, as monthly reports in different scientific
Journals; but of necessity in an insulated form, and without the connexion and the illustrations which it has now
been my endeavour to bestow upon them. They are intended to form (in a *Second Volume*) the basis of such a
methodical account as I have hinted at, of the climate of
London: or rather of that district in which the Metropolis
with its suburban branches, have during the last ten years,
been rapidly extending.

Of the Calendar and Arrangement.

In introducing to the reader's notice this collection of
observations, I ought in the first place to account for the
peculiarities of the arrangement. I had given them, from
the first, to the press, not as usual in calendar months, but
in periods of a lunar revolution. In so doing I had two
objects in view.

In the first place I obtained an earlier insertion in the
periodical publications (which come out on the first of
each month) than would have been possible, had I carried
them up to the close of the preceding month : the difference, as the reader will perceive, is on an average two
weeks in my favour, though at the expense to the pub-
lisher of inserting one table more in the course of two
years. Secondly, and what more induced me, my attention had been for some years called to the question, so
much agitated among meteorologists, whether, and in
what way, the relative positions of the moon in the different parts of her complex orbit, influence the state of
our atmosphere. I thought the most convenient way of
investigating this subject, and which might bring out,
even unexpectedly, facts capable of deciding the question,
would be to digest my results in lunar periods at once.
I lost by this means the facility of having them compared

monthly with those of other observers : and I obtained in
return, sufficient materials for deciding in the affirmative
the first part of the question above-mentioned, as well as
for throwing, possibly, some light on the second; which,
however, is one of too great difficulty to be treated with
much success, except by *combined* and *extensive* observations.

Having adopted, at the commencement, a period beginning at the *new moon*, the first three years will be
found thus arranged. From the close of the year 1809,
however, I preferred the *last quarter;* as by setting out
from this, the phases of new and full moon appear in
the midst of their respective moieties of the observations.

One other circumstance I may here notice. I have pre-
fixed to the calendar names of the months their numerical
designations, and where I write myself, I use these in pre-
ference. This is the phraseology of the Society of Friends
to which I belong, and is from principle, as well as edu-
cation, a part of my form of Christianity. The reader
who may not approve of this peculiarity, will therefore
be pleased to tolerate it : which he will do the more will-
ingly, on finding that it interferes but little with his
convenience in reference.

The following is the method I use in noting my observations. About *nine* in the morning, I make the round
of the instruments (the *situation* of each of these will
be noticed, in treating of it in its place); I find it the
securest way to do this with the slate and pencil in hand.
The direction of the wind, for the past twenty-four hours,
usually from memory, but with due reference to the present posture of the vane, is noted for the *first* column.

The actual place of the quicksilver in the *barometer*,
with the place it has moved from, as indicated by the
hand index, are put down for the second and third
columns; and for the fourth and fifth the situation of the

capable of receiving a fir pole: the latter should be well painted before fixing, and the vane put on and adjusted after it is fixed.

In this stage of the business the workman will require a compass, if letters are as usual attached to the vane; and he must observe, that when the needle, by moving the compass round, is made to point to the *variation* north (at present 24° west of north), the north on the card of the compass will be the point for the north of the vane. The exact variation is now annually inserted in the Philosophical Transactions, and other publications.

The vane in the figure is on the scale of an inch to a foot: the section, in which the spindle, the flint, and the guides are represented, is drawn two inches to the foot: the spindle and tube are stout enough for a much larger vane: the branches carrying the ball are seen edgewise.

Of the Barometer.

On the construction and uses of this instrument much has been written which need not here be repeated. In applying it to the purpose of the meteorologist, it is the due attention of the observer to the changes it undergoes, rather than the perfection of the instrument itself, that serves to promote science. Yet as the mean of a given number of observations at any place is applicable to other not unimportant ends; and as these means must for the most part differ by but small quantities, it is desirable that every barometer, from which we are about to take the pains to register a series of changes, should have the previous labour bestowed upon it of adjustment to a fixed standard, which is probably done in but few instances at present. As to corrections for the slight variations of temperature which take place in an inhabited apartment, I have not thought it needful to enter upon them. These niceties appear to belong to a more advanced state of the science; and there are other sources of discordance, at present more obvious, in the generally varying and imperfect construction of the instrument.

I have employed the same barometer in all my observations at home: but in consequence of slight accidents, it has been more than once re-filled, in doing which I have endeavoured to restore, as nearly as possible, the former adjustment: the amount of the error of this will therefore be discoverable at any time. It is on the wheel or siphon construction, made by *Haas* about the year 1796. The scale extends through a space of 18 inches, and the workmanship is delicate; the weight which rests on the quicksilver preponderating by but a few grains, so that the *radius*, or hand, makes a sensible libration on suddenly opening or shutting the door of the room, as well as during the passage of the strong gusts in a storm of wind. I find an advantage in having it fixed in a place by which, when at home, I pass frequently in the course of the day: for as often as I perceive, by the divergence of the radius from an *index*, which is also made to traverse the circle, that there has been a movement, I adjust the latter to the place, and thus secure the extremes. At the hour of observation, if the movement has been wholly in ascent or descent, the extremes are found, the one in the highest or lowest point of yesterday, the other in the present place of the radius: but if there has been a change of *direction* to any extent worth notice, the index (which is never put back again in this case, but left in the place) points out one of the extremes to be noted, and the radius the other. It is rare that such changes happen in the night, or more than once in the

its power. The sinking of the index, which defeats the
whole purpose of the invention, I conceive will thus be
effectually prevented. At any rate, stiffer springs and a
larger magnet must be resorted to, to regain its reputation.

3. There is yet a third defect. At the first approach
of cold weather, a small bubble of air sometimes makes
its appearance in the spirit, which, increasing and getting
into the tube, at length occupies several degrees of the
scale ; but without disturbing the results in proportion.
I have got rid of this, by first cooling the instrument, so as
to bring the air back into the spirit tube, and then making
the bubble move to and fro in the spirit before the fire :
the pressure caused by the expansion of the included air
and vapour, soon drives it back into the spirit ; but it is
subject to re-appear. The *radical cure* is, to break the
point of the upper bulb, and let in the atmospheric air :
very little waste of the spirit will ensue in a long time, if
the point be fine : if it be too solid, the glass may be detached from the frame, and the point cautiously drawn
out before breaking, by heating it at the flame of a spirit
lamp, and applying another piece of glass when it is hot
enough to adhere. But as this requires a practised hand,
the workman should leave the upper bulb with a point
fine enough to be easily nipped off, should it be found
needful. Spirit may be introduced into this upper bulb
at any time, by first warming, and then cooling it with
the aperture immersed.

4. The descending scale is often, and indeed generally,
graduated further than the lower end of the index is capable of following it. The workman may avoid this, by
sparing a few degrees at the ascending extremity of the
scale. Here he need not, *for our climate*, ever go beyond
110°—while, at the cold extreme, the index should be
capable of showing 10° below *zero*, which will suffice in
all cases.

c

5. A Six's thermometer should be *mounted* (at about five feet from the ground), on a fixed support, not hung up free, as the present construction indicates, and liable to swing and strike with violent gusts of wind. A mahogany frame, or back, may be provided with two projecting brackets, into which the box scale may enter by a pivot at each end. By this means the thermometer may be set facing that part of the general north exposure, where, from the disposition of the surrounding objects, the heat may have the freest *radiation to the open sky;* a point which late discoveries show to be important: and in situations where it is inconvenient to go in front, the scale will admit of being turned towards the observer.

To the above mounting, it will in some situations be proper to add a small shelter above the instrument, which shall suffice to keep off direct showers at least—and, at a suitable distance on the west side, a shade moveable on hinges, to be interposed, in the heat of summer only, between the instrument and the rays of the afternoon sun.

As the *position* of the thermometers which I have used varied in the different stations, it will be most convenient to describe the position, when treating of the mean temperature at each.

Of the Hygrometer.

In determining to substitute for the daily amount of evaporation, in one of my columns, the degrees of moisture indicated by the hygrometer, I had no hesitation in making choice of that of *De Luc*. This instrument is capable of bearing, with little injury, a constant exposure to the air abroad: I have accordingly kept it hung up, near the thermometer, in a small tin frame, the sides and bottom of which are of open work, with a glass in front: so that the whole instrument is visible, and the air freely admitted, while the rain, and the fingers of the curious, are excluded.

My instrument consists, essentially, of a very slender strip of *whalebone*, which, having been cut out of the piece *across the grain*, and reduced by scraping to the requisite thinness, with a length of about $3\frac{1}{2}$ inches, is so mounted in a brass frame, with a counteracting spring of wire, as to move an index round a circular scale of three inches circumference. The shortening of the strip of whalebone by dryness, and the lengthening by moisture, while the spring keeps it extended, respectively carry the index towards the extremes of the scale. The *moist* extreme, which the inventor fixed at 100°, is now and then attained, in winter, in the natural state of the air: but the *dry*, never.

The latter is accordingly fixed by a method, in which the buyer may at any time prove for himself, whether the instrument has had, or retains, its proper adjustment. For this purpose, a few ounces of fresh-burned quicklime are to be put into a dry wide-mouthed stopper bottle, with sufficient room above for the hygrometer to rest upon the lime without soiling the whalebone. The instrument being placed, the stopper is to be put in, and, for greater security, closed round with putty. In 48 hours, or thereabouts, it will be seen whether the index, under these circumstances, will pass to *zero*, which point it ought not at any rate to exceed. The hygrometer is next to be exposed for a few hours, under a close glass to the vapour of water, at a mean temperature, and if convenient, in a mean state of the barometer; but without being immersed in the water. The index, if right at the other extreme, should now go to 100°.

Should it prove out of its place, yet with the proper range, there is at the bottom of the frame a small screw, by turning which it may be adjusted: but should the range prove several degrees too large, or too small, either the parts of the instrument have not been duly proportioned by the maker, or it has suffered by use or accident.

This hygrometer shows the effect of moisture on whalebone to be precisely the same as on a deal board: which, as every one knows, will swell, or extend itself in breadth, in a moist atmosphere, and contract again as it dries. Now the board, after a few summers and winters, loses this property, or becomes *seasoned:* and there is no doubt that the same effect must be produced in time on the whalebone. But the texture of the latter substance is so greatly superior in hardness and firmness to deal, that it is probable few single observers will wear out their instruments. If deterioration, however, be dreaded, it may be indefinitely put off, by exposing the hygrometer only at the time of observation: the few minutes taken up in observing and noting the other instruments, may often, though they will not always, suffice to make it take the degree of moisture present in the air. And for such a mode of observing, the delicate *hair* hygrometer of Saussure, of which I have had as yet but little experience, will be preferred by accurate meteorologists. It is necessary to observe, that though graduated alike, these two instruments do not range together, and their results must not be entered in the same column.

I have caused a workman to add to the hygrometer of De Luc, a pair of detached indexes, to be moved by the one in connexion with the whalebone; in such a manner as to show the greatest degrees of moisture and dryness, which take place in a given interval: but I have not yet had opportunity to prove how far they will answer in practice.

Of the Gauges for Rain and Evaporation.

These are treated of together, as being connected in the most essential part,—the graduated measure for the water.

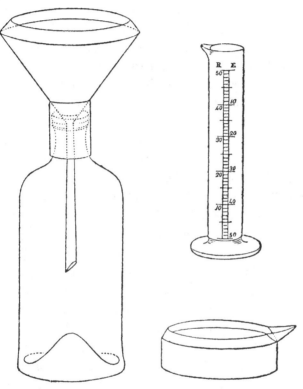

The *rain-gauge* consists of three pieces, a funnel, a bottle, and the measure. The *funnel* is most conveniently made of five inches opening, and of the form represented in the figure: the mouth-piece of brass, turned in a lathe, the remainder of tinned copper. It has two necks: the inner and longer one widening a little downwards, enters deep into the bottle, and conveys the rain: the outer neck is soldered on the cone of the funnel, having no opening into the latter: it serves the necessary purpose of preventing the entrance of water from the outside; and

by resting on the shoulder of the bottle, it gives steadi-
ness to the funnel.

As to the *bottle*, a common wine-quart will contain
from two to two and a half inches of rain on this funnel;
but it is better to use a three-pint bottle (technically
termed a *Winchester quart*), which has the proportions
given in the figure. For an unusual fall of rain may
happen, when a previous quantity has not been measured
out : and it is on such occasions that we would wish,
more especially, to be certain of the amount.

A cylindrical glass of the depth of eight inches, exclusive
of its foot, and $1\frac{1}{3}$ inch in diameter, serves to make the
measure. It is graduated into parts, each of which is
equal in capacity to the depth of $\frac{1}{100}$ of an inch on the
area of the mouth of the funnel. A glass of the above
size will measure out fifty such parts, or half an inch at
once. The graduation is conducted on the principle
(which is a medium between calculation and experiment)
that a cylinder of water at a mean temperature, an inch
deep, and five inches in diameter, weighs 10 ounces *troy.*
The hundredth part of this, or 48 grains, is accordingly
taken for the graduating quantity, and the scale is formed
by successive additions, at each of which the surface is
marked. Considering the nature of this operation, which
scarcely admits of our going to fractions of a grain, I
suppose the above standard to be sufficiently correct. I
have been accustomed to *etch* the scale on the glass with
fluoric acid, but it is more conspicuous when engraved at
the glass-cutter's wheel. Previously to sending it for this
purpose, the whole scale should be traced, either on a
strip of paper pasted on before it is divided, or in oil
paint on the glass itself. A diamond, or steel point, may
be used for engraving the scale, in default of other means.

Although I recommend these dimensions as convenient,
and have had them executed in different instances for
others, I have hitherto used a gauge, the funnel of which

has eight inches aperture, and the measure is graduated by the quantity of 124 grains, the bottle being large in proportion.

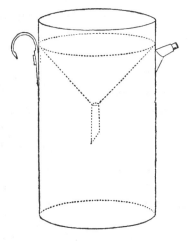

For tropical climates, and in cases where a large bottle is found inconvenient, the whole recipient part may be of tinned copper, the *rim* excepted, which is still to be of turned brass. On this construction, a moveable funnel may be let in, so as to rest below the rim and prevent evaporation ; a spout, with a small aperture, should also be provided at the side, both for the convenience of emptying the water into the measure, and to permit the air, on occasion, to pass out freely.   (See the figure above.)

The *position* which, since the year 1811, I have preferred for the rain-gauge, is to sink it into the ground, bringing the mouth of the funnel nearly to the level of the turf; which should be kept cut, so as to leave a clear space of an inch or two around.   In winter, when snow may be expected, it is proper to raise it a few inches.   A thick sheet of snow is apt to have a large depression above the funnel, the surface of which, slightly thawed and frozen again, has, more than once, collected and sent into my gauge a redundancy of water.   On the subject of different products from different situations of the gauge, the reader may consult the appendix to Table lxiv. in this volume.

The graduated measure for the rain, being numbered on the opposite side of the scale downward, serves also to ascertain the *evaporation*. For this purpose, a cylindrical tinned copper vessel is employed, of five inches diameter within, furnished with a rim to prevent spilling, in which is a lip, set on clear of the cylinder. Two measures, or an inch of water being poured in, fills two thirds of the cylinder: the vessel is then placed near the ground, in a situation where it may be sheltered from rain, and have the sun's rays without reflection. At the end of 24 hours, or a longer period, extending to a week if desirable, but regulated by the season of the year, the water being returned into the measure, the quantity which has evaporated may be read off, and the vessel replenished. For warmer climates, or longer periods of observation, the depth of the vessel may be increased, and a greater number of measures put in. See fig. p. xxi.

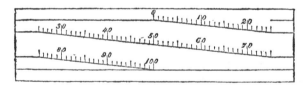

Where the *evaporation* alone is in question, and the observer wishes to ascertain it daily, without trouble, the following contrivance may be used.—On a plate of glass, 6 inches long, and an inch and a half wide, a line is to be drawn, near and parallel to one side, to serve for a base. From this a diagonal scale, etched with fluoric acid, is to be carried up, ascending at the rate of *one* inch in *ten;* so that the tenths of an inch into which it is ultimately divided, shall rise in progression just $\frac{1}{100}$ of an inch above each other. The glass being now fixed perpendicularly on its edge, in a vessel of the proper capacity and depth (if this be *square* it may be set in diagonally, and supported by the angles) a little water is first to be put in ; the surface of which is to be brought, by adjusting the position of the vessel, to range with the horizontal line at the bottom of the scale. This adjustment made, more water is to be added, up to the line which cuts the division at *zero*. Then, in proportion as the surface is lowered by evaporation, it will cut the several divisions in succession, indicating at sight the effect to the 100th of an inch.

Lastly, for delicate occasional observations on *rain, dew,* or *evaporation,* I have an instrument which will indicate either to the 1000th of an inch, and which I likewise find useful in graduating other gauges. This instrument, a figure of which is given on the succeeding page, I shall now describe.

A funnel like that of the rain-gauge, but with an upright cylindrical rim, five inches in diameter, terminates in a glass tube 20 inches long, and of half an inch calibre; having at bottom a good stop-cock. The tube is graduated on the principle of the glass measure above described : but the divisions are here wide enough to admit of decimal subdivisions. When the instrument is used *for evaporation,* the tube is first to be filled to the *zero at top;* a bottle of water is then to be added, so that the surface may stand at a proper height in the cylinder during the experiment; at the close of which, the same quantity of water being returned into the bottle, the deficiency will appear in the tube. When for *dew,* or for *rain,* in minute quantities, or at short intervals, water is to be introduced up to the *zero at bottom,* and the inside of the funnel moistened with a sponge at the outset; the difference in volume caused by change of temperature, must in these delicate experiments be obviated, or allowed for. This instrument requires likewise a support to keep it upright and steady in use.

d

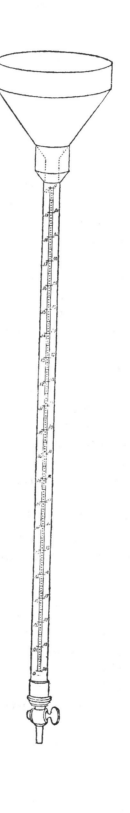

Of the Vapour-point.

As I have two or three times in the course of my observations made mention of the *vapour-point*, I may here
explain, that by this term is meant the degree of Fahrenheit's thermometer, at which a body colder than the air
(such as a glass of pump water) will no longer collect
moisture from it. For this experiment (which was first
introduced by Dalton, and is a useful one in studying
the subject of *rain*, though seldom performed) the liquid
in the glass should be cold enough, either naturally, or by
artificial refrigeration, to ensure the effect above-mentioned:
then, as the temperature of the glass slowly increases by
the contact of the moist air, it is to be repeatedly dried
with a clean cloth, till the dew no longer re-appears on
its surface; at which moment a delicate thermometer,
previously immersed in the liquid within, gives the vapour-
point, or the lowest temperature at which vapour can subsist in the present state of the atmosphere. See the notes
to table i. lxvii. lxxxiii.

Of the Cyanometer.

One of these instruments having been put into my
hands by Professor Pictet, at Geneva, in the summer of
1816, I brought it home to make trial of its use: but
the almost continual recurrence of turbid skies since that
period, has nearly defeated my purpose hitherto. I shall,
however, describe and figure it for the reader's information. The figure on the succeeding page is drawn one
fourth of the actual dimensions. I have not attempted to
express more than the general outline.

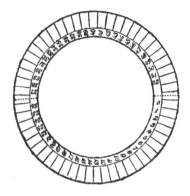

The cyanometer of Saussure is, in effect, a circle of
small pieces of paper tinted with blue, and pasted on a
card, which is open in the middle, and folds in two, with
the patterns inward. They are numbered from 0 to 52 : the
last is of the colour of solid indigo, that is, nearly black;
and the colours lighten gradually through the whole
series, till, at 0, nothing remains but the white paper.
The colour goes quite to the outer edge; but on the inner,
a space is left for the number.

Its use is, to assist the judgment in determining the
degree of intensity of the blue colour of the sky; which
varies greatly in different seasons, and still more in different climates, and at different elevations in the atmosphere. For this purpose it is held up in such a direction
that, while a full light falls on the pattern, the sky may
be seen at the same time; and the card is turned till the
sameness, or near approximation, of the tint of some
number is decided on; which is then set down for the
colour of the sky.

This invention is chiefly useful to the traveller, who,
in ascending mountains, and in changing climates, meets
with a range of colour to which a single situation scarcely
affords a parallel. About half the range of the scale may be
found, probably, in our own skies. That they do not
attain to the intensity of those on the more elevated parts

of the Continent, is manifest from the surprise with which our travellers view for the first time the blue rivers and lakes of those countries. They forget that they are in fact contemplating, in a natural cyanometer, a phenomenon to which, by its gradual approach, the eye had already become accustomed, when turned to the vault of the heavens.

Simple as this little instrument appears, I have great doubt whether our workmen, who may attempt it, can give it any improvement; save, perhaps, by securing the back with morocco leather, and providing a case. The form and size were certainly adopted after mature reflection, and different trials. Those who incline to exercise their ingenuity on this subject, may do it with more promise, by trying different combinations or thicknesses of *blue glass*, to be viewed against a ground of dead white, in the manner above-mentioned.

Of the Electrical Apparatus.

There are several ways in which the electricity of the atmosphere may be investigated, as

1. By *small instruments* managed by the hand, with some of which the most minute quantities of it may be detected. Of these I have had very little experience, and shall therefore omit to treat of them.

2. By the *insulated rod*, or conductor. In the first three years of my observations, the reader will find pretty frequent notice of the state of the natural electricity, obtained in this way. My apparatus was constructed on the plan of that described in the Philo. Trans. vol. lxxxii. part ii. by John Read, who, in the years 1790 and 1791, accumulated a valuable mass of facts on this subject, which he has there reported. I made, however, the following variations in the apparatus: the *conductor*

itself was a single taper rod of iron, which by a screw at
the bottom entered firmly into a brass cap, cemented upon
a glass pillar; the latter standing *free* in a socket of wood.
By means of a stout glass tube, and other defences of
glass and cork, the rod passed up through an angle of my
observatory, rising seven or eight feet above the roof, which
height I found sufficient. I added to the part which communicated with the ground, a stout *arm*, which, turning on a
joint, could at any time be shut to the cap of the pillar,
so as to make the whole a conductor to the earth : and in
this state I commonly left it, when out of use. When I
removed my residence, I took down this apparatus, and
have not erected it at Tottenham. Hence the deficiency
of electrical observations in the latter years.

3. By the *insulated kite*. This subject is well handled
by Cavallo, in the second volume of his Treatise on
Electricity. The few experiments I have made with the
kite amply confirm to me his opinion, that the metallic
thread is the *conductor* in this case, and the kite its
support only. I have likewise discovered what seems a
material improvement. Instead of twisting together the
conducting and supporting threads, as heretofore, I leave
the kite with its string precisely in the state in which
boys raise it : so that it may be flown in a pretty high
wind, and carry out plenty of line of sufficient strength ;
or in gentle gales, a lighter string. The conducting thread,
which may now consist of the fine lace-thread before
intertwisted with the common string, is tied to a loop in
the latter, two or three yards below the kite : so that
while the kite is set up by two persons, a third lets out
the lace from an insulated reel; keeping always to
leeward, that the *stress* may be wholly on the common
string. When this is let out to the desired length, it is
likewise insulated, by attaching a loop tied in it to a hook
fixed on a glass handle. By this management, the con-

ducting thread hangs as nearly perpendicular as the wind may permit, and is in little danger of breaking.

Those who attempt to use this apparatus should have skill enough to know when the clouds actually portend a discharge to the earth: at which times alone I believe it to be attended with any danger. It is the most likely method I know of, to obtain satisfactory observations on the electricity given to the air by the passage of *clouds*. That of rain, hail, or snow, is certainly best collected by the insulated rod.

4. The variable electricity of the atmosphere has been found to affect considerably the action of *De Luc's electrical column.* As some instances of this kind occur in the observations of my friend Thomas Forster, inserted in this work, I shall follow his description of the instrument. It is composed of a great number of small circular and very thin plates, of silver, of paper, and of zinc, alternately placed on each other, and pressed closely together, so as to form a column. One end of the column thus arranged is observed to become permanently electrified *plus*, the other *minus*: a *bell* is connected with each, and a small ball of metal is suspended between the bells. The whole is enclosed in glass and *insulated*. From the tendency of the electricity to become equalized, while it is continually renewed, the metallic clapper keeps passing to and fro between the bells. And the varieties in the kind of pulsation produced by this means, with its occasional intervals, or even cessations for a considerable time, constitute the effects to be studied in connexion with the other phenomena of the season.

Of the Notes and Miscellaneous Matter.

The notes appended to my earlier tables of observations were published in an incomplete state: the confined in-

terest of the subject at that period, and some uncertainty
as to the probable reception of the terms used to designate
the clouds, made me sparing of them for two or three years.
On concluding to insert the suppressed notes in this work,
I found that, from the mode of reference by letters instead
of dates, it would be no easy task to incorporate them
with the rest. They are therefore left in the less convenient form of *additions*.

The *nomenclature of clouds*, to which I have just alluded, having now grown into pretty general use among
the meteorologists of our own country, I shall not need to
introduce it to the reader as a novelty. In the second
volume the terms for the clouds, which are used in the
notes, will be fully explained. In the mean time the
reader, who may not have acquired them, will find the
system in its original state in Tilloch's Philosophical
Magazine, vol. xvi. xvii. and (with some changes not
affecting the nomenclature) in Rees's Cyclopædia, article
Cloud; as also in Nicholson's Philosophical Journal, vol.
xxx. It has been abridged and reported in several other
publications: and, in the supplement now publishing to
the Encyclopædia Britannica, with the addition of a set
of new terms for the modifications, intended for the use
of English readers. I mention these in order to have the
opportunity of saying that I do not adopt them. The
names for the clouds which I deduced from the Latin are
but seven in number, and very easy to remember: they
were intended as *arbitrary terms* for the *structure* of
clouds, and the meaning of each was carefully fixed by
a definition: the observer having once made himself
master of this, was able to apply the term with correct-
ness, after a little experience, to the subject under all its
varieties of form, colour, or position. The new names,
if meant for another set of arbitrary terms, are super-
fluous: if intended to convey in themselves an explanation

in English, they fail in this, by applying only to some part or circumstance of the definition; the *whole* of which must be kept in view to study the subject with success. To take for an example the first of the modifications—the term *Cirrus* very readily takes an abstract meaning, equally applicable to the rectilinear as to the flexuous forms of the subject. But the name of *Curl-cloud* will not, without some violence to it's *obvious* sense, acquire this more extensive one; and will, therefore, be apt to mislead the learner, rather than forward his progress. Others of these names are as devoid of a meaning obvious to the English learner as the Latin terms themselves.

But the principal objection to English, or any other local terms, remains to be stated. They take away from the nomenclature its present advantage of constituting, as far as it goes, an universal language, by means of which the intelligent of every country may convey to each other their ideas without the necessity of translation. And the more this facility of communication can be increased, by our adopting by consent uniform modes, terms, and measures for our observations, the sooner we shall arrive at a knowledge of the phenomena of the atmosphere in all parts of the globe, and carry the science to some degree of perfection. What would geography have been at this moment, had such descriptions of boundaries as those we meet with in the book of Numbers, or in Joshua, never given place to the universal language of maps and globes?

The *miscellaneous extracts* might have been made more copious than they will be found on the whole, but I avoided taking them from sources where they were already easily accessible to the reader. They are inserted for different purposes: some, to explain the cause of

appearances recorded in the notes, as in the case of
distant thunder storms : others as being supposed to have
a more remote connexion with my own observations:
lastly, not a few to serve as examples of meteorological
and physical phenomena, to which I might have occasion to advert hereafter. The portion which I derived from the public papers was selected from materials
obtained in the following manner :—In the daily paper
taken for use at our laboratory, a line is drawn by my
desire, in the margin of all such passages as have any relation to the weather, or to physical phenomena. Thus,
though absent, I secure them for future notice. In turning over the file at intervals I extract what suits my
purpose, annexing the true or *most probable* date, and
retrenching what is extraneous or irrelevant in matter or
phrase.

If the conductors of our London papers intend (as
without doubt they do) that the many notices of the
weather and phenomena with which they favour us,
should promote science as well as gratify curiosity, they
will not be displeased at my requesting them to give us
these *with a precise date.* A provincial editor naturally
reverts to the day of the week on which a thing happened, and the date of *his* paper settles that of the event.
But when the latter comes to be transplanted into a subsequent London paper, sometimes with, sometimes without,
the corrective addition to the day, of *se'nnight,* or *fortnight,* the uncertainty is often such as to render it useless
to the accurate collector. The day of the month being
adhered to in all cases, would obviate the inconveni-
ence.

The *language* of these accounts is also commonly
vague and unphilosophical: a hard gale of wind is too
often " a tremendous hurricane," and frost and floods,

hail and thunder, are too frequently stated to have been the most severe or destructive " in the memory of the oldest persons living!" Reporters of unusual phenomena should be cautioned to describe, as accurately as they can, in the first place *what they saw*, and in the true *order of time:* after which, if necessary, they may draw comparisons, and give vent to their feelings. I am, however, indebted to the public papers for several circumstantial reports, evidently communicated by men of science; with which the reader may find it not unpleasant to relieve his attention, in looking through this volume for dry facts and coincidences; and from some of which he may possibly be induced to draw the conclusion, that the milder uniformity and tameness of our own climate, are, at least, equally desirable with the more splendid and various, but destructive, phenomena of other zones.

Some disadvantage, I am sensible, attends the publication of this first volume, while the second is as yet not nearly ready to appear. The reader may possibly be disposed to compare this collection of facts at the first view to the parts of a dissected map, turned out before him without the accompaniment of a design, by the aid of which he might put them together. But in this I have followed the natural course of the labour imposed on me by the undertaking. To make out of the detached portions of observations which had been separately formed, *a whole* was the first thing to be done. And this was most easily accomplished by putting them at once to the press, and supplying what was necessary to complete the series, as the work proceeded. They were thus brought into a fit state to be perused for the purpose of theory and deduction, in which considerable progress is already made. Until the results can be laid before the public, what is now produced may serve the purpose of reference, and occasional information on various points of the history of our climate, for the series of years comprehended in the tables. Persons moderately conversant with natural philosophy, will know how to make use of them in this way: still more, they who have been accustomed to make similar observations for themselves. There are indeed but few at present who can be said to *study* the subject, compared with the number of amateurs in chemistry, in astronomy, in electricity, &c. Yet it is one with which gentlemen possessing the requisite information, together with domestic habits, might very agreeably fill up a portion of their daily leisure. There is nothing splendid or amusing to be met with in the outset of such a course: but, I believe, that even in more attractive pursuits, the pleasure of study resolves itself sooner or later into the feeling of the gradual acquisition of knowledge, the perception of the relations, agreements, and differences of facts, and their orderly arrangement in the mind. Now, in no one department of natural knowledge is the field less trodden, or the opportunity for a successful exertion of the judgment in establishing general principles greater, than in meteorology in its present state. There is no subject on which the learned and the unlearned are more ready to converse, and to hazard an opinion, than on the weather—and none on which they are more frequently mistaken! This alone may serve to show that we are in want of more *data*, of a greater store of facts, on which to found a theory that might guide us to more certain conclusions; and facts will certainly multiply together with observers. He who wishes to study astronomy (the most perfect, perhaps, because

the most ancient of the sciences), must begin, I imagine,
where the Chaldeans began, though with so much better
means before him:—he must remark for himself in the
heavens, the actual courses of the planets, and the most
obvious points in the construction of our own system.
So, to become qualified to reason on the variations of our
own climate, we should begin by making ourselves familiar with their extent and progress, as marked by the
common instruments, and the common natural indications:
for which purpose such a model as the present volume
may be found very serviceable. A moderate knowledge
of the phenomena acquired in this way, will naturally
excite a desire to become acquainted also with their
causes, and eventually, with the principles of the science.
These have been ably investigated in parts by several
writers: in our own language, by Franklin, Cavallo, Kirwan, Dalton, Marshall, Wells: in French, by Saussure,
De Luc, Cotte, Bertholon: a work by Beccaria is extant
in an English translation; and there are many detached
extracts of the opinions of foreign authors, as well as
essays of minor bulk, dispersed in the Philosophical Transactions, and other periodical publications. Dr. Thomson
has given a good summary of meteorological facts in
the former editions of his System of Chemistry; and
Dr. Robertson has done the same in a separate work.
But we are not as yet possessed of a general elementary
treatise, displaying in a sufficiently familiar manner the
present extent of the science; which from this cause
appears more confined and imperfect than it really is. In
the early part of this introduction, I made mention of
some lectures on meteorology, which I had a view of
publishing. These are necessarily a more remote object
than the completion of the present work. I must, therefore, entreat the patience of my friends in regard to

these : for while I have long owed such a work to the
public, the materials have been accumulating, and the
labour of selecting and arranging in a small compass,
what may be deemed fittest for the purpose, is likely to
be yet considerable.

ERRATA ET CORRIGENDA.

Table 1. line 15—Barometer 39·83 read 29·83.
   11. in the references—*e* read *c*.
   23, 24. The lunar phasis omitted : which, with several inaccuracies dis-
         covered in the placing of these, will be rectified by a table of them
         all.
   25. line 24 in the reference *b* read *l*.
   50. *omitted* November 18, SW—29·25—29·17—51°—45°.
   51. line 14—The maximum and minimum of pressure appear to be re-
         versed.
   52. at bottom ; minimum of temperature 53° read 33°.
   58. in the notes, fourth page, line 6 from bottom ; *burnt* read *burst.*
   64. in the notes, fifth page, line 10 ; No. 1, read No. 2.
   102. in the notes, second page, line 6 from bottom : *burn* read *burnt.*

Meteorological Observations

MADE AT

PLAISTOW, NEAR LONDON,

IN THE YEARS

1806, 1807, 1808, 1809,

ARRANGED IN LUNAR PERIODS.

(First published Monthly in the Athenæum.)

TABLE I.

	1806.	Wind.	Pressure. Max.	Min.	Temp. Max.	Min.	Evap.	Rain, &c.
11th Mo.	New M. a. Nov. 10	N	30·10	30·10	48°	40°	—	
	11	SW	30·10	29·98	50	33	4	
	a. 12	SW	—	—	47	34	—	
	13	SW	30·18	29·98	57	48	4	
	14	SW	29·98	29·86	53	48	3	1
	15	SW	29·86	29·70	53	37	6	1
	16	SW	29·88	29·70	48	39	10	
	b. 17	SW	29·82	29·70	53	43	8	
	1st Q. 18	SW	29·70	29·58	55	47	11	4
	d. 19	SW	29·25	29·01	51	35	7	63
	20	SW	29·34	29·04	47	35	15	
	b. c. 21	SW	29·33	29·00	45	31	8	2
	b. c. 22	W	29·63	29·33	43	27	4	
	b. c. 23	SW	29·83	29·46	51	33	2	2
	24	SW	39·83	29·65	56	52	8	
	25	SW	29·46	29·26	57	47	8	3
	Full M.b. 26	SW	29·88	29·46	50	39	4	18
	27	NW	30·01	29·90	56	41	0	17
	28	SW	30·04	29·64	60	54	8	2
	e. 29	SW	29·69	29·53	58	36	20	
	30	W	29·69	29·13	45	33	6	4
12th Mo.	f. Dec. 1	SW	29·13	28·63	50	38	8	30
	L. Q. g. 2	W	29·59	28·63	42	38	8	1
	3	NW	29·67	29·63	43	31	3	2
	4	W	29·67	29·45	50	38	2	2
	5	SW	29·45	29·27	54	48	20	
	6	SW	29·30	29·25	58	33	12	
	7	W	29·03	29·00	47	38	7	6
	8	Var.	29·28	29·00	45	36	0	17
	9	NW	29·42	29·33	44	33	3	9
			30·10	28·63	60	27	1·99	1·84

NOTES.—*a.* Misty, the trees dripping without rain.

b. Lunar halo; very brilliant on the nights of 22 and 26. Two series of the prismatic colours were distinguished.

c. Hoar frost. Some hail on the 21st.

d. The barometer appeared to have descended still lower. Pretty strong signs of *positive* electricity the 18th p. m. after which very stormy from midnight to sunrise

e. Wind NW; at 8 p. m. slight shower from a mass of clouds passing in the S, in which it is said to have lightened. On examination a changeable electricity was found in the insulated conductor.

f. Very stormy night.

g. Loud thunder about 3 a. m. a *positive* electricity after sunrise. Soon after 2 p. m. a squall from the NW, with snow and finally rain. The conductor was highly charged *pos.* giving dense sparks. In proportion as the nimbus passed off to the SE this went off and a weaker *negative* charge succeeded.

Additional Notes from the MS. Register.—Eleventh Mo. 27. I found the vapour point about noon but one degree below the actual temp. of the air : hence no evaporation. 30. Much wind.

Twelfth Mo. 1. The barometer took to rising again at sunrise, at 2 p. m. a squall coming on, the electricity became *negative.*

RESULTS.

The prevailing wind has been decidedly SW; often amounting to a gale, more especially in the intervals between midnight and sunrise.

Mean barometrical pressure 29·54 in.; the column almost constantly in motion.

Mean temperature, by Six's thermom. 44·6°.

Total evaporation 1·99 in. This is probably greater than the evaporation at the surface of the earth, the gauge being at an elevation of 30 feet.

Total of rain 1.84 in. being a small amount for the season, though showers have been frequent. Rain noted as fallen by day 0·54, by night 1·30 in.

TABLE II.

	1806.		Wind.	Pressure. Max.	Pressure. Min.	Temp. Max.	Temp. Min.	Evap.	Rain, &c.
12th Mo.	New M.	Dec. 10	SW	29·33	29·33	43°	37°	2	1
	a.	11	SW	——	——	49	39		4
	a.	12	SW	29·36	29·03	55	40		3
	a.	13	SW	29·35	28·98	55	44		12
	a.	14	SW	29·55	29·27	52	40		51
		15	SW	29·76	29·60	53	38	10	5
		16	SW	29·60	29·56	55	52	7	
		17	SW	29·65	29·56	55	48	8	9
	1st Q.	18	SW	29·56	29·51	50	44	3	12
		19	S	39·61	29·45	52	36	5	
	b.	20	S	29·25	29·19	51	37	9	6
	b.	21	Var.	29·66	29·43	49	32	2	28
	b.	22	SW	29·82	29·63	55	49	10	
		23	SW	29·93	29·82	56	53	7	1
	b.	24	Var.	30·35	30·12	56	33	2	2
	Full M.	25	SW	30·17	30·00	55	38	9	1
	c.	26	SW	30·19	30·10	44	42	4	
		27	SW	30·10	29·85	51	42	4	
		28	SW	29·85	29·50	52	42	8	
		29	S	29·49	29·43	51	38	4	2
		30	Var.	30·00	29·43	53	40	4	3
1807.	*d.*	31	N	30·48	30·00	41	26	—	
1st Mo.	L. Q. *d.* Jan.	1	N	30·60	30·48	38	27	—	
	d.	2	W	30·60	30·33	30	28	—	
	e.	3	W	30·25	30·20	38	30	4	
		4	N	30·46	30·25	40	33	3	
		5	NW	30·47	30·45	38	22	—	
		6	W	30·45	30·33	33	25	2	
	f.	7	W	30·33	30·08	39	23	0	
				30·60	28·98	56	22	1·07	1·40

NOTES.—*a.* Very stormy nights. Suspecting too great an effect of the wind on the water in the evaporation gauge, these four results are omitted, and precaution taken to increase the shelter.

b. Lunar halo; succeeded (as usual) by wind, on 23, 24, 25.

c. The highest tide that has occurred, it is said, during fifteen years. In the last spring-tides, the addition of heavy rain caused a pretty considerable overflow of the River Lea. The present elevation seems to be independent of rain in these parts.

d. Hoar frost. 31. Bright day quite to sunset: 1. misty: 2. very misty and much rime. This gradation towards moisture is against the continuance of frost.

e. At 10 a. m. wind W; the melted rime descended in showers from the trees. There was enough on the rain gauge to make 0·02 in. when melted. Such an amount of rain would have wetted the ground considerably ; which however is dry, save under the trees. Rime is a peculiar *cristallization* from a freezing mist. Hence its abundance on surfaces opposed to the wind.

f. At sunrise, an unusually ruddy sky.

Additional Notes, &c.—20. Fine day. *Cirrus* passing to *Cirrocumulus*, &c. rather stormy evening. 21. Clear evening, wind NW, having gone round by E. 24. At noon, a sudden change of wind to N with a shower. 31. *Cirrus* at sunset. The catkins of the filberts expand prematurely.

" On the 25th Dec. a hedge sparrow's nest was taken at Doveridge, Derbysh. with four eggs; and near Warwick, a green linnet's, with two eggs. It is worthy of remark, that the heat was the same on the 24th Dec. as on the 24th June last, on both those days the thermometer being nearly 60°."— (ATHENÆUM.)

RESULTS.

The south-west wind, which had so long reigned, yielded, just at the close of the year, to the north and west. Some frost ensued, which however had not the characters of permanence, being neither ushered in by driven snows nor accompanied with a dry and serene atmosphere. Mean height of the barometer 29·84 in. Mean temperature 42·53°.

Evaporation in 25 days 1·07 in. Rain 1·40, of which by day 0·67, by night 0·58 in.

The effects of the late high winter temperature on vegetation must have been obvious to every one who has seen the country. To the very close of the year the grass continued to grow, the daisies to enamel the turf, and many of the inmates of our gardens (native and exotic) to thrive and blossom. Even hyacinth bulbs, left in the open beds, shot up and flowered. Ten years ago winter came on six weeks earlier, and with considerable severity.

TABLE III.

1807.				Wind.	Pressure. Max.	Min.	Temp. Max.	Min.	Evap.	Rain, &c.
1st Mo.	New M.	Jan.	8	NE	30·08	29·78	42°	26°	—	
			9	S	29·88	29·76	46	36	3	
			10	E	30·20	29·88	42	27	—	
	a.		11	W	30·20	30·08	38	29	—	
			12	SW	30·08	29·66	45	38	4	
	b.		13	NW	29·97	29·66	44	23	4	—
	c.		14	W	30·03	29·97	35	13	—	—
	d.		15	SW	29·97	29·66	47	15	4	9
	1st Q.		16	NW	29·90	29·87	51	41	4	
			17	SW	29·87	29·52	51	37	3	
			18	W	29·63	29·50	42	27	—	
	e.		19	Var.	29·50	29·00	45	28	—	18
			20	SW	29·05	28·98	40	27	—	
	f.		21	NE	29·20	28·80	41	27	—	2
			22	E	29·54	29·04	38	31	—	17
			23	NW	30·08	29·54	41	28	8	
	Full M. e.		24	W	30·43	30·08	43	23	—	
	e. g.		25	SW	30·49	30·43	27	25	—	
	e.		26	SW	30·45	30·43	41	27	4	2
			27	NW	30·50	30·45	44	31	—	
	e.		28	NW	30·54	30·40	43	28	—	
			29		30·40	30·10	42	28	6	
	L. Q.		30	W	30·10	29·84	38	33	3	
	h.		31	W	——	29·84	42	28	—	—
2d Mo.		Feb.	1	N	29·84	29·57	35	18	—	—
			2	S	29·57	28·90	39	26	8	39
			3	W	——	——	39	28	—	—
			4	SW	29·24	28·90	40	28	7	9
			5	W	29·33	29·24	43	29	—	
			6		29·65	29·33	45	31	9	2
					30·54	28·80	51	13	0·76	0·98

NOTES.—*a.* Misty.

b. A squall with rain at 2 p. m.

c. Snowy, p. m.

d. A faint but large solar halo for two or three hours. After sunset the wind rose, and the night was stormy with rain.

e. Hoar frost, and on the 26th, much rime with a thaw after it.

f. A very damp chilling atmosphere, there being a prodigious quantity of cloud, but as yet no wind or rain answerable to the great depression of the barometer.

g. Very foggy evening. It appears from the papers that a great storm of wind occurred about Exeter on the 22d.

h. This day, about noon, the haze which had long occupied the higher atmosphere, became arranged in broad parallel bars of *Cirri*, extending NE and SW beyond the horizon. In the latter quarter some appearance of denser clouds forming below the haze. The evening proved calm, with sunshine; but there followed in the night a brisk gale with heavy rain and snow, from the north. The latter fell also, but in small quantities, on the 2d, 3d, and 4th inst.

RESULTS.

Prevailing Winds Westerly.

Mean elevation of barometer .... 29·78 in.
Mean temperature...................... 34·75°
Evaporation .... 0·76
Rain .... 0·98 in.

The character of this period has been on the whole frosty, the temperature having usually fallen at night below 32°, though the vaporous state of the atmosphere has given occasion to frequent remissions with rain. Of this there fell by day 0·17, by night 0·31: the remainder was not divided.

The gravity of the air has undergone some notable changes. From the 1st to the 24th ult. the result of its frequent variations was a loss on the whole equal to 1½ in. of quicksilver. In four days, from the 20th, this quantity was restored, the rapid movement necessary to this effect being chiefly felt, as it seems, on the south coast. The increase was retained just four days; and in four days more, the whole, except the weight of two-tenths of an inch in the barometer, was again parted with. These effects resemble the increase, continuance, and subsiding of an inundation; and possibly the cause of this sudden swell may have been the meeting of simultaneous opposing currents from the north and south, which, after taking some time to unite and assume a new direction, passed off to the eastward.

TABLE IV.

	1807.		Wind.	Pressure. Max.	Pressure. Min.	Temp. Max.	Temp. Min.	Evap.	Rain, &c.
2d Mo.	New M. Feb.	7	NW	29·85	29·67	43°	28°	6	6
		8	SW	29·58	29·50	48	36	4	10
	a.	9	SW	29·53	29·28	52	37	11	4
		10	W	29·77	29·65	51	39	10	
		11	W	29·98	29·77	57	49	11	
		12	SW	30·15	30·15	57	38	11	
		13	SW	30·18	30·15	53	38	5	
		14	Var.	30·15	30·03	54	31	6	
1st Q.	b.	15	S	30·03	29·87	54	37	7	
	c.	16	SW	29·72	29·64	55	37	3	
	d.	17	Var.	29·49	29·33	45	24	—	
		18	N	30·06	29·49	29	23	—	
	e.	19	Var.	30·18	30·11	35	22	—	
		20	SW	30·11	29·73	46	34	17	10
		21	W	29·72	29·67	46	35	10	
Full M.		22	W	29·69	29·57	49	32	8	1
	f.	23	NW	29·88	29·69	44	23	6	
	g.	24	W	29·69	29·48	51	34	8	11
		25	SW	29·68	29·47	56	34	5	2
		26	NW	29·92	29·68	39	25	7	
	h.	27	N	30·38	29·92	37	26	2	
	h.	28	N	30·56	30·38	38	27	5	1
3d Mo.	L. Q. March	1	NE	30·56	30·50	42	31	4	
		2	NE	30·50	30·17	44	35	5	1
		3	Var.	30·17	29·84	44	31	5	
		4	N	29·84	29·83	43	23	—	
	h.	5	N	———	———	33	19	—	
	h.	6	W	29·86	29·66	38	23	10	11
	h.	7	N	29·97	29·77	40	23	—	
		8	W	29·77	29·30	45	31	8	5
				30·56	29·28	57	19	1·74	0·62

Notes.—*a.* A gale of wind.

b. A very fine day. A thermometer against a wall facing the south rose to 87°. About sunset the clouds came down as in summer evenings before thunder.

c. A very large and distinct lunar halo.

d. Stormy indications at sunset, followed by a severe and destructive gale from the NE which was not over till the next night; much lightning between 3 and 4 a. m. the 18th; abundance of snow fell on the more hilly parts of the country, though but a moderate quantity **here.**

1

e. The whole hemisphere very red for some time after sunset, which we ascribed to the reflection of light from elevated *Cirri.* Our Manchester correspondent, however, states the same phenomenon at the same time as an *Aurora Borealis.* Additional communications, decisive of this point will be acceptable. The phenomenon was repeated on the 21st, which, with the preceding and following night, was windy.

f. The *Cirrus* cloud continues to fill the higher atmosphere, and is now collected into parallel bars extending E and W beyond the horizon.

g. Hoar frost, with a strong positive electricity, which was found again at sunset in some rain, mixed with sleet.

h. Small quantities of snow at intervals; the atmosphere sensibly tending to a state of greater dryness.

Additional Notes, &c.—Second Mo. 7. At sunset, the shadows of large *Cumuli* were projected on the haze above them, and there was a bank of *Cirrus* and *Cirrostratus* in the W. 8. *Cirrostratus* at sunset, finely divided and wavy. 17. At sunset, a bank of *Cirrostrati* obliquely crossing some *Cirri* which pointed northward. 21. Very red *Cirri* at sunset. 26. A complete evaporation of the denser clouds in the evening. 27. A brilliant morning.

Third Mo. 7. *Cumulus* and *Cumulostratus,* which nearly evaporated at sunset leaving a red haze. 8. Overcast for rain *.

RESULTS.

Prevailing Winds Westerly.

Mean elevation of the barometer........ 29·86 in.
Mean temperature.................... 38·21°
Evaporation .... 1·74 in.
Rain, &c..... .... 0·62 in.

Character of the period, frosty, with a dry atmosphere for the most part.

Rain noted by day 0·10, by night 0·25 in.

* An immense ball of fire was observed at Glasgow on Friday night, the 6th of March, fifteen minutes before ten o'clock, directing its course (seemingly) from SE to NW, with a very long streaming tail, and, towards the end, of a sparkling appearance. (PHILO. MAGAZINE.)

TABLE V.

	1807.		Wind.	Pressure. Max.	Pressure. Min.	Temp. Max.	Temp. Min.	Evap.	Rain, &c.
3d Mo.	New M.	March 9	NE	29·93	29·30	39°	29°	9	9
	a.	10	NE	———	———	38	28	11	
	a.	11	NE	30·20	29·93	37	31	16	
		12	NE	30·17	30·13	42	34	16	
	a.	13	NE	30·1?	30·06	45	31	11	
	a.	14	N	30·06	29·92	39	26	3	
	a.	15	NW	29·98	29·84	40	18	7	
		16	NW	29·40	29·33	37	26	—	
1st Q.	b.	17	NW	29·43	29·28	42	28	14	1
		18	SW	29·60	29·28	52	28	14	
		19	SW	29·97	29·60	49	26	12	
	b.	20	W	30·14	29·97	51	31	9	
	b.	21	Var.	30·33	30·14	60	36	4	
		22	E	30·44	30·33	54	31	15	
Full M.		23	E	30·41	30·33	42	28	—	
		24	NE	30·41	30·29	42	28	—	
	c.	25	NE	30·29	30·18	44	33	52	
	a.	26	N	30·18	29·79	42	35	5	2
	a.	27	N	29·79	29·70	43	35	7	
	a.	28	NE	29·69	29·68	42	33	7	
	a.	29	NE	———	———	45	28	—	
L. Q.	a.	30	NE	29·78	29·63	42	29	—	
	a.	31	Var.	29·47	29·43	48	29	22	33
4th Mo.		April 1	NE	29·68	29·47	48	31	4	
	a.	2	N	29·77	29·68	34	27	0	4
		3	N	30·06	29·77	42	22	—	
		4	SW	30·18	29·93	45	23	13	
		5	SW	———	———	50	24	—	
		6	SW	30·02	30·00	50	24	14	
	d.	7	Var.	30·15	30·00	57	34	7	
				30·44	29·28	60	18	2·72	0·49

NOTES.—a. Snow in very small quantities till the 31st, when it was more plentiful, and followed by rain. On the 2d the trees and shrubs were much loaded all the forenoon, and it lay in the drifts several days.

b. Lunar halo.

c. About 8 p. m. a steady light in the NW, probably *Aurora Borealis;* it disappeared gradually after a flash of lightning in the SE. There were clouds in each quarter, but of no great density. *Dies albo calculo notandus!* not, however, on account of the weather, but for the accomplishment of a long deferred national act of justice in the abolition of the Slave Trade.

d. A fine spring day. Bats began their usual excursions in the twilight. Considering that these animals pursue small insects by the sense of hearing alone, their agility is surprising; and the silent motion of their leathern wings is admirably adapted to the purpose.

Additional Notes, &c.—Third Mo. 10. Frequent slight showers of granular snow, (or the *nuclei* of hailstones.) 15. A little opake hail. 18. a. m. *Cumulus* clouds with a veil of *Cirrus* above: windy, with a little rain at sunset. 21. A light breeze varying from W by N to E. 22. A steady breeze E. 25. Idem NE. 26 to 31. *Cumulostratus* daily, with a clear sky above, and now and then, granular snow.

RESULTS.

Prevailing Winds between North and East.

Mean elevation of barometer .... 29·94 in.
Mean temperature............. .... 36·28°
Evaporation .... 2·72 in.
Rain and snow .... 0·49 in.

Character of the period dry, frosty, and cloudy, with a high degree of transparency in the lower air. Rain, &c. by day 0·03, by night 0·42.

TABLE VI.

	1807.		Wind.	Pressure. Max.	Pressure. Min.	Temp. Max.	Temp. Min.	Evap.	Rain, &c.
4th Mo.	New M.	April 8	SW	30·23	30·15	63°	30°	5	
	a.	9	S	30·23	29·98	61	39	19	
	b.	10	SW	—	—	55	43	—	
		11	SW	29·98	29·30	57	45	32	
	c.	12	SW	29·30	29·25	59	42	23	
	d.	13	Var.	29·28	29·22	59	33	10	
		14	Var.	29·28	29·25	61	44	7	16
1st Q.	e.	15	NE	29·39	29·28	51	39	10	
		16	Var.	29·66	29·39	43	37	10	
	f.	17	N	29·78	29·77	42	25	—	—
		18	N	29·87	29·78	46	28	11	—
	g.	19	NE	30·12	29·87	50	29	11	1
		20	NE	30·16	30·12	45	30	12	
		21	SE	30·12	29·95	51	29	7	
Full M.		22	S	29·89	29·88	51	41	9	
		23	SW	—	—	62	43	—	
		24	Var.	30·08	29·89	60	47	—	
		25	E	30·17	30·03	69	40	34	
	h.	26	E	30·10	30·06	68	41	25	
	i.	27	SW	30·03	29·98	80	52	35	
		28	W	30·03	30·02	70	49	13	
L. Q.	k.	29	NE	30·02	29·88	72	46	17	4
		30	E	29·88	29·75	73	49	32	
5th Mo.		May 1	E	29·75	29·70	78	50	24	
	l.	2	Var.	29·67	29·63	79	49	21	
		3	Var.	29·63	29·60	71	49	16	22
		4	E	29·60	29·34	58	50	5	15
		5	SE	29·34	28·98	71	51	12	30
		6	SW	29·09	28·90	61	49	21	1
				30·23	28·90	80	25	4·21	0·85

NOTES.—*a.* A serene atmosphere, with a strong *positive* electricity the whole day. The pith-balls of the conductor diverged half an inch.

b. Windy, with a little rain, which was *negative*.

c. Barometer stationary about twenty hours, the air *positive*.

d. About 9 p.m. a meteor passed from the zenith to the south; there fell much dew in the night, and rain followed in some quantity.

e. A single swallow on the wing.

f. Several more of these harbingers of warm weather, which, however, met with a most inhospitable reception in a storm of snow and sleet, continuing most part of the day.

g. p. m. Many distinct *Nimbi* traversing the country in different quarters, and discharging showers of *hail*, which was highly charged with electricity. One of these, being carefully examined throughout, presented the following phenomena. While the cloud was on the horizon in the NE and the shower behind it, the pith-balls of the insulated conductor remained in contact. When the extremity of the upper surface of the inverted cone of cloud had arrived in the zenith, they opened *negative*, and diverged slowly to full two inches, at which time pretty strong sparks were drawn from the conductor. During the remainder of the approach of the shower, they gradually closed again. At the moment when the latter began to touch the observatory they opened *positive*, diverged more speedily, and the apparatus gave strong sparks for a considerable time, *positive*. As the cloud drew off to the SW, this charge gradually ceased, and the balls opened again *neg.* diverging gradually as before, then converging, and lastly were left a little charged *pos.* The reader who is conversant in electrical phenomena will see in all this the natural effects of the high positive charge in the column of falling hail, which might be six or seven miles in diameter, and which appeared to be surrounded with a negative *area*, extending into the dry atmosphere about three miles in every direction. Could the descent of the electric fluid have been rendered as obvious to our senses over the whole tract, as was that of the hail, its conductor, we should have pronounced it a shower of fire rather than of ice; for the latter, when melted into the rain gauge, made no more than one hundreth of an inch along with several previous showers.

h. First notes of the cuckow.

i. About 9 p. m. a sudden shower, which gave to the conductor a strong *negative* charge continuing some time after it. The air before was *positive*.

k. A mist from the Thames.

l. After repeated indications of strong electricity in the clouds for some days past, thunder was heard at intervals, in a *Nimbus* situated in the W and NW. Signs of negative electricity followed, for a few minutes only, when the edge of this cloud approached us. Soon after a breeze coming on from SW, this, with other clouds of the same kind, which had formed in the E, S, and SW, drew off to the northward, where they remained visible on the horizon till late at night, the lightning playing among them almost incessantly. Much rain followed on the ensuing days, which was several times examined, and found *positively* charged.

Additional Notes, &c —Fourth Mo. 9. Very serene and warm, *Cirri* pointing N, and *Cirrostratus* evening; the wind brisk. 10. *Cirrostratus* in abund•

ance the whole day. 12. At sunset the *Cumuli* evaporated, *Cirrus* and *Cirrostratus*, with much dew succeeded.

Fifth Mo. 5. Rain the whole evening at first non-electric, then strongly positive at intervals.

THUNDER STORMS.

At Poulton-in-the-Fylde, Lancashire, and the neighbourhood, on Thursday the 30th of April, and the two following days, there was the most tremendous thunder and lightning ever remembered by the oldest persons; on Friday, particularly so. As a girl, aged thirteen, was returning from school, in Poulton, about seven in the evening, she was struck dead within half a mile of the town: her bonnet, cloak, stockings, and shoes, were burnt or torn in pieces, several parts of which were carried into the hedge; she had a gallon of rum in a stone bottle, wrapt up in her apron, which there is no doubt exploded, as several pieces were found at a considerable distance from the spot. Two sheep were killed near Poulton, and the ground near where they lay was perforated in several places, and burnt—Great Marton Mill had three of its sails shivered in pieces, and the top set on fire; a large iron chain, which draws up the corn, was melted to a rod of iron, and as the bottom did not reach to the floor, considerable damage was done below it, such as tearing up the boards, &c.—On the same day, there was a dreadful thunder storm at Preston, attended with vivid and continued lightning, a fall of rain so heavy as to be compared to the setting in of the rainy season in Africa, and hail so large, that some of the stones measured three inches in circumference; it broke windows and sky-lights innumerable.—The storm was also severely felt at Lancaster, York, and many other places.—At Bakewell, in Derbyshire, hailstones fell, intensely frozen, from two to four inches round, and many windows were broken.

The inhabitants of Silkstone, near Penistone, Yorkshire, were visited by one of the most alarming phenomena ever remembered. The clouds had portended rain, but none had then fallen there, when suddenly a torrent of water deluged the town, which is situate in a valley, and several persons were unfortunately drowned. The greatest transition from cold to heat ever remembered had been observed in the last week in April, and the above inundation was occasioned by a mass of clouds, during the thunder storm, bursting in a field in the township of Bradfield, the waters taking their course down the Rivelin and Loxley, and thence into the Dun, which became suddenly swollen. Near Doncaster it is said to have risen nine feet in the space of an hour and a half. A great number of windows were broken, at Heckleton, during a severe hail storm the same evening. Pieces of ice of an oblong form exceeding five inches were picked up.—(ATHENÆUM.)

Process employed in the Maçonnais of France to avert Hail and dissipate Storms.
Mag. Encyclopedique, T. 2. p. 5.

This process, which is now universal in the part of France named in the title, was originally introduced by the Marquis of Cheviers, a naval officer; retired on his estate at Vaurenard, about 35 years ago, who having recollected to have seen the explosion of guns resorted to at sea in order to disperse stormy clouds, resolved to attempt a similar method to dissipate the hail storms, whose ravages he had often witnessed. For this purpose he made use of boxes of gunpowder, which he caused to be fired from the heights on the approach of a storm; this had the happiest effect, and he continued till his death to preserve his lands from the ravages of hail storms, while the neighbouring villages frequently experienced their baneful effects. He consumed annually between 200 aud 300lb. of mining powder. The inhabitants of the communes where the estate of the Marquis was situated, convinced of the excellence of the practice, from the experience of a great number of years, continued to employ it. Their example was followed by the surrounding communes; and the practice gaining ground, is at this moment in use in the communes of Vaurenard, Iger, and many others. The size of the powder boxes, their charge, and the number of times they fire them off, vary according to circumstances, and the position of the places. In the commune of Fleury they use a mortar which carries a pound of powder at a charge; and it is generally upon the heights, and before the clouds have had time to accumulate, that they make the explosions, which they continue until the stormy clouds are entirely dissipated. The annual consumption of gunpowder for this purpose, from the magazine at Mâcon, is from 1300 to 1600lb.—(ATHENÆUM.)

RESULTS.

Winds variable. Mean elevation of barometer 29·73 in. Its movements offer nothing remarkable, the rain having been preceded as usual by continued depressions of the column.

Mean temperature.............51·12°
Evaporation.................. 4·21 in.
Rain, &c..................... 0·85.

The most prominent feature in this period is the almost constant strong electrisation of the atmosphere, which terminated, though not in this part of the country, in violent thunder storms. Rain by day 0·30. By night 0·54.

TABLE VII.

	1807.	Wind.	Pressure, Max.	Pressure, Min.	Temp. Max.	Temp. Min.	Evap.	Rain, &c.
5th Mo.	New M. *a. b.* May 7	SW	29·58	29·13	58°	44°	15	2
	c. 8	SW	29·58	29·27	57	49	11	24
	9	Var.	29·30	29·23	58	46	7	11
	b. 10	Var.	29·80	29·30	51	41	7	19
	d. 11	Var.	29·80	29·56	59	48	6	33
	e. 12	SW	29·44	29·38	59	52	13	—
	13	Var.	29·70	29·44	64	46	11	37
1st Q.	*f.* 14	SW	29·72	29·60	56	49	4	30
	g. 15	SW	29·79	29·72	64	45	10	12
	16	SW	—	—	—	—	—	
	17				—	—	—	
	18		30·30	29·79	69	41	53	
	19	NE	30·30	30·18	59	44	14	
	20	NE	30·18	30·02	62	45	16	
Full M.	21	NE	—	—	—	—	—	
	22	NE	30·13	30·10	68	42	61	
	23	E	30·10	29·93	75	49	32	
	24	E	29·93	29·70	82	52	28	
	h. 25	Var.			85	56	—	
	i. 26	SW	29·72	29·66	70	54	60	
	27				63	46	—	
	k. 28		—	—	—	—	—	—
L. Q.	*l.* 29	NE			63	39	—	—
	m. 30	E	29·75	29·62	55	42	48	60
	31	E	29·80	29·78	54	46	13	—
6th Mo.	*b.* June 1	S	29·83	29·78	63	45	7	21
	n. b. 2	Var.	29·89	29·88	68	49	9	1
	3	NW	30·06	29·89	70	47	12	
	4	NW	30·06	30·03	66	47	10	
	5	Var.	30·03	29·88	68	52	11	23
			30·30	29·13	85	39	4·58	2·73

NOTES.—*a.* Strong *positive* electricity in the intervals of the showers. A pair of pith-balls, loaded with lead, so as to weigh ten grains, and suspended from the conductor by threads of the length of seven inches, exhibited a curious phenomenon. Besides the waving motion, during their divergence, which is not unusual, there was a sensible impulse of the fluid downward upon the balls, causing the threads to quiver incessantly like an insect's wing.

b. Rainbow.

c. Much wind at SW. *Negative* electricity, from a *Nimbus* going by in the S.

d. At 6 p.m. changeable electricity, from a *Nimbus* in the N, in which it thundered.

e. A strong wind at SW with much *scud*,* the rain strongly *positive*.

f. Almost incessant rain, which was void of all signs of electricity.

g. Rain still non-electric, and the air strongly and variably charged in the fair intervals.

h. Lightning in the W.

i. After a constant exhibition of the *Cirrus* cloud for several days past, with much dew, the latter deposition is suspended, and the sky overcast and threatening.

k. Hoar frost this morning, and a *Stratus* on the river and meadows after sunset.

l. Very stormy night; the newly expanded foliage suffered much.

m. Rain the whole day.

n. Strongly positive atmosphere.

Additional Notes, &c.—Fifth Mo. 7. The bow appeared very bright and perfect twice; at 4 and at 7 p.m. 26. For five days past the *Cirrus* cloud has appeared over the whole sky, with the points directed upward.

RESULTS.

Winds Variable.

Mean elevation of barometer .... 29·78 in.

Mean temperature................ ....55·41°

Evaporation .... 4·58 in.

Rain .... 2·73.

There have been almost continual indications of an active state of the atmospheric electricity; a result which seems naturally allied to the variable state of the currents, and a much greater deposition of water than is usual at this season. Rain noted as by day 1·38, by night 0·56.

* Loose shapeless portions of cloud moving swiftly in the lower current of the atmosphere.

TABLE VIII.

	1807.		Wind.	Pressure. Max.	Pressure. Min.	Temp. Max.	Temp. Min.	Evap.	Rain, &c.
6th Mo.	New M.	June 6	Var.	29·88	29·58	62°	48°	4	91
	a.	7	SW	29·57	29·55	67	51	·6	2
	b.	8	SW	29·68	29·57	68	52	19	
		9	SW	29·93	29·68	68	45	23	
	c.	10	SW	29·95	29·92	65	57	16	6
	c. d.	11	SW	30·10	29·90	70	44	16	
	c. e.	12	W	30·15	30·10	69	44	17	
	1st Q.	13	W	30·10	30·05	69	46	20	
		14	SW	30·05	29·90	70	46	13	
	f.	15	Var.	29·90	29·62	78	48	21	
	g. h.	16	W	—	—	74	46	—	
	g.	17	NW	29·95	29·62	68	49	44	
		18	NW	30·16	29·95	67	43	14	
		19	NW	30·17	30·16	66	42	12	
	Full M.	20	NW	30·28	30·09	69	47	16	
	i.	21	NW	30·18	30·10	72	53	16	
		22	NW	30·02	30·00	74	53	—	
	k.	23	Var.	30·00	29·92	67	51	35	
		24	Var.	29·92	29·76	71	47	21	
		25	NE	29·81	29·76	75	54	—	
		26	NW	29·88	29·76	79	54	36	
	l.	27	NW	29·92	29·90	74	54	17	
	L. Q.	28	NW	29·90	29·86	69	43	14	
	m.	29	NE	29·86	29·81	69	46	13	
		30	N	29·81	29·79	65	52	10	
7th Mo.	n.	July 1	W	29·85	29·81	65	48	9	
	i.	2	SW	29·81	29·81	71	54	26	
		3	W	29·86	29·81	64	54	11	
		4	Var.	29·93	29·86	66	45	8	
				30·28	29·55	79	42	4·57	0·99

NOTES.—*a.* Thunder at intervals.

b. Some lightning in the evening.

c. c. c. Much wind by night.

d. Slight solar halo.

e. Hoar frost.

f. A beautiful display of the *Cirrus* cloud all day.

g. g. Some indications of a tendency to thunder. The rod was charged *negative*, but the clouds at length passed off to NE.

i. i. Brisk winds.

k. The wind went from NW by E to SW.

l. Clouds highly coloured at sunset. The wind, as usual after this appearance, veered eastward.

m. A *Stratus* on the river and meadows this morning, owing, probably, to the sudden depression of the nocturnal temperature.

n. Abundance of *honey dew* *, on the elms and limes, as well as on the fruit trees.

Additional Notes, &c.—Sixth Mo. 4. The clouds, which have become daily more dense, now show for thunder: there are wanting however the concurrent indications of active electricity and a temperature above 80°. A few drops of rain to-day were wholly unelectrified.

" Lisbon June 12, 1807.

" On the 6th inst. at ten minutes before four in the afternoon, a dreadful shock of an earthquake was felt here, which lasted about eight seconds; it was more severe than the great one in 1755, but, thank God, not so fatal, a few lives only are lost, but a number of houses damaged."

Another account states that the shock took place about four o'clock in the afternoon, and lasted about twelve seconds. The shock was so severe, that several houses were much damaged, and the city thrown into the greatest confusion. It was not known that any lives were lost, but several had their arms and legs broken, &c. by jumping out at the windows, under the apprehension of the houses falling upon them. The shock was also felt at St. Ubes, Oporto, and generally throughout Portugal. It was felt on board the Lively frigate, then about eighty leagues off the rock of Lisbon.—(Public Papers.)

RESULTS.

Winds Variable.

Mean elevation of barometer .... 29·89 in.

Mean temperature .... 59°

Evaporation .... 4·57 in.

Rain............................... 0·99 in.

Character of the period fair and dry. There has been a remarkable approximation to the mean, both in the extremes of pressure and temperature, and in the daily rate of evaporation. The electricity has been nearly quiescent. Rain by day 0·35, by night 0·58 in.

---

* Some notice of the origin of this sweet fluid, which has abounded so as to drop from the trees, of late, may be interesting to most readers. It is well known to naturalists, that it neither falls on the leaves from the air, as it name implies, nor transpires from them; but that it is excerned by insects of the genus *Aphis*, which inhabit the under sides of leaves, and shed this liquor on the surface of those below. This little creature seems to subsist by drawing the juices from the sap vessels, and, by a peculiarity of constitution, rejects the richer saccharine part, which afterwards affords nourishment to many other insects. Ants are so fond of it, that a whole colony may be found travelling to the highest branches of a tree in search of it; and we have seen them seize the clear drop, while yet attached to the body of the Aphis, which, although defenceless, was not molested further by these predacious wanderers.

TABLE IX.

1807.			Wind.	Pressure. Max.	Min.	Temp. Max.	Min.	Evap.	Rain, &c.
7th Mo.	New M.	July 5	E	29·93	29·87	73°	48°	—	
		6	NE	30·08	29·93	68	43	45	
		7	E	30·19	30·08	63	39	18	
		8	N	30·19	30·16	70	41	20	
	a.	9	Var.	30·16	29·98	77	52	16	
		10	NW	29·98	29·68	81	54	14	
	b.	11	SW	29·72	29·58	78	56	—	1
1st Q.	c.	12	SW	29·78	29·72	79	59	53	1
		13	SW	—	—	—	—	—	—
	d.	14	SW	29·81	29·78	77	58	40	—
	d.	15	SW	29·80	29·78	73	55	14	3
		16	W	29·90	29·80	73	57	12	
		17	W	29·95	29·87	76	60	18	
		18	W	29·99	29·97	76	50	19	
	Full M.	19	SW	—	—	79	60	—	
		20	W	29·81	29·78	80	54	44	
		21	Var.	—	—	79	56	—	
		22	Var.	29·75	29·64	87	62	54	
		23	SW	29·64	29·58	81	62	—	
	d.	24	SW	29·68	29·60	77	57	—	—
		25	SW	29·78	29·69	82	59	—	
		26	SW	29·82	29·70	77	55	82	1
	L. Q.	27	SW	29·91	29·83	76	51	—	
		28	W	29·92	29·78	81	55	—	
		29	SW	29·78	29·58	81	63	—	
		30	SW	29·63	29·58	72	60	90	25
		31	SW	29·66	29·55	71	55	—	
8th Mo.		Aug. 1	S	29·80	29·66	76	47	—	
		2	SW	29·77	29·75	78	55	51	7
				30·19	29·55	87	39	5·90	0·38

NOTES.—*a*. The cockchaffer begins to appear in great numbers. A cat is observed to pursue these insects every evening, feasting on them with avidity.

b. After considerable appearances of a strong electricity in the clouds, which yet did not affect the insulated rod, several *Nimbi* formed suddenly about 5 p. m. from one of which fell a shower in large *warm* drops strongly *positive*. Some thunder and lightning followed, during which the rod was highly charged, but the kinds were not noticed. The wind by the vane continuing strong at E, the lower clouds moved directly N, and the sky cleared over head: but

NOTES, &c. *(continued)*.

there remained a bank of clouds but little elevated above the N horizon, in which till eleven p. m. were the usual appearances of a thunder storm in action.—(*From MS. Reg.*)

c. Thunder and lightning in the night, with a little rain. An ox was struck dead in the level opposite Woolwich, and found at 5 a. m. the 13th, with the hair singed and the body beginning to putrefy.

d. Slight showers.

RESULTS.

Prevailing Winds West and South-west.

Mean elevation of the barometer........ 29·81 in.
Mean temperature.................... 66·08°
Rain............................... 0·38 in.
Evaporation .... 5·90 in.

Rain by night 0·01, by day 0·35 in.

TABLE X.

	1807.			Wind.	Pressure.		Temp.		Evap.	Rain, &c.
					Max.	Min.	Max.	Min.		
8th Mo.	New M.	Aug.	3	SW	29·86	29·75	72°	50°	—	
			4	W	29·90	29·86	71	53	—	
			5	SW	29·89	29·78	72	57	—	
			6	SW	29·78	29·75	75	58	98	1
			7	NW	29·82	29·78	71	57	—	
			8	Var.	29·93	29·82	65	55	—	14
			9	NE	29·96	29·93	71	51	13	
	1st Q.		10	NE	29·93	29·88	68	49	—	
	a.		11	SE	29·88	29·64	75	56	38	2
			12	W	29·64	29·62	79	62	—	
	b.		13	SE	29·65	29·50	81	61	—	
	c.		14	NW	29·85	29·50	72	57	—	93
			15	W	30·00	29·85	71	60	—	
	d.		16	NW	30·15	30·08	75	58	74	
			17	E	30·12	30·05	79	54	—	
	Full M. e.		18	E	30·05	29·89	79	56	—	
	e.		19	SE	29·89	29·85	79	59	—	
	f.		20	NW	29·85	29·81	75	60	56	1
			21	E	29·80	29·78	78	59	—	
			22	E	29·78	29·76	82	60	—	
			23	SW	29·78	29·75	81	63	55	3
			24	W	29·81	29·76	71	58	—	
			25	W	29·90	99·81	76	56	—	
	L. Q.		26	NW	29·91	29·90	79	51	—	27
			27		29·90	29·61	—	—	47	1
			28	SW	29·83	29·64	79	56	39	
			29	SW	29·86	29·83	74	49	26	
			30	W	29·97	29·86	70	46	19	
			31	NW	30·12	29·97	61	51	5	5
9th Mo.		Sept.	1	W	30·15	30·12	69	45	11	
					30·15	29·50	82	45	4·81	1·47

Notes.—*a.* Rainbow; some lightning in the evening.

b. Much lightning in the night.

c. Very heavy rain a. m.

d. A *Stratus* on the marshes.

e. Foggy morning.

f. Some thunder p. m.; almost constant lightning in the evening.

ACCIDENTS BY LIGHTNING.

At Stockport, on the 26th, I examined a cotton manufactory, in which a fire had occurred by lightning the night before. It had fallen on the gable end of the building, about seventy feet from the ground, directed apparently to this spot by a packing press, containing much iron-work. In its way it had shivered some slates on the roof, and forced out a quantity of bricks from the wall over a window, which was also shattered. The press was much burned, being apparently set on fire all over by the iron. This effect was most conspicuous in a drawer containing loose iron-work, and some lead, which latter was found to have been melted. The fire having been promptly extinguished, there was little damage, save the loss of some cotton.

The affixing of proper conductors to such manufactories, containing immense quantities of detached iron and brass-work, and crowded with people, cannot be too often recommended, especially such as stand alone and on elevated ground.

I conversed here also with the survivor of two men (George Bradbury and Peter Sidebotham) who were struck, near this spot, about a month before. They had been at work in a garden, and had retired for shelter from a heavy rain to a little summer-house on the bank of the Mersey, which rises here thirty or forty feet from the water. They sat back to back on a chest, with the door shut. Bradbury says he neither *saw, heard, nor felt* any thing, but on reviving found himself extended on the floor, together with his lifeless companion.

There was in the middle of the roof of this small building a lead gutter, having no metallic communication with the ground, and not more than three or four feet above their heads. A piece of fir, which was laid from wall to wall, about a foot below the gutter, was split by the stroke. The panes of the window were nearly all broken, and the glass thrown *outward*. There was a strong sulphureous smell in the place for some time. It appeared to me that the stroke had passed from the gutter on the body of Sidebotham, and from thence through the thigh of Bradbury, on which the effects yet remained, viz. an ulcer on the outer part a little above the knee, another on the inner, and a third on the calf of the leg, which by bending the knee might be brought into contact with the second. The use of the limb seemed at first to be taken away, but in these parts there presently arose hard swellings which ulcerated.

Bradbury's account of the swiftness of the effect, preventing sensation, is undoubtedly correct; and it is not possible to imagine an easier death, however terrible from its suddenness, than must have befallen his companion.

RESULTS.

Prevailing Winds Westerly.

Mean height of barometer .... 29·85 in.
Mean temperature............ . ... 64·96°
Rain.... .. .. 1·47 in.
Evaporation .... 4·81 in.

Above 1·10 in. of rain appears to have fallen by day, the remainder by day and night.

TABLE XI.

1807.			Wind.	Pressure. Max.	Pressure. Min.	Temp. Max.	Temp. Min.	Evap.	Rain, &c.
9th Mo.	New M. Sept.	2	W	30·15	30·07	70°	44°	15	
		3	NW	30·07	29·88	72	49	20	
		4	SW	29·88	29·70	69	59	18	
	d.	5	W	29·70	29·37	72	55	—	
	d.	6	SW	29·62	29·36	62	45	—	
	a.	7	W	29·90	29·62	68	37	91	
	1st Q.	8	W	29·92	29·65	59	45	—	
		9	W	29·83	29·47	66	49	—	
		10	NW	30·03	29·83	55	39	21	65
		11	W	29·90	29·80	58	41	—	
	a.	12	W	29·91	29·79	61	33	31	
	a. b.	13	N	29·98	29·91	54	26	10	
		14	N	29·98	29·92	58	41	12	
		15	NW	29·92	29·92	59	37	9	
	Full M.	16	NW	29·92	29·88	58	38	9	
		17	N	29·88	29·87	57	30	9	
	d.	18	NE	29·85	29·79	54	43	24	
	d.	19	NE	30·16	29·85	61	35	18	
	d.	20	W	30·16	30·01	59	51	—	
	d.	21	SW	30·01	29·72	61	54	25	7
		22	SW	29·72	29·56	60	54	6	24
	d.	23	S	29·56	29·49	61	54	8	10
	L. Q.	24	S	29·52	29·41	66	47	13	
	d.	25	Var.	29·45	29·37	63	46	8	23
	d.	26	SW	29·49	29·45	66	47	—	13
	d.	27	SW	29·46	29·42	66	52	30	5
	d.	28	Var.	29·91	29·46	60	41	11	
		29	E	29·91	29·45	61	45	—	
	e.	30	W	30·16	29·46	59	39	18	
				30·16	29·36	72	26	4·06	1·47

NOTES.—a. Hoar frost.

b. This night put a period to the growth of the more tender vegetables, as potatoes, kidney beans, &c. where they lay exposed. The
cucumbers were quite killed. This very low temperature (for the season) was confined to a small height, and was detected by exposing the
thermometer horizontally four inches above the turf. In its usual position, six feet from the ground, it had not descended on some previous nights to 32°, though spiculæ of ice had been formed on the
grass.

c. " From 7 to 8 p. m. I observed a star in the NW, about 25 or 30°

above the horizon, apparently a comet, as it has a luminous brush or tail pointing southward, the extremity being raised above the horizontal line." I owe this observation on the comet, probably one of the earliest made in this part of the kingdom, to my friend and assistant in Meteorological Observations, John Gibson.

   d. Strong winds.

RESULTS.

Prevailing Winds Westerly.

Mean height of barometer .... 29·76 in.
Thermometer .... 52·94°
Evaporation .... 4·06 in.
Rain .... 1·47 in.

Of which by day 0·10, by night 0·49, the remainder by day and night.

It is remarkable that the rain of this period agrees exactly in quantity with that of the last. The warm and dry weather terminated, not as usual in thunder showers, but by a sudden and continued depression of temperature; which, with the arid state of the country, constituted a sort of premature winter. The strong southerly winds which blew during the latter half of the period, brought, however, a seasonable supply of moisture, and vegetation has since revived.

It appears by the Register of my friend Thomas Hanson, the results of which appeared monthly in the Athenæum, that there fell in this month at *Manchester* 6¼ in. of rain, the winds having been chiefly W and SW, the mean of barometer 29·62 in. of thermometer 51·73°, the latter probably too high.

TABLE XII.

	1807.			Wind.	Pressure.		Temp.		Evap.	Rain, &c.
					Max.	Min.	Max.	Min.		
10th Mo.	New M.	Oct.	1	S	30·16	30·02	60°	36°	18	
	a.		2	W	30·03	29·98	65	51	20	13
	a.		3	SW	30·11	30·03	65	51	11	
	a.		4	S	30·06	30·02	67	48	3	
	a.		5	W	30·02	29·93	65	53	2	1
			6	W	29·96	29·95	61	50	5	
	b.		7		29·95	29·75	65	54	—	
	1st Q.		8	W	30·10	29·75	61	48	45	
			9	W	30·10	30·07	59	52	—	1
			10	SW	30·12	30·07	65	54	14	
			11	SW	30·11	29·98	61	55	—	
			12	W	30·12	29·98	62	56	—	
			13	Var.	30·16	30·12	65	54	26	
	c.		14	SW	30·16	30·01	69	54	10	
	d.		15	SW	30·11	30·01	65	48	9	
	Full M.		16	SW	30·06	29·99	58	52	12	
	e.		17	W	29·99	29·85	63	54	—	
	d.		18	NW	30·24	29·88	62	36	24	
	d.		19	Var.	30·25	29·92	61	43	8	
			20	S	29·92	29·42	65	52	15	
	b.		21	Var.	29·46	29·44	61	53	15	22
	f.		22	NW	29·44	29·03	63	41	4	30
			23	N	29·30	29·20	50	33	5	
	L. Q.		24	SE	29·54	29·30	56	36	5	11
			25	E	29·64	29·54	56	49	9	11
			26	E	29.62	29·48	53	35	—	—
			27	N	29·83	29·62	52	35	6	24
			28	SW	29·85	29·64	51	38	9	
	f.		29	SW	29·63	29·54	54	34	—	4
			30	W	29·68	29·45	53	39	11	
			31	N	29·91	29·68	48	35	12	
					30·25	29·03	69	33	3·08	1·17

NOTES.—*a.* Calm clear days, with *Gossamer* and *Stratus* by night.
b. b. Windy.
c. The swallows were not seen after this day.
d. d. d. Much dew.
e. A very large and perfect lunar halo.
f. f. Stormy nights.

Additional Notes, &c.—Tenth Mo. 1. This morning's tide (at the Laboratory, Stratford, situate on the River Lea,) was remarkably high, the water rose 18 inches above the usual height of spring tides. The wind yesterday morning was very strong at the W and NW, the barometer rising 7 tenths of an inch in 24 hours. 27. Rainy morning, preceded by a thick fog.

From the Journals of the French National Institution, " M. Messier has collected all the particulars he could of the violent thunder storm that occurred at Paris, on the 21st October, 1807, and the extraordinary gale of wind that came on the next day. In his Journal, which he has kept for 50 years, he finds nothing equal to them: but on the 3d November following, there was a storm as violent, during which the church of Montevilliers was struck by lightning."

Barometer at Paris on the 19th at noon 30·29 inches (English), on the 22d 29·31, on the 23d in the morning 29·12 : this was a very wet day there, with the wind strong at S ; while with us, *with the same state of the barometer*, it was N and fair. A lower state of both barometer and thermometer ensued, with them as with us, for some time after.

Rain in this month at Manchester, 2·37 in.   Mean of barometer 29·65 in.

Hanson.

RESULTS.

Winds Variable.

Mean height of barometer ... ... 29·84 in.
Thermometer........................ 53°
Evaporation .... 3·08 in.
Rain .... 1·17 in.

Of which there fell by night 0·38 inches, the remainder was not divided.

TABLE XIII.

1807.			Wind.	Pressure. Max.	Min.	Temp. Max.	Min.	Evap.	Rain, &c.
11th Mo.		Nov. 1		29·90	29·66	55°	43°	—	
	a.	2	SW	29·66	29·20	53	39	—	
		3	SW	29·30	29·21	47	37	36	17
		4	W	29·56	29·30	49	33	17	
	b.	5	SW	29·43	29·31	52	36	7	
		6	W	29·48	29·20	49	41	13	47
1st Q.		7	SW	29·26	29·14	50	41	9	34
		8		29·35	29·16	46	32	—	
		9		29·47	29·35	44	33	—	
		10		29·40	29·01	47	33	—	
	c.	11	NW	29·81	29·03	42	28	28	61
	c.	12	NW	30·01	29·81	37	29	4	
		13	NW	30·04	29·99	37	31	1	4
	d.	14	N	30·03	29·98	41	30	2	1
Full M.		15	NE	29·98	29·90	44	35	11	
		16	NE	29·91	29·90	43	39	8	6
		17	Var.	29·90	29·60	43	37	—	
		18	Var.	29·60	29·44	40	30	4	
	e.	19	Var.	29·44	28·69	45	32	—	49
	f.	20	SW	29·32	28·68	41	30	19	
		21	SW	29·54	29·32	37	25	—	
L. Q.	g.	22	E	29·54	29·17	45	30	5	48
		23	S	29·19	29·06	48	31	4	16
		24	SW	29·42	29·19	37	32	—	
		25	W	29·48	29·42	42	29	11	
		26	W	29·61	29·48	38	27	1	
		27	NW	29·82	29·61	33	22	—	
		28	NW	29·82	29·65	31	23	5	
				30·04	29·01	55	22	1·85	2·83

NOTES.—a. A very stormy night. About 3 a. m. a sudden violent gust, with hail, after which the wind fell.

b. Hoar frost.

c. c. A little snow. The water in the evaporation gauge having been frozen, a small quantity of salt was now introduced to prevent its forming a solid mass.

d. Misty a. m.

e. Snow, in considerable quantity, for the season; which was dissolved by rain in the course of the day.

f. A smart shower of hail, followed by heavy rain, a. m.; fair p. m.; wind very high the whole day.

g. The earth was much hardened this morning by the frost, though the depression of temperature below 32° was not considerable. This is to be ascribed to the previous cooling it received from the melted snow.

RESULTS.

Mean height of barometer .... 29·47 in.
Temperature .... 37·92°
Evaporation.. .... 1·85 in.
Rain, &c....... .... 2·83 in.

Of which by night 1·59 in.   By day 0·16 in.

The barometer has ranged for the most part below the mean, though the middle of the month is distinguished by a pretty bold curve in elevation. The fore part of the period was almost constantly windy; the latter very cold, with hoar frosts; the greatest depression of temperature being after sunrise.

The mean of the barometer at *Manchester* for this month, (of which our period takes in 28 days) is 29·46 inches, the mean temperature 38·20.—*Hanson.* The extremes being *52°* and *19°*, (each 3° lower than ours), I suspect the mean here stated to be 3° too high, for want of a sufficient proportion of the lowest observations. The *rain*, as before, greatly exceeds ours, being very nearly 5 inches.

TABLE XIV.

	1807.		Wind.	Pressure. Max.	Min.	Temp. Max.	Min.	Evap.	Rain, &c.
11th Mo.	New M. *a.*	Nov. 29	N	29·72	29·65	33°	30°	—	—
	b.	30	N	29·95	29·72	36	30	1	
12th Mo.		Dec. 1	Var.	29·95	29·84	39	32	1	
	c.	2	SW	29·91	29·84	44	29	4	
		3		29·96	29·87	42	32	—	
	c.	4	W	29·87	29·70	47	38	—	
1st Q.	*c.*	5	SW	29·70	29·46	47	35	—	—
	a.	6	W	29·55	29·48	41	27	10	
	d.	7	N	29·59	29·55	35	20	—	
	e.	8	W	29·76	29·50	33	19	8	16
		9	NW	30·01	21·76	34	21	—	
	b.	10	NW	30·01	30·00	32	17	4	
	f. b.	11	NW	30·20	30·01	38	32	—	
	b.	12	W	30·25	30·20	44	31	—	—
Full M.	*b.*	13	W	30·24	30·13	43	36	6	
	g.	14	Var.	30·20	30·16	44	29	2	
	b.	15	W	30·16	30·06	43	34	2	
		16	Var.	30·07	30·06	37	30	0	
	b.	17	E	30·06	30·02	36	32	—	
	b.	18	E	30·09	30·00	37	33	3	
	b.	19	SE	30·30	30·09	39	28	0	
	b.	20	Var.	30·41	30·30	31	28	0	
	b.	21	NW	30·40	30·29	30	26	0	
L. Q.	*h.*	22	W	30·29	30·20	36	28	0	
	b.	23	SW	30·20	30·18	36	31	—	
		24	SW	30·18	30·12	39	35	4	3
		25	SW	30·12	29·93	48	39	4	
	c.	26	SW	29·93	29·72	54	45	—	
		27	SW	29·95	29·71	46	33	10	2
	i.	28	SW	29·71	29·55	48	43	9	
				30·41	29·46	54	17	0·68	0·21

NOTES.—*a.* Snow at intervals.

b. Misty.

c. Windy.

d. a. m. hoar frost. p. m. large lunar halo.

e. A fall of snow, unusually deep for this part of the island: during it the thermometer was three times noted in gradual ascent, *viz.* at 10 a. m. 19°; at 3 p. m. 22°; at 9 p. m. 26°: it amounted on the plain to about 6 inches: to the small depth which it exhibits in the rain-gauge should be added near as much more, for loss by evaporation, and spilling, before the column was melted down into the funnel.

f. A little snow very early, followed by a thaw and cloudy weather.

g. Hoar frost, which speedily went off: clear morning.

h. The melting rime fell in showers from the trees, and the air cleared up.

i. The clouds highly coloured at sunrise, indicating wind; the evening proved stormy.

METEOROLITES IN NORTH AMERICA.

Greenfield, Massachusetts, December 19, 1807.—On Monday morning last, the 14th inst. in the vicinity of this place, several bodies of stone were discovered, which appear to have descended from the regions above.—Several pieces of this stone were shown me by different persons, by whom the fact was so well attested as to make it impossible altogether to disbelieve it. But being resolved to get the best evidence of such an extraordinary occurrence, which the nature of the case could admit, I devoted this day, in company with the Rev. Mr. Holly, in visiting the different places where the stones had fallen.

The first place we visited is about three miles and a half in a north-easterly direction from my house, in a lot firmly covered with grass, about 25 rods from the house of Elijah Seeley. The breach here made in the ground was about four feet diameter, and nearly the same depth, in a rather sloping direction, which was occasioned by the stone striking a shelly rock and glancing. The rock on which the stone fell was much shattered, and the stone itself very much broken, the largest pieces weighing not more than six or eight pounds; the quantity altogether about a bushel. A quart or two of these fragments we gathered here; the greater part having been previously carried away by the inhabitants. By the fall and glancing of the stone, the dirt and sod were strewed two or three rods round the breach, and several pieces of sod carried before the fragments to the lowest depth to which they sunk into the earth, and were removed by myself. Mr. Seeley and his wife say that just after day-light, they saw vivid flashes of light in rapid succession for five or six seconds; and in about a minute afterwards it was followed with a dreadful explosion resembling three cannons fired in quick succession, ending in a cracking rumbling noise; that about ten o'clock the same morning, going into this lot, he discovered the breach in the ground above described, and conceiving it to have been caused by something discharged from above, at the time he heard the explosion, he called his wife out to witness the facts; and in the course of the day it was visited by all the neighbourhood. Mr. Seeley and his wife are sober, discreet, and intelligent persons, implicitly to be relied on.

The next place we went to view was about four miles N E from the first, in the court yard of Mr. William Prince, a respectable and wealthy farmer. The court yard is a grass plot, smooth as a carpet, and firmly trodden. Here we found a hole, about the size of a post hole, two feet two inches in depth, from which had been taken, on the evening of the aforesaid Monday, a stone weighing thirty-five pounds, the texture and appearance of which resembled exactly the one which fell at Seeley's. Mr. Prince and his wife and sons (men grown) give the same account of the flashes and explosion as was given above, with this further particular, that they heard, about a minute after the explosion, the fall of the stone, at the noise of which they were much alarmed, but could not discover the cause. After it was light Mr. Prince went out, and passing across the door yard discovered a hole in the grass plot, only twenty-seven feet from his house. The ground appeared fresh broken, and no dirt thrown out. He looked into it, but could see nothing, and no further discovery was made until evening, when his sons returned from a town meeting at Weston, where they heard of the fall of the stone at Seeley's. This induced them to examine further the hole in the court yard. On hauling out the dirt which lay loosely over the stone, they soon discovered it, and took it out entire, except some small pieces, that were broken off by stones in the ground. We examined this hole, and found the sod and grass as in the other case, driven before the stone to the bottom of the hole, which we took up, with pieces of the stone that had not before been found. The hole was perpendicular in the earth, and in diameter no larger than the stone. A Mr. David Hubbell, a man of undoubted veracity, was passing in the street about 25 rods from this stone when it fell, who saw a ball of fire, emitting sparks, with a tail about four feet long, shoot across the horizon in a southerly direction, and in about a minute or two afterwards he heard the explosion, which he described as the others had done; and a minute or two after that, he heard a loud whistling through the air, which made a noise like a hurricane. The same appearance and explosions were witnessed by Judge Wheeler and Russel Tomlinson, who were ten miles distant from each other; men of great candour and careful observation; but being distant from the places where these stones fell, neither of them heard the whistling just mentioned. The largest piece of the stone taken out of Prince's yard, which remains entire, weighs about eleven pounds, and is now in my possession; the rest of it has been broken into small parts and scattered amongst the inhabitants.

The third and last place where these stones have been discovered to have fallen, is about five miles north-east of Mr. Prince's, and seven below Newtown, near the turnpike road which leads from thence to Bridgeport.

The stone which fell there was small, and falling upon the top of a rock, that projected two feet above the ground, was dashed into small fragments, none of which weighed more than four or five ounces; and it was judged the whole of these collected would not more than have filled a quart measure.—This stone fell about thirty yards from the house of Mr. Merwin Burr; he and his wife being up, they both ran to the door the moment they discovered the flashes of light, and in a minute or two heard the same explosion as heretofore described; and in about a minute after the explosion they heard something fall near by them which made quite a loud report; and in quick succession three or four other noises at greater distance, which they thought to be something falling in a swamp, 20 or 30 rods in front of the house. Mr. Burr took a candle, and with his wife went out immediately, to see if they could find any thing in the direction where they heard the loudest report, but found nothing; when it grew lighter, and before sunrise, Mr. Burr went again, and found the fragments of the stone which had been dashed to pieces on the rock, and which have precisely the same appearance as those found at the two former places; the swamp, being full of water, has not been explored. Besides the large piece, I have many smaller ones collected at these three different places by Mr. Holly and myself, exactly resembling each other; and from the mouths of all these witnesses I have named (except Mr. Burr, who happened to be from home) we have taken the facts just as I have here related them. The ball of fire and explosion were witnessed by hundreds in this and the neighbouring towns; and I myself, as I was returning from New-York in the stage, a little on this side of Rye, at the same hour in the morning, saw vivid flashes of light, which lasted four or five seconds of time; and though the curtains were down, the stage was perfectly illuminated, but we heard no report.

This stone, I presume, possesses considerable iron, as it is strongly attracted by the needle; and I judge it is one fourth heavier than the common granite. The outside is covered with a smooth, glazed, sooty crust, as thick as foolscap, looking like the back of a chimney; and the broken surfaces are of a blueish lead colour, the whole appearing to have undergone the action of intense heat.— The body of stone which fell at Seeley's must have weighed more than one hundred pounds.—*J. Bronson.*

Other accounts are given of the storm, and of the meteor here noticed, in communications from other parts of the country; they are however too much alike to be inserted here.—(ATHENÆUM.)

1

RESULTS.

Prevailing Winds Westerly.

Mean height of barometer . . . . 29·80 in.
Thermometer. . . . . 36·26°
Evaporation . . . . 0·68 in.
Snow and rain. . . . . 0·21 in.

Character of this period frosty, and, for the most part, calm; the atmosphere turbid and almost constantly depositing moisture.

At *Manchester*, there fell in this month 2·62 inches of rain, the mean of the barometer being 29·65 inches.—*Hanson.*

Extracts from a Paper by M. Cotte, in the Journal de Physique, tom. lxviii. p. 331, &c. inserted for the purpose of comparison with the corresponding dates in this Register.

1807.—*Feb.* 18. A dreadful gale with much snow on the French coast of the channel, in Flanders, Picardy, and Normandy; at Montmorency, near Paris, stormy, with snow and hail, the barometer exhibiting great variation. *March* 8. Barometer very low at Montmorency, a violent storm of wind at Mende, (in the S of France), and on *March* 30, an earthquake in the department of Puy-de-Dôme. *April* 13. Much damage by hail, (grêle desastreux) with thunder, in the Haute Garonne, S of France. Barometer low at Montmorency. Frost and snow in this month so far south as Naples, after which in France, as here, there followed excessive heat about the close of the month. *May* 26. Great hail at Tonneins. 29. Hurricane about Dijon, and in the departments to the N E: no wind near Paris, but the barometer low. *June* 6. Earthquake at Lisbon. 17. Thunder storm with very large hail at Antwerp. *July, August.* An extremely hot and dry season in France. *Sept.* 4, 5. Earthquake at Genoa and Naples. 7, 30. Violent storms of wind at the Hague: much wind at Paris. *Oct.* 1. An earthquake, with a storm of wind and thunder at Vienna, the barometer very low. Barometer high at Paris, with calm weather. *Nov.* 3. A great storm of wind and thunder at Montivillois, Lower Seine. 23. A furious hurricane with much rain at Bordeaux: for a week before and after which, there were frequent shocks of an earthquake at Algiers. Barometer low during this time at Paris.

TABLE XV.

	1807.	Wind.	Pressure. Max.	Min.	Temp. Max.	Min.	Evap.	Rain, &c.
12th Mo.	New M. *a.* Dec. 29	SW	29·56	29·37	53°	38°	—	—
	a. 30	SW	29·37	29·32	48	35	16	
1808.	*a.* 31	SW	29·32	29·24	46	42	—	
1ts Mo.	Jan. 1	S	29·24	28·93	47	39	22	28
	2	SW	29·23	28·96	44	31	4	37
	b. 3	SW	29·95	28·96	39	31	—	
	c. b. 4	SW	29·95	29·82	45	32	5	4
1st Q.	5	SW	30·22	29·95	48	39	—	
	b. 6	SW	30·42	30·22	47	39	—	
	7	SW	30·50	30·42	47	41	8	3
	8	SW	30·51	30·50	44	38	—	—
	f. 9	SW	30·50	30·41	45	38	—	
	10	SW	30·41	29·97	48	42	—	
	b. 11	NW	29·98	29·95	48	32	—	
	b. 12	NW	29.95	29·81	37	31	—	
Full M. *d.*	13	W	29·81	29·21	45	37	16	
	a. 14	SW	29·98	29·16	46	29	7	2
	b. 15	NW	30·16	29·98	33	23	—	
	16	NW	30·36	30·16	31	23	—	
	17	NE	30·51	30·36	34	21	—	
	18	N	30·51	30·26	35	24	—	
	19	SW	30·26	29·66	40	34	13	2
L. Q. *e.*	20	NW	29·82	29·66	40	20	—	
	e. 21	NW	30·19	29·82	27	12	—	
	g. 22	SW	30·19	30·05	31	14	—	—
	23	SW	30·05	29·87	38	30	9	1
	f. 24	Var.	29·87	29·45	37	31	—	
	f. 25	Var.	29·45	29·25	34	24	—	
	g. b. 26	NW	29·51	29·25	33	24	5	
	g. a. 27	SW	29·37	29·25	46	27	—	—
	h. 28		29·51	29·25	43	33	8	22
			30·51	28·93	53	12	1·13	0·99

NOTES.—*a.* Strong winds: the night of 31 stormy, and that of 14 tempestuous.

b. Fine clear days.

c. a. m. (and 26th after sunset) a little opake hail.

d. Cirrostratus highly coloured at sunset.

e. Snow at different times.

f. Misty weather.

g. Hoar frost.

h. A smart shower mixed with hail at night.

On the 23d, at 10 a.m. a faint but nearly perfect rainbow appeared under singular circumstances. The night had been clear and frosty, and at this time a few light clouds had began to appear in different quarters, but none over the place of the bow, nor was the falling mist that afforded it of sufficient density to obscure the sky. Precipitation after this went on rapidly ; in 30 or forty minutes the sky was overcast, and before noon it rained a little. On the 26th, in the midst of calm sunshine, wind NW, a patch of *Cirrostratus*, in the form of a *cyma recta*. The wind changed to SW, and blew pretty strong on the 27th, with rain.

Additional Notes, &c.—First Mo. 6. The morning was beautifully clear, with *Cirrocumulus* and *Cirrostratus*, stretching from N to S. 14. The day was very stormy: all night the wind blew quite a tempest, the barometer rising rapidly. The weather appears to have been equally tempestuous this day at Plymouth. 23. After a little rain, a rapid thaw, continuing through the following day.

RESULTS.

Prevailing Winds Westerly and chiefly South-west.

Mean elevation of barometer . . . . 29·82 in.
Thermometer . . . . 36·98°
Evaporation . . . . 1·13 in.
Snow and rain . . . . 0·99 in.

Of which by day 0·59, by night 0·28 in.

TABLE XVI.

1808.		Wind.	Pressure. Max.	Pressure. Min.	Temp. Max.	Temp. Min.	Evap.	Rain, &c.
1st Mo.	New M. *a.* Jan. 29		29·70	29·51	42°	37°	—	
	30	S	29·74	29·62	50	40	12	9
	31	SW	29·83	29·74	51	41	—	
2d Mo.	*a.* Feb. 1		29·81	29·63	52	47	19	2
	b. 2		29·74	29·56	51	37	—	—
	c. 3		30·25	29·74	43	30	13	1
	d. 4	W	30·31	30·17	42	30	10	
1st Q.	*e.* 5	SW	30·17	29·94	49	35	5	
	e. 6	W	29·94	29·79	50	44	14	3
	f. 7	W	29·86	29·79	48	37	8	28
	8	W	29·79	29·74	39	29	—	—
	9	NW	29·99	29·79	38	27	—	
	10	NW	30·10	29·95	34	25	—	
	g. 11	SW	29·95	29·20	47	29	—	—
	h. 12	Var.	29·69	29·24	31	23	—	—
Full M.	13	N	30·01	29·69	30	19	—	
	i. 14	N	30·18	30·01	29	17	—	
	g. 15	SW	30·01	29·89	37	23	21	29
	16	NW	30·02	29·89	39	26	—	
	17	W	30·02	29·99	41	30	5	
	18	NW	30·22	30·02	43	30	4	
	k. 19	NE	30·49	30·22	42	26	—	
L. Q.	*k. l.* 20	E	30·53	30·49	39	24	—	
	k. 21	NE	30·53	30·50	39	29	—	
	k. 22	NE	30·50	30·45	40	32	—	
	23	NE	30·45	30·40	38	32	32	1
	24	NE	30·71	30·45	39	28	—	
	25	E	30·71	30·61	37	24	20	
			30·71	29·20	52	17	1·63	0·73

Notes.—*a.* At 7 h. 20 m. p. m. a brilliant meteor in the east. It was a small bright blue mass of light, with a short red train. It appeared suddenly, at a moderate elevation, and after descending a few degrees, with a small tendency northward, became extinct.

b. a. m. stormy with rain.

c. The skylark sings. About noon a warm electrical shower, giving plenty of small sparks from the rod.

d. Hoar frost, a. m. Electric signs again, from a *Nimbus* passing in the south. The blackbird sings. A small bright lunar halo.

e. e. Strong winds.

f. Lunar halo, consisting of a coloured circle near the moon, and a pale one much more distant.

g. Snow at intervals.

h. The wind, which was full south in the night with rain, blew furiously from the north all day, with a continued fall of snow as fine as dust, which was at length drifted to the depth, in some places, of three or four feet.

i. The sky this evening being uncommonly serene, I suspected a strong evaporation of the snow to be going on. A Six's thermometer was therefore placed horizontally upon it, in a place exposed to the wind. It descended to 6·5°, or 10·5° lower than the standard instrument at five feet elevation. The morning of 15th was cloudy, with rime on the trees.

k. Hoar frost.

l. The clouds beautifully coloured at sunset.

Additional Notes, &c.—Second Mo. 3. The clouds to-day and yesterday, in the intervals of the showers, assumed the bold and compound structure usual in summer.

RESULTS.

Winds Westerly to the last quarter—afterwards Easterly.

Mean height of barometer.............. 30·02 in.
Temperature .... 35·91°
Evaporation .... 1·63 in.
Rain and snow..................... 0·73 in.

Character of this period, variable and frosty.

Rain by night 0·28: remainder by day and night.

" *Manchester.* On Thursday the 11th Feb. it was remarkable tempestuous, particularly in the evening, with a great fall of snow, which was much drifted. The temperature was under the freezing point; and the barometer, during the day, fell four-tenths of an inch. On the 21st, about seven o'clock in the evening, a luminous body was seen descending in a SW direction.—*Hanson*."

TABLE XVII.

1808.			Wind.	Pressure. Max.	Pressure. Min.	Temp. Max.	Temp. Min.	Evap.	Rain, &c.
2d Mo.	New M.	Feb. 26	Var.	30·61	30·43	39°	30°	9	
		27	W	30·43	30·41	48	33	6	
		28	NW	30·41	30·14	50	39	—	
		29	NW	30·31	30·14	51	43	9	4
3d Mo.		March 1	NW	30·36	30·31	52	45	4	
		2	NW	30·35	30·35	51	42	6	
		3	NW	30·43	30·34	48	30	5	
	a.	4	E	30·46	30·43	54	35	7	
1st Q.	b.	5	E	30·46	30·46	46	30	11	
	b.	6	NE	30·46	30·37	46	31	19	
		7	NE	30·37	30·33	42	32	22	
		8	NE	30·36	30·31	41	31	14	
	c.	9	E	30·36	30·35	43	32	11	1
	c.	10	E	30·42	30·36	40	30	17	
	c.	11	NE	30·40	30·32	40	32	13	
Full M.		12	NE	30·32	30·22	40	35	5	
	e.	13	E	30·22	30·14	43	35	11	
	d.	14	NE	30·14	30·02	42	34	13	
		15	NE	30·02	30·00	45	32	13	
	b.	16	NE	30·02	30·02	44	27	—	
	d.	17	NE	30·02	30·00	37	28	—	
	d.	18	NE	30·00	29·79	35	29	—	
L. Q.	d.	19	SE	29·79	29·55	42	32	36	7
	f.	20	E	29·55	29·70	45	37	0	7
	f.	21	NE	29·90	29·70	41	36	6	
		22	NE	29·96	29·90	43	28	13	
	b.	23	NE	29·93	29·93	41	31	15	
	g.	24	E	29·95	29·91	39	18	—	6
		25	NE	29·95	29·92	43	23	12	
				30·61	29·55	52	18	2·77	0·25

NOTES.—a. Very misty a. m. with hoar frost. Wind W.

b. Hoar frost.

c. Lunar halo.

d. Snow in small quantities.

e. A little rain a. m. At 2 p. m. a slight shower of hail.

f. Misty a. m.

g. A smart shower of snow, covering the ground in a few minutes.

RESULTS.

Prevailing Winds Easterly.

Mean height of barometer . . . . 30·16 in.

Temperature . . . . 38·11°

Evaporation . . . . 2·77 in.

Rain and snow . . . . 0·25 in.

The atmosphere during this period has preserved, with remarkable uniformity, the character peculiar to the season ; dry, dense, and clear below, though mostly cloudy above. From the rules laid down by Kirwan, the probability seems to be as five to one in favour of a dry summer. Rain noted by day 0·04, by night 0·08 in.

The rules above alluded to are stated to have been deduced from a series of observations made from 1677 to 1789, and are as follows:

1. When there has been *no storm* before or after the vernal equinox, the ensuing summer is generally *dry*, at least five times in six.

2. When a storm happens from an easterly point on the 19th, 20th, or 21st of March, the succeeding summer is *dry*, four times in five.

3. When a storm arises on the 25th, 26th, or 27th of March, and not before, in any point, the succeeding summer is generally *dry*, four times in five.

4. If there should be a storm at SW, or WSW, on the 19th, 20th, or 22d of March, the succeeding summer is generally *wet*, five times in six.

" *Dry* summers (this philosopher states) are the consequence of *uniform* winds, from whatever quarter they may blow ; as *wet* summers are of their *variation*, particular if in opposite directions." Again, " Southerly winds are most frequently accompanied with rain in most parts of Europe at least, and probably in most parts of our hemisphere ; but northerly and easterly, with clear, dry, and serene weather." And it seems reasonable to suppose that the wind which is to prevail during the summer, may most frequently set in with the vernal equinox.

TABLE XVIII.

1808.		Wind.	Pressure. Max.	Pressure. Min.	Temp. Max.	Temp. Min.	Evap.	Rain, &c.
3d Mo.	a. March 26	NE	30·05	29·95	47°	34°	6	
New M.	27	NE	30·06	30·05	38	32	—	
	28	NE	30·14	30·06	40	27	—	
	29	NE	30·12	30·05	45	28	—	
	30	NE	30·05	29·98	39	32	—	
	31	N	29·98	29·78	41	25	64	
4th Mo.	April 1	NW	29·84	29·67	41	22	—	
	2	NW	29·91	29·84	42	27	—	
b.	3	SW	29·91	29·72	51	37	34	
1st Q. c.	4	SW	29·72	29·37	53	44	13	6
c.	5	SW	29·58	29·26	53	48	—	—
c.	6	SW	29·79	29·58	54	50	27	7
c.	7	SW	29·80	29·63	56	38	8	1
d.	8	NW	30·24	29·80	52	33	9	6
	9	NW	30·24	30·24	51	38	9	
Full M. c.	10	N	30·24	30·24	55	45	10	
c.	11	W	30·24	30·05	54	42	12	—
c.	12	NW	30·05	30·29	51	36	8	2
e.	13	NW	30·29	30·19	60	39	—	
e.	14	W	30·19	30·05	66	39	30	
	15	N	30·06	30·00	63	31	17	
f.	16	Var.	30·13	30·06	52	30	12	
L. Q.	17	Var.	30·11	30·04	49	29	12	1
g.	18	Var.	30·04	29·61	46	32	11	
c. h.	19	Var.	29·62	29·57	46	25	17	2
c.	20	SW	29·57	29·14	49	36	—	59
	21	SW	29·44	29·09	47	34	20	
c.	22	W	29·41	29·40	54	36	14	7
	23	Var.	29·58	29·41	48	35	5	11
	24	N	29·87	29·58	47	33	8	7
			30·24	29·09	66	22	3·46	1·09

Notes.—*a.* Snow.

b. The atmosphere, which has been long serene, begins to grow turbid.

c. Windy, sometimes amounting to a gale.

d. A shower at sunset, with *negative* electricity.

e. Misty morning.

f. The swallows made their first appearance this morning.

g. a. m. a strong *positive* electricity in the clear atmosphere.

h. p. m. much snow, which gave *positive* signs till nearly over, and then *negative*. The rain from the 17th to the end was always found

NOTES, &c. *(continued)*.

electrified variably, often so as to afford moderate sparks. All the showers were mixed with hail. On the 23d, p. m. after a sudden jet of fire from the insulated to the uninsulated part of the conductor, there followed thunder pretty far to the east and south-east. The cuckow had been heard this morning.

Additional Notes, &c.—Fourth Mo. 15. *Cirrostratus* in large patches, which came down towards evening with an appearance as for thunder, the air being somewhat electric. 19. The flakes of snow this afternoon were unusually large. 20. Heavy rain in the evening.

RESULTS.

Wind and Character Variable. Hoar Frost has been frequent.

Mean height of barometer.............. 29·86 in.
Temperature ... . 41·82°
Evaporation (in 29 days) .... 3·40 in.
Rain, &c.......................... 1·09 in.

Rain noted as fallen by day 0·20, by night 0·72 in.

TABLE XIX.

1808.		Wind.	Pressure. Max.	Pressure. Mib.	Temp. Max.	Temp. Min.	Evap.	Rain, &c.
4th Mo.	New M. *a.* April 25	NW	29·91	29·87	43°	33°	5	6
	26	N	29·91	29·90	47	34	8	41
	27	N	29·90	29·87	42	36	6	
	28	N	29·90	29·87	46	35	9	
	29	W	29·90	29·88	45	32	10	
	30	Var.	29·95	29·90	52	39	8	
5th Mo.	*b.* May 1	Var.	29·93	29·91	65	38	13	
	c. 2	E	29·91	29·81	66	50	32	
1st Q.	3	E	29·81	29·77	79	45	28	
	d. 4	NE	29·80	29·77	80	44	30	
	e. d. 5	Var.	29·77	29·75	72	53	37	
	f. e. d. 6	S	29·75	29·59	78	52	21	
	g. 7	SE	29·59	29·52	75	51	24	3
	h. 8	Var.	29·54	29·52	63	45	11	7
	9	SW	29·77	29·52	57	44	10	12
Full M.	10	SW	29·90	29·71	62	48	16	2
	i. 11	SW	30·13	29·90	64	46	12	3
	12	SW	30·24	30·13	64	53	12	
	k. 13	SW	30·24	30·15	73	55	18	
	14	W	30·15	30·11	78	54	21	
	15	S	30·11	29·95	83	60	33	
	16	SW	30·01	29·95	79	55	33	
L. Q.	17	Var.	30·04	29·98	80	51	22	
	18	NE	30·21	30 04	61	37	—	
	19	NE	30·21	30·02	63	38	24	
	20	Var.	30·02	29·77	66	52	—	—
	21	N	29·97	29·58	66	50	27	40
	22		29·62	29·54	69	51	10	30
	23		29·80	29·62	66	46	10	11
			30·24	29·52	83	32	4·90	1·55

NOTES.—*a.* The nightingale sings.

b. Negative electricity pretty strong the whole day.

c. Lightning about sunset.

d. Much dew.

e. Lunar halo.

f. The evening twilight at nine was brilliant in the highest degree, casting a strong shadow into the light of the moon, though the latter was near the meridian.

g. a. m. rainbow, with little appearance of cloud. About 5 p. m. wind E, heavy showers with lightning to the west, and the bow again twice. Sparks of *negative* electricity from the extremity of one shower. After this, from 8 to 10 p.m. wind SE. A thunder storm passed in the E from S to N. Three distinct *Nimbi* were perceptible on the horizon, illuminated by continual discharges, the sound of which did not reach us: (nor did their electricity in the least affect the insulated conductor.—*From MS. Reg.*)

h. At 8. a.m. a steady rain, non-electric; then showers, with a changeable electricity. The master of a small vessel, whom I met with on the 9th, informed me that, being in the Channel, about 90 miles E of my residence, he had the above storm for several hours as far to the *west* of him. He also was nearly out of hearing of the thunder, but had a dry squall at 9 p.m. so violent as to oblige him to strike all his canvas. This storm, therefore, though it extended far to the north, passed us in a column not exceeding 30 or 40 miles in width, following the coast and the hills.

i. Squalls with rain: a fine bow p.m.

k. Hazy atmosphere, with abundance of the *Cirrostratus*.

Additional Notes, &c.—Fifth Mo. 1. The pith-balls of the insulated conductor diverged one-eighth of an inch all day: there were few clouds, chiefly nascent *Cumuli.* 2. A smart breeze and clear sky till afternoon, when a very few drops fell. 5. The wind veered from E by S. 8. The last shower gave sparks *negative,* the charge continuing for some time after it: at length after a sudden *positive* charge the air began to clear. 11. One of these *Nimbi* gave during its approach first a weak *positive,* then a strong *negative* charge: the rain, when it came, was first weakly *positive,* then non-electric: the weight of it fell to the NE of us. 14. a.m. *Cirrocumulus.*

The following account of a new Volcano in one of the Azores has lately been given to the public by the American Consul at St. Antonio.

Fayal, (Azores) June 25, 1808.

" A phenomenon has occurred here not unusual in former ages, but of which there has been no example of late years; it was well calculated to inspire terror, and has been attended with the destruction of lives and property. On Sunday, the 1st of May, at one p.m. walking in the balcony of my house at St. Antonio, I heard noises like the report of heavy cannon at a distance, and concluded there was some sea engagement in the vicinity of the island. But soon after, casting my eyes towards the island of St. George, ten leagues distant, I perceived a dense column of smoke rising to an immense height; it was soon judged that a volcano had burst out about the centre of that island, and this was rendered certain when night came on, the fire exhibiting an awful appearance. Being desirous of viewing this wonderful exertion of nature, I embarked on the 3d of May, accompanied by the British Consul, and ten other gentlemen, for St. George; we ran over in five hours, and arrived at Vellas, the principal town, at eleven a.m. We found the poor inhabitants perfectly panic struck,

and wholly given up to religious ceremonies and devotion. We learned that the
fire of the first of May had broken out in a ditch, in the midst of fertile pastures, three leagues SE of Vellas, and had immediately formed a crater, in
size about twenty-four acres. In two days it had thrown out cinders or small
pumice stones, that a strong NE wind had propelled southerly; and which, independent of the mass accumulated round the crater, had covered the earth
from one foot to four feet in depth, half a league in width, and three leagues in
length; then, passing the channel five leagues, had done some injury to the
east point of Pico. The fire of this large crater had nearly subsided, but in
the evening preceding our arrival, another small crater had opened, one league
north of the large one, and only two leagues from Vellas. After taking some
refreshment, we visited the second crater, the sulphureous smoke of which,
driven southerly, rendered it impracticable to attempt approaching the large
one. When we came within a mile of the crater, we found the earth rent in
every direction, and, as we approached nearer, some of the chasms were six
feet wide: by leaping over some of these chasms, and making windings to avoid
the larger ones, we at length arrived within two hundred yards of the spot; and
saw it, in the middle of a pasture, distinctly at intervals, when the thick smoke
which swept the earth lighted up a little. The mouth of it was only about
fifty yards in circumference; the fire seemed struggling for vent; the force with
which a pale blue flame issued forth resembled a powerful steam-engine, mul-
tiplied a hundred fold; the noise was deafening: the earth where we stood had
a tremulous motion, the whole island seemed convulsed, horrid bellowings were
occasionally heard from the bowels of the earth, and earthquakes were frequent.
After remaining here about ten minutes we returned to town; the inhabitants
had mostly quitted their houses, and remained in the open air, or under tents.
We passed the night at Vellas, and the next morning went by water to Ursulina, a small sea-port town, two leagues south of Vellas, and viewed that part
of the country covered with the cinders before-mentioned, and which have turned
the most valuable vineyards in the island into a frightful desert. On the same
day (the 4th of May) we returned to Fayal, and on the 5th and succeeding
days, from twelve to fifteen small volcanos broke out in the fields we had tra-
versed on the 3d, from the chasms before described, and threw out a quantity
of lava, which travelled on slowly towards Vellas. The fire of those small
craters subsided, and the lava ceased running about the 11th of May; on which
day the large volcano, that had laid dormant for nine days, burst forth again
like a roaring lion, with horrid belchings, distinctly heard at twelve leagues
distance, throwing up prodigious large stones and an immense quantity of
lava, illuminating at night the whole island. This continued with tremendous
force until the 5th of June, exhibiting the awful yet magnificent spectacle of a
perfect river of fire (distinctly seen from Fayal!) running into the sea. On
that day (the 5th) we experienced that its force began to fail, and in a few
days after it ceased entirely. The distance of the crater from the sea is about
four miles, and its elevation about 3,500 feet. The lava inundated and swept
away the town of Ursulina, and country-houses and cottages adjacent, as well
as the farm-houses, throughout its course. It, as usual, gave timely notice of

its approach, and most of the inhabitants fled; some few, however, remained in the vicinity of it too long, endeavouring to save their furniture and effects, and were scalded by flashes of steam, which, without injuring their clothes, took off not only their skin but their flesh. About sixty persons were thus miserably scalded, some of whom died on the spot, or in a few days after. Numbers of cattle shared the same fate. The Judge and principal inhabitants left the island very early. The consternation and anxiety were for some days so great among the people, that even their domestic concerns were abandoned, and, amidst plenty, they were in danger of starving. Supplies of ready-baked bread were sent from hence to their relief, and large boats were sent to bring away the inhabitants, who had lost their dwellings. In short, the island, heretofore rich in cattle, corn, and wine, is nearly ruined, and a scene of greater desolation and distress has seldom been witnessed in any country."—(ATHENÆUM).

RESULTS.

Winds Variable.

Mean height of barometer .... 29·87 in.
Temperature .... 55·18°
Evaporation .... 4·90 in.
Rain, &c.......................... 1·55 in.

The current month has been highly favourable to vegetation, which had been long retarded by the low temperature. Rain noted by day 0·52, by night 0·07 in.

TABLE XX.

1808.			Wind.	Pressure.		Temp.		Evap.	Rain, &c.
				Max.	Min.	Max.	Min.		
5th Mo.		May 24		29·90	29·80	66°	49°	—	
	New M.	25		29·88	29·70	77	56	—	
		26		29·70	29·65	65	52	—	—
	a.	27	SW	29·94	29·70	63	47	38	20
		28	SW	30·15	29·94	68	44	16	
	b.	29	SW	30·17	30·15	68	59	9	
		30	SW	30·15	29·97	76	52	16	12
		31	Var.	29·97	29·78	87	59	16	
6th Mo.	1st Q. June	1	SW	29·99	29·79	63	44	8	3
		2	SW	29·98	29·88	70	47	20	
		3	Var.	29·88	29·68	65	47	—	3
		4	Var.	29·68	29·65	71	49	33	
		5	Var.	29·68	29·65	66	44	15	
		6	W	29·80	29·68	58	46	—	
		7	W	29·82	29·77	68	48	17	
	Full M.	8	W	29·77	29·65	65	51	16	2
		9		29·80	29·64	58	48	—	—
		10	NW	30·01	29·80	63	47	15	4
	c.	11	N	30·15	30·00	72	47	14	24
	d.	12	Var.	30·15	30·11	70	45	8	
	d.	13	SW	30·12	30·02	70	49	11	
		14		30·02	29·85	70	54	19	2
	L. Q.	15	NW	30·07	29·85	65	44	12	
	e.	16	NW	30·10	30·07	65	54	12	
		17	W	30·07	30·00	66	58	16	
		18	W	30·04	30·00	76	54	10	
		19	NW	30·04	30·00	76	54	11	
		20	Var.	30·00	29·95	72	56	14	
		21	E	29·95	29·78	71	56	14	
		22	SW	29·78	29·69	68	54	13	39
	f.	23	Var.	29·83	29·70	68	44	17	
				30·17	29·64	87	44	3·90	1·09

Notes.—*a.* Windy.

b. Much dew.

c. p. m. thunder in the NW.

d. Hazy atmosphere a. m.

e. No dew this morning.

f. The cuckow has not yet left us: the nightingale is still heard, but less constantly.   An extensive *Stratus* on the meadows this evening.

Additional Notes, &c.—The *Cirrostratus* in parallel waved bars made its appearance on the 26th of Fifth month: which day in other respects (a somewhat

2

low barometer excepted) indicated nothing against fair weather: on the 10th of Sixth month, the same prognostic was observed, with the barometer several tenths higher. In each case rain followed in some quantity. The rain on the 22d was preceded by a regular gradation of clouds, during twenty-four hours, from the *Cirrus* to a completely full sky; and it was followed by the *Stratus*. 23. a. m. *Cumulus* with *Cirrus*; the lower air remarkably clear.

Of the rain in this period 0·12 in. is noted as having fallen by night.

RESULTS.

Prevailing Winds Westerly.

Mean height of barometer............29·89 in.
Thermometer.......................59·41°
Evaporation........................ 3·90 in.
Rain, &c........................... 1·09 in.

This spring, though very late in its commencement, is remarkable for having proceeded through a period of more than seven weeks without a single frosty night, at a season when they are almost always experienced. Fruit of almost every kind has consequently succeeded well; apples in particular, which seldom amount to a crop in these parts, promise to be abundant.

TABLE XXI.

1808.			Wind.	Pressure. Max.	Min.	Temp. Max.	Min.	Evap.	Rain, &c.
6th Mo.	New M.	June 24	Var.	29·98	29·83	74°	48°	15	
		25	NE	30·03	29·98	75	47	11	6
		26	E	30·04	30·03	75	54	17	
		27	N	30·08	30·04	67	52	—	
		28	NE	30·13	30·05	61	52	23	
		29	NE	30·24	30·11	75	50	20	
		30	NE	30·24	30·15	70	49	18	
7th Mo.	1st Q.	July 1	NE	30·15	30·05	70	48	20	
		2	N	30·05	29·99	68	52	17	
		3	NE	29·99	29·97	69	44	12	17
	a.	4	N	29·97	29·88	70	47	19	
	b.	5	NW	30·08	29·95	66	47	15	
		6	W	30·12	30·08	67	56	—	
Full M.		7	W	30·08	30·01	75	52	33	
	b.	8	SW	30·01	29·95	75	59	24	—
	c.	9	NE	30·01	29·97	73	57	9	—
		10	SW	30·13	30·01	76	58	—	
	d.	11	SW	30·12	30·08	83	59	38	—
	e.	12	S	30·08	29·96	92	63	30	
	f.	13	S	29·96	29·93	96 !	60	35	
	g.	14	Var.	29·96	29·92	94	63	31	
L. Q.	h.	15	NE	29·92	29·91	81	62	27	—
	i.	16	Var.	29·97	29·91	88	59	20	—
	k.	17	NW	30·03	29·97	83	57	16	
	l.	18	E	29·97	29·90	86	54	25	
	l.	19	SW	29·90	29·75	86	59	25	
	m.	20	SW	29·78	29·70	75	55	17	
		21	SW	29·71	29·70	78	56	21	
		22	S	29·76	29·71	79	59	16	02
				30·24	29·70	96	44	5·54	0·25

Notes.—*a.* Misty, a. m.

b. b. Rather windy; a slight shower the 8th at night.

c. A little rain from the N a. m. and at sunset cloudy, with temperature 67°.

d. a. m. a few drops of rain, and much dew at night.

e. A very fine day. The *Cirrus* cloud only appeared. Temperature at 10½ a. m. 86°: the maximum 92° was about 2 p. m. with a fine breeze, so that the heat was not oppressive. Evaporation from a vessel on the ground between 1 and 2 p. m. 0·04 in.; between 4 and 8 p. m.

about the same quantity; in two hours after sunset there evaporated no sensible quantity of water, and dew fell.   About 11½ p. m. a bright small meteor passed from SW descending to W.

f. Dew on the grass.   Temperature at 9 a. m. 84°.   The intense heat of the *maximum* lasted nearly three hours, till about 4 p. m.   At 6 p. m. temperature 90°; after which it declined rapidly.   The thermometer is defended from the sun by a thick laurel tree, and subject to no other reflection than that of the grass-plot round it.   That at the laboratory rose (out of reach of the sun's rays) to 98·5°; but it is contiguous to a large building, and might be affected by a current of heated air from the roof or walls.   Another, at Plashet, a mile and a half eastward, indicated 96° as the *maximum* under the shade of a house.   Evaporation *in the shade* from 9 to 2, 0·1 in.; from 2 to 7, 0·19 in.   The sky was clear till near sunset; then appeared some haze, indicating the fall of dew, in the SE, and a few traces of thunder clouds in the NW.   The *vapour point* about 2 p. m. was 65°.

g. At 2½ a. m. several birds were singing by moonlight; the lark on wing and the chimney-swallow were distinguished.   The cuckow is said also to have been heard.   Sudden strong gusts of wind occurred between 6 and 7 a. m., and 2 and 3 p. m.   Some lightning in the W at night.

h. Dew on the grass: a fine breeze from ENE a. m.   Much lightning in the W this night: a few drops of rain p. m.

i. No dew: at sunset a smart breeze from SW, with lightning in the NE, and a few drops of rain.

k. Dew; little wind; evening twilight very brilliant, and the clouds highly coloured.

l. Dew.

m. A fine breeze from SSW.

Additional Notes, &c.—Sixth Mo.   29. *Cirri* p. m. pointing downward and to NE: at sunset a beautiful rose-coloured haze on the W horizon.

Seventh Mo.   2. At sunset a low group of *Cirrostratus* in the NW.   4. *Cumulostratus* p. m. and much *Cirrostratus* at sunset.   15. The evaporation in a sheltered spot near the ground was only 0·11, and on the 16th only 0·09 in.

Rain by day 0·06, by night 0·17 in.

RESULTS.

Prevailing Winds Northerly.

Mean elevation of barometer . . . . 29·97 in.

Temperature . . . . 65·60°

Evaporation . . . . 5·54 in.

Rain, &c. . . . . 0·25 in.

There has been very little indication of electricity by the rod; a few thunder clouds have appeared at intervals, which at no time threatened a discharge. The lightning we have perceived belonged to storms too distant for the clouds composing them to be visible above our horizon.

To the Editor of the Athenæum.—Sir, your correspondent L. H. takes notice that, on the 14th of July last, at half-past two in the morning, several birds were heard singing, and the lark and the chimney-swallow were observed upon the wing. Give me leave, therefore, to inform him, through the channel of your useful and entertaining miscellany, that on the 12th of the same month, at a quarter before three a. m. I heard the song thrush, *turdus musicus,* as well as the blackbird, *turdus merula,* at Hackney, in Middlesex. I remain, Sir, yours, &c.—S. R.

Gloucestershire.—On the night of Friday the 15th July, after several days of uncommon and oppressive heat, the city of Gloucester experienced a storm of thunder and lightning, which extended many miles round, and exceeded in awful phenomena any one remembered for many years past. Unlike the tempest of the milder zones, the thunder was remarked to roll in one continuous roar, for upwards of an hour and a half, during which time and long afterwards, the flashes of lightning followed each other in rapid and uninterrupted succession. But the most tremendous circumstance of this storm was the destructive *hail shower* which accompanied its progress. It may be doubted, however, whether such a name be applicable; for the masses of ice which fell on the places where the tempest most fiercely raged, bore no resemblance to hailstones in magnitude or formation, most of them being of a very irregular shape, broad, flat, and ragged, and many measuring from three to nine inches in circumference. They appeared like fragments of a vast plate of ice, broken into small masses in its descent towards the earth.—The storm rose in the south-west, and spreading to the north-west, gradually died away in the north-east, from which quarter it was opposed for nearly its whole duration by a strong breeze, particularly hollow and mournful in its sound. The damage done in different parts of this and other counties is very considerable. A fire ball burst in the College Green, carrying away one of the pinnacles on the west end of the cathedral, two cows were killed in a field at Sneedham Green, ten couple of ducks in the fold yard of a gentleman near Cheltenham, and twelve at Upton, near this city. A summer-house on an eminence, belonging to Edward Sheppard, Esq. of Uley, was entirely burnt. At Tewkesbury, many windows that lay in the direction of the storm have been broken, as well as the glass of the hot-houses, &c. in the gardens. At Tetbury, and in its neighbourhood, some houses were set on fire, and many panes of glass broken. Upwards of 600 panes were broken, in the house and garden of Mr. Cave. At Frenchhay, near Bristol, the orchards are stripped of their fruit, and the gardens of their plants. The greatest part of the windows, on the south side of the Broadway are broken, and the tops of the beans in the same neighbourhood have been cut off. The row of trees before Mr. Tucker's house at Moore-end, was so broken, that the leaves and small branches lay in the road a foot deep. At Newton, Corston, and Kelston, most of the windows that lay in the direction of the storm have been broken. The plantations and shrubberies of Mr. Langton, of Newton, were covered with leaves and branches of trees, and the pines, and other fruit in his hot-houses entirely destroyed. In the south and west fronts of Mr. Jollyfe's mansion at Amerdown, not a pane escaped, and the ground was even the next morning covered with the ice that fell. At Radstock, several fields of corn are nearly destroyed, the stalks being mostly cut off in the middle by the masses of ice. At Writhlington, near Radstock, very great damage has also been done to the growing crops. A boy belonging to Mr. Harding, of Keysham, was struck down by the lightning, and his recovery was for some time doubtful: a sheep which was near him was killed. All the glass in the gardens, and a great part in the house of Mills Park, were destroyed, and a great number of pine-apples, grapes, fruit trees, &c. were cut to pieces. A valuable horse belonging to Mr. Hyatt, of Shepton Mallett, was struck dead in the field. William Simkins, jun., mowing with two other men at Kilmarton, Wilts, was killed by the lightning, which also struck down his companions, but they recovered after a short time. In many parishes near Monmouth, roofs of houses, barns, and sky-lights were destroyed by the hail, boughs of trees were cut asunder, and the apples and pears scattered in such quantities that they might have been raked together in heaps. In the park of the Earl of Digby, near Sherborne, the limbs of a large oak tree were shivered in pieces, while the middle or heart was left standing; two sheep were killed under another tree. A flash of lightning struck the back part of a house in the lower town of Bridgenorth; carrying down a large proportion of the chimney, it descended into a bed-room over the kitchen, demolished the windows, and three doors in the room, and forced down the whole front of a large closet, splintering the wood in all directions; three children in bed escaped unhurt. The storm appeared to have spent its fury between Piper's Inn and Ashcot. Here, as well as at Glastonbury, the corn was laid flat, the roads inundated, the apple trees stripped not only of their fruit, but their very leaves, and almost every pane of glass in the village of Ashcot was broken.—(ATHENÆUM).

Thermometer at Paris, (reduced to Fahrenheit's Scale) with the Wind and State of the Sky.

1808.	Max.	Min.	Wind.	
7th Mo. July 10	82·6	55·1	NW	Cloudy.
11	86·5	59·5	E	Very fine: hazy at noon.
12	90·2	65·7	SE	Fine, with light clouds.
13	93·8	61·0	SE	Very fine day.
14	95·0	66·8	S	The same: hazy at noon.
15	97·2	70·0	SE	The same.
16	89·3	68·7	NW	Thick haze and clouds.
17	81·5	62·7	W	Cloudy.
18	93·8	74·0	Var.	Much cloud.
19	95·5	66·3	NE & SW	Cloudy: rain in the evening.
20	79·6	64·4	W	Cloudy, with rain.
21	76·1	61·7	S	The same.
22	79·6	61·2	SW & SE	Cloudy: fine at intervals.

TABLE XXII.

1808.			Wind.	Pressure, Max.	Min.	Temp. Max.	Min.	Evap.	Rain, &c.
7th Mo.	New M.	July 23	SE	29·76	29·74	85°	59°	19	01
		24	S	29·74	29·70	80	59	10	19
		25	NW	29·75	29·71	70	58	05	33
		26	NE	29·76	29·75	72	56	10	06
		27	E	29·75	29·44	75	58	10	1·88
		28	W	29·59	29·43	67	60	14	29
		29	SW	29·68	29·59	78	60	10	
	1st Q.	30	SW	29·75	29·68	76	55	14	—
		31	E	29·68	29·57	80	61	09	42
8th Mo.		Aug. 1	SW	29.74	29·52	72	57	11	14
		2	NW	30·06	29·74	73	56	18	
		3	W	30·06	30·04	65	54	08	
		4	SW	30·04	29·84	80	50	19	
		5	SW	29·84	29·70	80	61	17	
	Full M.	6	SW	29·78	29·70	77	59	19	—
		7	SW	29·79	29·76	75	59	09	89
		8	SW	29·79	29·56	75	59	16	04
		9	NE	29·69	29·54	72	56	07	75
		10	NW	29·69	29·67	70	59	11	
		11	W	29·73	29·66	74	50	09	
		12	SW	29·73	29·69	78	61	15	01
		13	SW	29·69	29·64	70	60	07	09
	L. Q.	14	SW	29·64	29·60	72	57	15	—
		15	W	29·80	29·64	71	53	16	—
		16	W	29·85	29·76	71	58	17	—
		17	NW	30·00	29·85	68	54	09	
		18	W	30·08	30·00	71	54	12	
		19	NW	30·14	30·08	71	48	07	
		20	Var.	30·16	30·14	76	53	10	
				30·16	29·43	85	48	3·53	5·10

NOTES *from MS. Register.*—Seventh Mo.   24. A shower about 1 p. m. followed by several others; some appearances for thunder in the evening.   25. Morning showery : at 2 p. m. a thunder storm, which lasted till three : after which the wind shifted to W: the lightning was not very vivid, but followed almost instantly by the reports.   26. A very showery day.   27. Heavy rain commenced between 7 and 8 p. m. and continued all night.   28. Very stormy, with almost unceasing rain : the evening cleared up.   30. Very fine day.   31. A thunder storm commenced about 4 p. m. lasting till nine, with continued but not heavy rain : thunder distant, and at last it lightned only.

Eighth Mo.   1. Showery : sun at intervals.   2. Fair : a fine breeze in the evening.   5. *Cirrus* clouds, very beautiful, in the evening.   7. A very large

1

lunar halo about 10 p. m. 8. A heavy storm of wind and rain from 4 to half-past 7 a. m. in which time 0·89 in. of rain fell. 9. A thunder storm p. m. with rain of an hour and a half's continuance. 20. A *Stratus* in the marshes at night.

RESULTS.

Prevailing Winds Westerly.

Mean height of barometer .... 29·76 in.

Thermometer......................... 65·30°

Evaporation .... 3·53 in.

Rain, &c............................. 5·10 in.

The amount of rain is greater by 1·88 in. than that found in the rain gauge at Plaistow. It was obtained in my absence from home by a temporary gauge fixed near the River Lea, at the laboratory, and *less elevated* by about 30 feet than the gauge at Plaistow.

Yorkshire.—At Eglestone, near Barnard Castle, on the 1st August, a most alarming storm of thunder and lightning came on in the afternoon, which caused the water in Blackstone-burn, Gate-beck, and Hill-beck, to overflow. (Some details of the destruction of mills, bridges, &c. follow). At Bedburn, *where not a drop of rain fell,* a dyeing mill was washed away by the flood.

Scotland.—The beautiful wooded bank, immediately opposite to Springfield paper-mill, lately gave way with a dreadful crash into the river Esk, which runs at the bottom, and so completely choaked it up, that not a drop of water passed for several hours. The bank, which is about 200 feet in height, had discovered symptoms of agitation on the preceding day, and, for about an hour before it gave way, the agitation was extremely violent, and the trees were seen falling in all directions; but, when it began to move in a body it was awfully grand and terrific, and the noise was equal to the loudest thunder. The slip is supposed to have been occasioned by water lodged in the bank, which had loosened it from the bed.—(ATHENÆUM.)

TABLE XXIII.

1808.		Wind.	Pressure. Max.	Pressure. Min.	Temp. Max.	Temp. Min.	Evap.	Rain, &c.
8th Mo.	Aug. 21	NW	30·15	30·11	73°	51°	7	
	22	NE	30·14	30·11	76	53	7	—
	23	NE	30·11	30·07	72	53	12	1
	24	NE	30·09	30·07	68	54	11	
	25	NE	30·07	29·83	72	50	10	
a.	26	SW	29·83	29·55	74	49	18	14
b.	27	NW	29·66	29·55	75	55	10	
	28	W	29·85	29·66	66	47	14	
	29	S	29·85	29·72	68	43	14	
	30	SE	29·72	29·58	73	59	13	6
c.	31	SW	29·58	29·50	70	58	14	11
9th Mo.	d. Sept. 1	SW	29·71	29·54	68	54	18	10
	2	SW	29·79	29·71	'67	55	13	18
	3	W	29·79	29·76	67	56	10	3
	4	W	29·78	29·76	66	48	9	—
c.	5	S	29·76	29·62	67	54	11	6
	6	SW	29·72	29·66	68	52	11	5
c.	7	SW	29·66	29·57	67	57	15	2
	8	SW	29·57	29·30	65	50	19	4
	9	S	29·30	29·28	64	54	7	47
e.	10	SE	29·40	29·30	63	52	5	12
	11	W	29·58	29·40	62	52	3	48
f.	12	SE	29·62	29·58	62	52	2	41
	13	Var.	29·71	29·62	67	53	6	20
g.	14	N	29·97	29·71	68	54	5	
g.	15	NE	30·23	29·97	71	54	17	
g.	16	NE	30·23	30·21	67	44	25	
g.	17	E	30·21	29·96	66	54	16	
	18	Var.			68	57	16	10
g.	19	SW			67	50	10	
			30·23	29·28	76	43	3·48	2·58

NOTES.—a. Misty morning.

b. A heavy shower between 7 and 8 a. m.

c. Much wind by night.

d. Squalls with rain.

e. Very wet a. m.; fair p. m. but distant thunder about five.

f. Lightning in the W after 10 p. m.

g. Fair days, with misty mornings, and abundance of dew.

RESULTS.

Winds Variable.

Mean height of barometer.............29·76 in.
Thermometer.......................60·34°
Evaporation .... 3·48 in.
Rain............................ 2·58 in.

Of which there was noted by day 0·78 in. by night 0·42 in.

Character of this period wet and changeable. The *Manchester Register* presents the following points of comparison during this period. Eighth month, 22, 23, 24, Wind there, NE; with us, SE: barometer steady with both. 25, 26, Both barometers descend, with a change to SW, followed by rain. 29, Wind S, with us; there, strong at W. 30, SE, with us; SW and W, strong, at Manchester. 31, At night, this strong current appears to have reached us. Ninth month, 3 to 8, Winds with us, W, S, and SW, and 0·2 of rain; with them, the same winds, but mixed with SE, and 0·87 of rain. 9 to 13, With them, wind E and NE, and 0·27 of rain: with us, the winds variable and mixed with SE, and 1·68 of rain. 14 to 17, With us, NE, N, and E, and fair: the same with them, except a mixture of SE, (now dry, having probably left its water with us). 18, Variable winds at each station, with about the same small quantity of rain. 19, Fair, with a westerly wind, at both stations.

TABLE XXIV.

1808.	Wind.	Pressure. Max.	Min.	Temp. Max.	Min.	Evap.	Rain, &c.
9th Mo. b. a. Sept. 20	W	30·29	30·20	68°	44°	9	—
b. a. 21	W	30·20	30·12	68	45	5	1
b. a. 22	Var.	30·12	29·60	67	51	—	23
23	N	29·90	29·60	59	43	18	1
24	NW	30·03	29·90	56	46	9	
25	Var.	30·04	29·90	60	41	5	
b. a. 26	Var.	29·90	29·90	64	47	8	
27		29·90	29·77	55	36	9	
c. 28		29·77	29·47	56	36	5	
29		29·49	29·47	58	38	4	
30		29·77	29·49	51	34	5	
10th Mo. Oct. 1		29·77	29·70	54	38	7	
2		29.80	29·70	54	34	3	
b. 3		30·04	29·80	55	46	4	
b. 4		30·05	30·04	56	41	2	
h. 5		30·05	29·90	62	46	4	40
d. 6	SW	30·03	30·00	65	48	7	
e. 7	W	30·00	29·31	57	46	12	19
f. 8	NW	29·70	29·31	52	41	14	1
9	NW	29·77	29·71	53	38	9	
a. 10	W	30·01	29·77	57	37	6	
a. 11	W	29·92	29·80	56	41	7	6
a. 12	NW	30·06	29·92	54	34	9	
13	NW	30·06	29·50	50	40	8	6
g. 14	SW	29·50	29·18	57	38	12	4
f. 15	W	29·52	29·18	48	42	4	3
f. 16	W	29·53	29·48	51	40	10	7
f. 17	NW	29·73	29·53	49	38	7	
f. 18	SW	29·73	29·50	54	38	11	3
		30·29	29·18	68	34	2·13	1·14

NOTES.—a. Much dew a. m. : the product in the gauge on the 21st was *dew*.

b. Misty.

c. Lunar halo at night.

d A *Stratus* on the marshes.

e. A stormy night.

f. Windy.

g. a. m. stormy with rain. Wind south.

h. This is the product of rain fallen at different intervals (not noted) since the 26th.

NOTES, &c. (continued).

Additional Notes, &c.—Tenth Mo. 3, 4. A *Stratus* at night. 13. A fine day; cloudy after sunset; the depression of the barometer was chiefly by night.

Greenock, Oct. 8.—It blew a heavy gale from SW all yesterday, but about midnight it suddenly chopped round to the NW, and has since continued to blow almost a hurricane.—(Pub. Ledger.)

Chester, October.—The tremendous storm on the 8th instant has done considerable damage to the shipping in the Mersey.—(Papers.)

Manchester, Oct. 8.—Very boisterous, with a great fall of rain, *viz.* 0·78 in. Wind NW. This was a very rainy *month* there, the total being stated at 5·32 in. Mean of barometer 29·49 in.—*Hanson.*

The 8th of tenth month appears to have been very wet at Paris, with the wind strong at W. Barometer about 29·5 in. English.

Fogs preceded this great flow to the eastward alike at Manchester, London, and Paris, and by nearly the same short interval at each station.

RESULTS.

Prevailing Winds Westerly.

Mean height of barometer . . . . 29·78 in.

thermometer . . . . 48·84°

Evaporation . . . . 2·13 in.

Rain . . . . 1·14 in.

Of which by night 0·52 in. remainder by day and night.

TABLE XXV.

	1808.	Wind.	Pressure Max.	Pressure Min.	Temp. Max.	Temp. Min.	Evap.	Rain, &c.
10th Mo.	Oct. 19	NW	29·68	29·50	49°	40°	8	
a.	20	NW	29·68	29·29	53	42	5	36
b.	21	W	29·55	29·48	54	38	10	
c.	22	W	29·79	29·55	51	34	8	
	23	W	29·55	29·30	52	42	10	42
d. c. b.	24	SW	29·69	29·55	54	38	11	
e.	25	SW	29·69	29·15	54	46	11	14
1st Q. b.	26	SW	29·40	29·15	56	42	13	
e. d.	27	SW	29·42	29·39	53	40	8	80
	28	SE	29·65	29·39	53	43	6	25
f.	29	S	30·05	29·65	54	40	3	21
	30	NE	30·33	30·05	54	43	3	
g.	31	NE	30·33	30·27	52	44	7	
11th Mo. b.	Nov. 1	NE	30·27	30·15	51	47	14	
	2	NE	30·15	30·06	54	47	13	
Full M.	3	NE	30·30	30·15	48	42	—	
h.	4	NE	30·30	29·91	50	39	30	
	5	Var.	29·92	29·88	47	25	—	—
i. c.	6	E	29·88	29·72	45	36	7	
	7	E	29·72	29·66	47	41	4	
g. h.	8	NE	29·66	29·57	52	43	3	8
	9	NE	29·75	29·57	53	42	2	
k.	10	NE	29·84	29·72	54	42	—	
L. Q. b.	11	NE	30·07	29·84	49	36	—	
	12	NE	30·13	30·10	46	37	31	
c.	13	NE	30·12	30·06	41	30	—	
	14	Var.	30·11	30·00	42	28	—	
c.	15	S	30·00	29·63	51	33	9	
e. b.	16	S	29·63	29·29	54	49	14	9
e.	17	S	29·29	28·81	56	44	7	12
			30·33	28·81	56	25	2·37	2·47

Notes.—*a.* Very dark and cloudy a. m. the wind rising.

b. Windy.

c. Hoar frost.

d. Hail.

e. Stormy nights.

f. Swallows seen for the last time.

g. Misty, the trees dripping.

h. Showers in the evening.

i. Much rime on the trees a. m.: the leaves fall abundantly; a large mulberry tree cast its whole foliage in an hour or two.

k. Much dew a, m.

l. No dew a. m. the sky being veiled with clouds. Some drizzling rain followed.

Additional Notes, &c.—Tenth Mo. 20. This nocturnal fall of the barometer was preceded by a sensible fluctuation: the tendency was first to rise. 27. *Nimbi,* well formed, at intervals through the day: at night, two small meteors almost at the same instant.

Eleventh Mo. 16. Numerous small fluctuations preceded this rapid fall of the barometer.

Deal, Oct. 24.—Last night and this morning it blew very hard from the SSW.

Torbay, Oct. 26.—We have had heavy gales from the SW for several days, accompanied with dreadful squalls and showers.—A similar account from Plymouth.

Cork.—A heavy gale on the night of Nov. 17, the wind SW, shifting to NW.—(PUBLIC PAPERS.)

Cotte.—Journ. de Physique, v. 68, p. 335, notes, under Oct. 26, an earthquake at Leghorn, and very stormy weather, with a low barometer at Montmorency.

RESULTS.

Winds Variable: the prevailing one NE, attended with dry weather. The South-west has been stormy and wet.

Mean height of barometer .... 29·76 in.
temperature................... 45·36°
Evaporation .... 2·37 in.
Rain .... 2·47 in.

Of which by night 2·08 in.

TABLE XXVI.

1808.	Wind.	Pressure. Max.	Pressure. Min.	Temp. Max.	Temp. Min.	Evap.	Rain, &c.
11th Mo. New M. a. Nov. 18	NW	29·30	28·72	49°	32°	9	36
19	SE	29·73	29·30	43	34	6	
b. 20	SW	29·77	29·74	54	38	5	
21	SW	30·20	29·74	56	38	10	
c. 22	SW	30·20	30·08	50	38	5	
23	NW	30·18	30·08	54	44	4	2
1st Q. 24	SW	30·18	29·97	50	43	—	40
d. 25	SW	29·98	29·95	53	44	9	
e. 26	W	29·95	29·47	56	47	2	2
e. 27		29·77	29·40	54	30	—	2
g.f. 28	NW	29·83	29·77	43	30	7	
h.f. 29	SW	29·77	29·10	50	34	3	46
30	W	29·35	29·10	48	35	8	
12th Mo. e. Dec. 1	SW	29·45	29·15	49	37	6	16
e. 2	SW	29·43	29·15	49	41	12	
Full M. 3	N	30·02	29·40	50	42	9	
4	SE	30·31	30·02	46	27	3	
f. 5	SW	30·31	29·97	52	30	4	—
e. 6	NW	29·86	29·72	53	33	11	3
k. e. 7	NW	29·93	29·86	41	36	9	
8	N	29·97	29·93	43	37	—	
f. 9	NW	30·07	29·97	42	32	7	
L. Q. f. 10	NW	30·30	30·07	41	31	1	
i. 11	W	30·30	30·27	40	30	1	
i. 12	NW	30·34	30·27	42	36	2	
13	W	30·38	30·36	39	31	0	
e. 14	N	30·36	30·08	42	34	2	5
e. 15	N	30·11	30·07	31	32	3	
l. 16	W	30·07	29·84	37	26	—	
		30·38	28·72	56	26	1·38	1·52

NOTES —a. Snow on the ground a. m. followed by clearer weather.

b. Much wind last night: very cloudy. The maximum of temp. obtained at 9 a. m.

c. Windy: maximum at 9 again.

d. The same phenomena repeated.

e. Windy.

f. Hoar frost.

g. Large lunar halo.

h. Stormy night: snow fell in a more elevated part of the county.

i. Misty air.

k. Some snow a. m.

l. A bright small meteor moving from S towards W, soon after
6 p. m.

Additional Notes, &c.—Twelfth Mo. 9. Ice on the water : probably rather
from the effect of evaporation than the temperature of the air.

Plymouth, Nov. 18.—A heavy gale at WNW.—(PAPERS).

As the fall of the barometer, terminating about this day, was a great and
continued one, it may be proper to note, that *Cotte,* (Journ. de Phys. *ubi antea*)
places under this date a storm with high tides on the coast of Normandy,
disastrous inundations in the department of La Lozere, &c. near the gulf of
Lyons, and much wind with a low barometer at Montmorency.

December 8.—A violent storm in the Channel, and on the coast of Holland.
But *little variation* in the barometer (though cloudy and windy) at Montmorency. *Cotte, (ubi supra).*

He notes likewise the following singular circumstances of temperature in this
month. On the 13th, thermometer at Altona, —14 Reaum. about *zero* of Fahrenheit, at Montmorency, 41° Fahrenheit : and on the 14th, 40° at Montmorency,
while it was only 35° at Venice and Naples !

Lunar Rainbow.—The following notice of this rare phenomenon has been
handed to me from my friend John Capper, of Stoke Newington. " On the 1st
of 12th month (Dec.) 1808, a little after 5 o'clock a. m. I observed a lunar
rainbow; the moon being near setting to the north of west. It was of unequal
brightness. In the most northerly part of it, near the earth, the colours were
very distinct, in the other parts they were scarcely distinguishable; and it had,
on the whole, the appearance of a white arch, at the usual distance from
which the second (or doubly reflected) bow was visible. Opposite the brightest
parts, at the northern end, it was very strong, and coloured, as the inner bow
in that place was. The white part did not last more than ten minutes; the
coloured part continued visible much longer."

RESULTS.

Winds Westerly.

Mean height of barometer . . . . 29·86 in.

temperature . . . . 41·01°

Evaporation . . . . 1·38 in.

Rain . . . . 1 52 in.

Of which noted by night 0 63 in.

TABLE XXVII.

1808.		Wind.	Pressure.		Temp.		Evap.	Rain, &c.
			Max.	Min.	Max.	Min.		
12th Mo.	New M. *a.b.* Dec.17	N	29·65	29·45	38°	24°	—	—
	c. 18	N	29·65	29·62	31	26	—	—
	19	N	29·71	29·68	32	25	—	—
	20	N	29·99	29·71	30	14	—	
	d. 21	SW	29·99	29·20	34	20	—	—
	22	Var.	29·42	29·20	33	27	—	—
	e. 23	E	29·45	29·39	31	28	—	—
1st Q.	24	Var.	29·61	29·56	32	25	—	
	f. 25	NW	29·55	29·50	30	21	—	—
	g. 26	E	29·55	29·49	33	23	—	—
	a. 27	E	29·60	29·55	36	34	34	14
	a. 28	E	29·55	29·52	38	36	—	—
	29	E	29·52	29·49	42	37	4	15
	30	E	29·60	29·50	44	38	5	
	h. 31	E	29·62	29·60	38	36	1	12
1st Mo.	Full M. 1809. Jan. 1	E	29·60	29·49	39	38	—	—
	i. 2	NE	29·49	29·35	40	30	—	—
	3	NE	29·65	29·49	30	28	—	—
	k. 4	E	29·70	29·65	33	30	—	—
	k. 5	SE	29·65	29·56	38	32	8	19
	k. 6	SE	29·56	29·43	48	36	6	24
	k. 7	SE	29·43	28·80	45	39	5	16
	k. 8	S	29·37	28·50	46	37	4	23
L. Q.	*l.* 9	S	29·25	29·20	44	36	1	39
	k. 10	NW	29·38	29·16	46	34	4	2
	m. 11	W	29·49	29·38	47	36	1	
	12	N		29·49	40	30	—	—
	13	N	29·82		37	28	—	—
	14		29·82	29·77	34	29	—	—
	n. 15	NE	30·08	29·77	29	26	12	20?
			30·08	28·50	48	14	0·85	1·84

Notes.—*a.* Misty.

b. About 2 p. m. a heavy squall from NW with rain, sleet, and snow, giving strong sparks from the rod : the night proved very stormy.

c. Clear; brisk wind; snow at intervals, and pretty much of it in the night. Wild geese migrate in large flocks.

d. Rime on the trees. *Minimum* of temperature at 9 a. m.

e. Snow at intervals for three days past, and much this night. In the day it fell very sparingly, and regularly crystallized in *stars*.

f. Small rain, freezing on the ground.

g. Clear morning ; sleet in the night.

h. Cloudy for three days past.

i. Snowy morning after a wet evening.

k. Cloudy and windy weather. On the 6th a meteor of moderate size, passing eastward.

l. Hoar frost: *Cirrostratus:* a *Nimbus* in the S. These successive indications were followed by steady rain.

m. A fine day. The whole level, bordering on the Lea from Stratford upwards is now, by the continued rains and swelling of the river, several feet under water.

n. Snow at intervals. The inundation has subsided.

Additional Notes, &c.—Twelfth Mo. 17. This night is described in the Papers as very stormy, with the following differences in the wind : Torbay and Deal, NW. Yarmouth, NNE. Sheerness, NE. Ramsgate, (with much snow) N.

17, 18, 19. A great snow in the Netherlands, and disasters by *avalanches* in Switzerland. *Cotte.*

Aberdeen, Dec. 31.—We have experienced a strong gale from ESE to SE, with squalls and rain, which has prevented all intercourse by shipping, there not being a single arrival or sailing for eight days.—(Pub. Ledger).

Scotland, December.—A *whale* ran itself on shore on the banks of the Frith, between Alloa and Cambus : where it was with difficulty killed by the country people. Considerable damage occurred in several places by the high wind, heavy rain, and floods; and in one instance by lightning, which perforated a large building in several places, as if it had been battered by cannon shot.— (Athenæum).

Plymouth Dock, Jan. 7.—Last night it blew a tremendous hurricane, with heavy rain till daylight this morning, when the weather moderated.—(Pub. Ledger).

RESULTS.

Prevailing Winds Easterly.

Mean height of barometer .... 29·52 in.

thermometer .... 33·68°

Evaporation .... 0·85 in.

Snow and rain .... 1·84 in.

Character cloudy and frosty, with frequent rain and snow. The barometer has departed little from the mean, save in one great *depression,* the crisis of which occurred about 2 p.m. the 8th; and though the subsequent elevation went on at the rate of more than 0·1 inch per hour, the wind was quite moderate. It is very rare to observe so steady a wind from the SE, as that which preceded this fall of the barometer.

Rain noted as by night 0·63 in.

TABLE XXVIII.

| | 1809. | | Wind. | Pressure. | | Temp. | | Evap. | Rain, &c. |
				Max.	Min.	Max.	Min.		
1st Mo.	New M.	Jan. 16	E	30·12	30·07	30°	25°	—	
		17	E	30·07	29·94	30	18	—	
		18	E	29·87	29·77	28	19	—	
		19	Var.	29·53	29·48	30	23	—	
		20	N	29·50	29·43	33	31	—	—
		21	Var.	29·54	29·14	35	31	—	—
		22	Var.	29·65	29·08	35	19	—	—
	1st Q.	23	Var.	29·73	29·44	33	27	—	—
		24	W	29·45	29·40	45	34	15	1·89
		25	SW	29·82	29·47	48	36	5	27
		26	SW	———	29·18	51	40	—	
		27	SW	29·55	———	52	46	18	
		28	S	29·55	29·22	56	47	14	
		29	W	29·36	28·93	53	43	15	1
		30	W	29·75	28·70	51	37	12	23
	Full M.	31	S	29·94	29·76	50	37	4	
2d Mo.		Feb. 1	SW	29·76	29·52	54	48	14	
		2	SW	29·52	29·20	51	49	—	—
		3	SW	29·40	29·20	55	44	19	12
		4	SW	29·45	29·37	51	41	7	7
		5	SW	29·43	29·37	51	41	8	23
		6	SW	29·93	29·30	51	38	10	12
		7	E	30·00	29·90	39	33	10	—
	L. Q.	8	E	29·90	29·40	44	33	5	25
		9	SE	29·31	29·29	54	44	9	1
		10	S	29·29	29·11	52	44	14	2
		11	SW	28·85	28·75	51	42	11	7
		12	SW	28·80	28·70	51	40	10	8
		13	SW	29·20	28·80	53	43	14	1
				30·12	28·70	56	18	2·14	3·38

The period I am now reporting is so extraordinary in its character, that I must exchange the usual form of notes for a continued narration.

The wind has been inconstant, though the greatest quantity of air has undoubtedly flowed from the SW. The movements of the barometer have been, in like manner, desultory, and the *mean* much lower than for a considerable time past: there have been, moreover, some great and pretty sudden depressions. The temperature (after the thaw) was very high for the season, and the evaporation and precipitation great.

The new moon was very conspicuous on the 17th, the whole disk appearing, well defined. A brilliant small meteor descended on the SE horizon about 6 p. m. On this and the preceding day the snow exhibited its beautiful blue and pink shades at sunset, and there was a strong evaporation from its surface. I found a circular area, of five inches diameter, to lose 150 grains troy, from sunset on the 15th to sunrise next morning, and about 50 grains more by the following sunset; the gauge being exposed to a smart breeze on the house top. The curious reader may hence compute for himself, the enormous quantity raised in those 24 hours, without any visible liquefaction, from an acre of snow: the effects of the load thus given to the air were soon perceptible. On the 18th, though the moon was still conspicuous, the horns of the crescent were obtuse. On the 19th appeared the *Cirrus* cloud, followed by the *Cirro-stratus.* In the afternoon a freezing shower from the eastward glazed the windows, encrusted the walls, and encased the trees, the garments of passengers, and (it is said) the very plumage of the birds with ice*. Its composition, which I examined on a sheet of paper, was no less curious than these effects. It consisted of hollow spherules of ice, filled with water; of transparent globules of hail; and of drops of water at the point of freezing, which became solid on touching the bodies they fell on. The thermometer exposed from the window indicated 30,5°. This shower was followed by a moderate fall of snow. From this time to the 24th we had variable winds and frequent falls of snow, which came down on the 22d in flakes as large as dollars, with sleet at intervals. On the 24th a steady rain from W decided for a thaw. This and the following night proved stormy: the melted snow and rain, making about two inches depth of water on the level, descended suddenly by the rivers, and the country was inundated to a greater extent than in the year 1795. The River Lea continued rising the whole of the 26th, remained stationary during the 27th, and returned into its bed in the course of the two following days. The various channels by which it intersects this part of the country were united in one current, above a mile in width, which flowed with great impetuosity, and did much damage. From breaches in the banks and mounds, the different *levels*, as they are termed, of embanked pasture land, were filled to the depth of eight or nine feet. The cattle, by great exertions, were preserved, being mostly in the stall; and the inhabitants, driven to their upper rooms, were relieved by boats plying under the windows. The Thames was so full during this time, that no tide was perceptible; happily, however, its bank suffered no injury; the evacuation of the water from the levels has in consequence proceeded with little interruption, and is now (23d of 2d month) pretty fully effected. No lives were lost in these parts. Several circumstances concurred to render this inundation less mischievous than it might have been, from the great depth of snow on the country. It was the time of *neap* tide; the wind blew strongly from the *westward*, urging the water *down* the Thames; to which add moonlight nights, and a temperate atmosphere, both very favourable to the poor, whose habitations were filled with water.

On the 28th appeared a lunar halo of the largest diameter. On the 29th, after a fine morning, the wind began to blow hard from the south, and during

* Birds thus disabled were seen lying on the ground in great numbers in different parts of the country. Nineteen rooks were taken up alive by one person at Castle Eaton Meadow, Wilts.—(PUB. LEDGER).

See on this subject WHITE, *Nat. Hist. of Selborne*, vol. ii. p. 303.

K

the whole night of the 30th it raged with excessive violence from the west, doing considerable damage. The barometer rose, during this hurricane, one-tenth of an inch per hour. The remainder of the moon was stormy and wet, and it closed with squally weather; which, with the frequent appearance of the rainbow, indicated the approach of a drier atmosphere, a change on few occasions within my recollection more desirable.

Additional Notes, &c.—Second Mo. 12. At 9 a. m. the barometer was rising, p. m. it fell again, rose in the evening, fell in the night, and next morning was again found rising: most of these oscillations were of course but small: they were followed by squalls of wind.

The Public Papers give numerous details of inundations consequent on the thaw of the 24th ultimo, which appear to have prevailed in low and level districts all along the east side of this island: but in no part with more serious destruction of property, public works, and the hopes of the husbandman, than in the Fens of Cambridgeshire: where by some accounts 60,000, by others above 150,000 acres of land have been laid under deep water, through an extent of 15 miles.

The following fact is worth preserving:—" About 500 sacks were filled with earth, and laid on the banks of the Old Bedford River, at various places, where the waters were then flowing over. This proved effectual in saving that part of the country from a general deluge."

RESULTS.

Prevailing Wind South-west.

Mean height of barometer.............29·44 in.
thermometer...........40·86°
Evaporation........................ 2·14 in.
Rain and snow .... 3·38 in.

Portsmouth, Jan. 29.—The whole of this day it has blown a heavy gale from WSW, which now (7 p. m.) continues with unabated fury. A subsequent account says,—" Upwards of five tons of lead, in three pieces, were blown off the roof of the Clock Storehouse, in Portsmouth Dock-yard. One piece, of about 30 cwt. and another of about 40 cwt. were carried, in different directions, near 40 yards; and the other piece was carried 25 yards. It being so extraordinary a circumstance, the pieces of the lead were weighed the next morning, and the distances they fell accurately measured, for the purpose of being registered at the yard."

Plymouth Dock, Feb. 8.—A most tremendous hurricane accompanied with a very heavy rain, commenced about 11 last night, and has continued ever since without the least intermission. We have had much thunder and lightning here from the late storms.—(Papers).

Earthquakes in Scotland.—Convulsions of the earth, resembling earthquakes, have lately been felt in various parts of the Highlands. One was felt at Dunning, in Perthshire, on the 18th of January, of which Mr. Peter Martin, surgeon, gives the following account :—He was on his way home, about two in the morning, when his attention was suddenly attracted by a seemingly sub-terraneous noise, and his horse immediately stopping, he perceived the sound to proceed from the north-west. After continuing for about half a minute, it became louder and louder, and apparently nearer, when, all on a sudden, the earth gave a perpendicular heave, and, with a tremulous waving motion, seemed to roll or move in a south-east direction. The noise was greater during the shock than before it, and for some seconds after it was so loud, that it made the circumjacent mountains re-echo with the sound, after which, in the course of about half a minute, it gradually died away. At this time the atmosphere was calm, dense, and cloudy, and for some hours before and after there was not the least motion in the air. Fahrenheit's thermometer, when examined, about half an hour after the shock, indicated a temperature of 15° below the freezing point of water. The preceding day had been calm and cloudy; thermometer 8 a.m. 14°, 8 p.m. 13°. The morning of the 18th was calm and cloudy, but the day broke up to sunshine; thermometer 8 a.m. 19°, 8 p.m. 16°. About the same time a similar shock was felt at the bridge of Allan, near Stirling, where it was so violent along the foot of the hills as to make the tables and chairs rattle. On the 9th of January, about half-past five in the morning, a smart shock was felt at the village of Comrie, near Crief, the noise attending which was loud and greatly prolonged. During the time of the shock the air was calm and serene. The moon shone bright, and the sky was soon afterwards covered with whitish clouds, moving rapidly from NW to SE. The following is the substance of a letter from Strontian, in the west of Argyleshire: "On Tuesday, the 31st of January, we distinctly felt *five* shocks of an earthquake. It extended over the neighbourhood, and was accompanied with a noise like distant thunder. On Wednesday there was another, on Saturday following there were two more, on Sunday two, and this day (Feb. 6) one. The first, on Saturday, was the most severe; every moveable in my house was displaced, and the building much shaken, but fortunately alarm was the only consequence, as I have heard of no accident. The shocks were distinctly felt by the miners below ground; they continued only for a few seconds, and have all taken place between five and seven in the evening."—(Athenæum).

TABLE XXIX.

	1809.		Wind.	Pressure. Max.	Pressure. Min.	Temp. Max.	Temp. Min.	Evap.	Rain, &c.
2d Mo.	New M.	Feb. 14	SW	29·67	29·20	52°	41°	9	—
	a.	15	SW	29·80	29·53	53	42	7	5
		16	S	29·56	29·52	52	46	8	3
		17	SW	29·80	29·77	55	45	11	—
	b.	18	SW	30·47	29·77	57	35	10	
	c.	19	SW	30·47	30·17	54	45	9	
		20	SW	29·99	29·90	51	38	10	—
	d.	21	NW	30·35	29·99	42	30	7	—
1st Q.		22	NW						—
		23	Var.	30·35		52	33	14	—
		24	SW	30·35	30·25	48	38	6	—
	e.	25	W	30·45	30·25	47	29	—	—
	f.	26	SW	30·45	30·37	50	33	10	
	g.	27	SW	30·37	30·33	53	32	4	
	g.	28	SW	30·33	30·24	53	35	4	2
3d Mo.	Full M.	March 1	Var.	30·39	30·24	51	38	3	11
		2	NE	30·45	30·37	47	32	9	
		3	NE	30·37	30·16	46	35	9	—
		4	Var.	30·18	30·13	48	39	4	—
		5	N	30·29	30·18	43	34	6	1
	h.	6	N	30·39	30·29	40	35	4	
	i.	7		30·49	30·39	40	33	0	
L. Q.		8	W	30·49	30·31	47	32	2	
	k.	9	W	30·31	30·23	55	39	5	
	l.	10	E	30·27	30·23	46	31	18	
	l.	11	NE	30·27	30·09	47	36	12	
	m.	12	NE	30·20	30·09	50	34	21	
	m.	13	NE	30·25	30·20	47	34	14	
	n.	14	NE	30·44	30·25	46	36	10	—
	n.	15	NW	30·35	30·24	48	33	7	
				30·49	29·20	57	29	2·33	0·22

NOTES.—a. Cloudy and windy, with showers.

b. The temperature higher at 9 a. m. than for the last 24 hours. Barometer wavering. The ground springs, or those of the superficial strata, begin to subside, so that it is now possible to dig without puddling the soil.

c. a. m. much dew. At sunset the *Cirrus* cloud, highly coloured, with a strong pink tinge in the horizon. Some wind in the night.

d. a. m. very cloudy; a squall with a little hail, then a fair day.

e. Slight showers, with a clear air the last four days. Lunar halo, small and coloured.

f. Hoar frost; very misty.

g. Still misty weather: coloured halo, as also the following night.

h. Slight showers and misty weather for several days past.

i. A very wet mist this night. The evaporation gauge received an increase of about one hundredth of an inch.

k. The atmosphere begins to clear below.

l. Fine clear weather: smart breeze from NE.

m. Much dew, a. m. windy.

n. Cloudy: a little rain from the N, a. m. the 14th.

Additional Notes, &c.—Second Mo. 14. During a heavy storm of wind, accompanied by thunder, rain, and hail, the house of a baker in New Navy Row, *Plymouth Dock*, was struck by lightning, and together with the two adjoing houses materially damaged. On the *same day*, at *Paris*, the barometer being just past the minimum for the month, the wind very strong at W, with much rain, there was thunder at 6 p. m. and a house near the city was struck by lightning. A portion of the bell-wire in this house was dispersed and fixed as a very extensive *stain* in the wall of the apartment. At the end of vol. 69, of the Journ. de Physique, is a curious engraving of this picture: exhibiting one portion of the iron thrown off each way, at right angles from the course of the wire, in mossy ramifications; and another above it more perfectly dissipated, in a figure resembling an electrified lock of hair in the midst of a cloud of smoke. The stain measured six feet by four.

Second Mo. 17. Temperature 55° at 9 a. m. 21—25. Occasional slight showers: a pretty clear air but chilly.

Third Mo. 11. *Cirrocumulus* a. m. succeeded p. m. by *Cirrostratus*.

Accidents by Lightning.—The following singular circumstance took place on board the Warren Hastings, recently launched at Portsmouth, while the vessel was moored at the Mother-bank:—The morning being fine, it was deemed necessary to get up the top-gallant masts, which occupied some hours. About three o'clock in the afternoon the atmosphere was overcast to the westward, and every appearance indicated the approach of a violent storm. Several alert sailors were sent aloft to strike the top-gallant masts as speedily as possible, but while lowering them the wind blew tremendously, and the rain fell in torrents, accompanied by heavy claps of thunder. In the midst of the confusion, occasioned by the storm, three distinct balls of fire were emitted from the heavens; one of them fell in the main-topmast cross-trees, killed a man on the spot, and set the main-mast on fire, which continued in a blaze for about five minutes, and then went out. The seamen both aloft and below were almost petrified with fear. At the first moment of returning recollection, a few of the hands ran up the shrouds to bring down their dead companion, when the second ball struck one of them, and he fell, as if shot by a cannon, upon the guard iron in the top, from which he bounded off, into the cross jack braces. Finding that he still survived, he was relieved from his perilous situation, and

2

brought upon deck with his arms much shattered and burnt. This poor fellow was expected to undergo immediate amputation, as the only means of saving his life. The third ball came in contact with a Chinese, killed him, and wounded the main-mast in several places; the force of the air, from the velocity of the ball, knocked down Mr. Lucas, the Chief Mate, who fell below, but was not much hurt. For some time after the storm subsided, a nauseous, sulphureous smell continued on board the ship. (Apparent date by the Papers, Feb. 14.—ATHENÆUM.)

RESULTS.

Winds variable; the Westerly have blown longest, and the Easterly with the greatest force; though neither have much exceeded a breeze.

Mean height of barometer ....30·17 in.
             thermometer ....42·46°
Evaporation........................ 2·33 in.
Rain .... 0·22 in.

Character of this moon quite the reverse of the last. Barometer remarkably high and steady: temperature rather low and uniform: frequent mists, dews, and hoar frost: winds moderate and drying, with very little return of water from the clouds; so that much of the period had rather an *Autumnal* aspect. It has been highly favourable to the agricultural business of the season.

Prognostics of Rain from the Habits of Animals.

I have no doubt there are some of these on which we may place considerable reliance, as indicative of a present tendency to rain. I should explain them simply by the supposition of certain feelings in the animal, induced by the present state of the air, and connected with its comfort or preservation.

The " Bucula cœlum suspiciens patulis captavit naribus auras " of Virgil must be familiar to classical readers. I once observed this action of snuffing the air, very decidedly, in a young heifer; and conceived it to be occasioned by the fresh smell of the turf, watered by some showers which had been falling to windward.

Walking to-day (Fourth Month, 6, 1809), I observed the ducks busy throwing the water over their backs with their heads. Having heard this action cited as portending a shower, I looked up for the cloud: but chancing to direct my view the wrong way, saw nothing like rain approaching, and rejected the prognostic as futile. I had not proceeded, however, ten yards from the spot ere some drops began to fall from the zenith, and, on a more attentive examination, I found the *Nimbus* behind me. In about an hour it set in for a wet afternoon.

Some snow and hail had fallen at intervals, but no rain, for a week past; and the ground was dusty.

The reason of this I take to be, that these fowls, though so much in the water, do not love to be wet to the skin; and therefore, when warned by the peculiar sensation preceding rain, they presently close their plumage, by throwing a sudden weight of water on their bodies in the direction of the growth of the feathers.

On another occasion I observed several of these fowls run suddenly out from under a shed, with signs of much gratulation, on the immediate approach of a heavy shower; yet before a single drop had fallen. This action might be connected with the expectation of a supply of food; insects being brought down from the air, and worms and snails out of their strong holds, on such occasions.

TABLE XXX.

1809.		Wind.	Pressure, Max.	Min.	Temp. Max.	Min.	Evap.	Rain, &c.
3d Mo.	New M. March 16	NW	30·24	30·18	52°	42°	8	
	a. 17	NW	30·18	30·15	56	43	6	
	18	NE	30·15	30·05	55	37	16	
	b. 19	E	30·00	29·97	52	42	14	2
	c. 20	Var.	30·12	29·97	53	38	4	
	21	E	30 13	30·08	53	39	3	
	f. d. 22	S	30·13	29·94	66	37	12	
	g. d. 23	Var.	29·94	29·64	64	47	21	1
1st Q.	b. 24	W	29·64	29·32	57	43	10	11
	25	SE	29·32	29·11	52	33	5	
	e. 26	S	29·30	29·11	55	31	7	
	i. 27	E	29.40	29·30	56	36	8	
	k. 28	NE	29·71	29·40	56	40	6	15
	29	N	29·82	29·71	47	32	3	
	30	NE	29·81	29·79	44	39	11	
	Full M. b. 31	NE	29·79	29·73	50	36	21	
4th Mo.	l. b. April 1	N	29·80	29·71	45	30	10	—
	m. 2	NW	29·90	29·78	45	27	—	—
	n. 3	NW	30·07	29·90	46	26	—	—
	n. 4	NW	30.29	30·07	43	24	—	—
	5	SW			44	26	—	—
L. Q.	6	SW	30·15		48	34	26	14
	d. 7	Var.	30·33	30·15	50	31	7	
	8	SW	30·33	30·12	49	34	13	
	9	SW	30·12	29·89	54	40	7	3
	10	NW	29·89	29·53	56	45	10	—
	11	N	29·77	29·37	53	30	7	15
	e. 12	SW	29·77	29·30	52	40	13	3
	o. 13	SW	29·30	29·09	53	37	15	10
			30·33	29·09	66	24	2·63	0·74

NOTES.—*a.* The bat and black beetle come abroad. The roads are become unpleasantly *dusty.*

b. Cloudy, windy.

c. c. Very misty mornings. The mist as it broke away on the 27th, exhibited a faint white *bow* in the NW.

d. Much dew.

e. Hoar frost.

f. About 7 p.m. a meteor appeared suddenly in the W, and, descending with a slight inclination to the NW, became extinct. This meteor must have been large, or but little elevated, as it was much more conspicuous in the twilight than the planet Venus.

g. a. m. Very clear atmosphere below: *Cirrus* cloud only above This soon became *Cirrocumulus,* and p. m. came down, as before thunder, with a dull red halo round the moon. A shower, attended (it is said) with lightning, between 1 and 2 a. m.

i. Coloured *Cirrus* at sunset, and faint large lunar halo at night.

k. A thunder storm passed from W to E in the N about 2 p. m.

l. Showers, mixed with hail p. m. At sunset an extensive *blush* of rose coloured haze, spreading on the face of an opaque twilight, the clouds beneath (mostly *Cirrostratus*) at the same time rapidly dispersing.

m. Showers of hail, or rather of those hard snow balls which form its basis. A huge *Nimbus,* affording these balls, mixed with snow, passed over about sunset: after which the haze, &c. of last evening were repeated.

n. n. Showers of the kind last mentioned. Many very distinct specimens of the *Nimbus* cloud.

o. Hail has appeared occasionally in the showers for the last few days. Stormy night.

Sussex, March 28.—Between one and two o'clock the town of Horsham was visited by a storm (of thunder and hail); the tempest appeared to run in a SW direction, with a thick and gloomy atmosphere, and after many awful flashes of lightning and tremendous explosions, produced *hail* with a degree of violence that dealt destruction to the windows, &c. (The damage done to the *gardens* more especially is here noted). The hail stones were from two to three inches in circumference, and from their uneven formation appeared like rugged pieces of ice, covering the streets nearly shoe deep. The storm, though so heavy, was limited chiefly to the town.

Manchester.—The mean temperature (April) less than last month; snow and hail have fallen in small quantities. On the 13th, there was a heavy shower of hail, with a high SW wind: the hail stones were very large; the temperature during the storm was lowered 6°.—*Hanson.*

Malton.—Character of the period (April) wet, stormy, and changeable, with much hail and snow.—*Stockton.*—(ATHENÆUM).

RESULTS.

Winds Variable.

Mean height of barometer.............29·81 in.

thermometer...........44·01°

Evaporation .... 2·63 in.

Rain............................... 0·74 in.

Character, fair and warm to the first quarter: then cold, cloudy, and unfavourable to vegetation. Rain by day 0·30, by night 0·17 in. remainder by day and night.

TABLE XXXI.

1809.			Wind.	Pressure.		Temp.		Evap.	Rain, &c.
				Max.	Min.	Max.	Min.		
4th Mo.	New M. a.	April 14	Var.	29·49	29·06	52°	39°	5	22
	b.	15	SW	29·49	29·07	52	39	6	26
	c.	16	W	29·11	29·87	59	39	15	66
	d.	17	NW	29·55	29·11	45	33	3	16
	e.	18	N	29·77	29·55	43	30	8	—
	f.	19	N	29·77	29·71	49	35	13	—
	g.	20	SE	29·71	29·55	46	36	5	—
	h.	21	NE	29·67	29·58	50	39	17	79
	1st Q. i.	22	NE	29·99	29·67	50	42	11	
	i.	23	E	30·29	29·99	46	41	5	
	i.	24	N	30·36	30·30	50	38	6	
		25	Var.	30·30	29·90	56	40	5	8
	k.	26	SW	29·90	29·69	53	47	3	12
		27	Var.	29·69	29·40	56	46	7	6
	l.	28	NE	29·64	29·40	56	40	9	11
		29	N	29·78	29·74	48	36	15	
	Full M. m.	30	NW	29·74	29·32	57	43	8	2
5th Mo.		May 1	NW	29·41	29·32	54	35	7	10
	n.	2	W	29·79	29·41	55	33	11	1
		3	SW			53	38	11	
		4	SW	29·89	29·86	56	45	—	—
	o.	5	Var.	30·22	29·89	57	34	23	3
	L. Q.	6	SW	30·31	30·22	63	47	12	
	p.	7	SW	30·32	30·28	68	45	13	
		8	NW	30·28	30·17	68	38	9	
		9	E	30·17	30·04	67	43	27	
		10	E	30·04	29·97	71	44	—	
		11	Var.	29·97	29·95	75	47	70	
		12	SW	29·95	29·93	78	46	25	
	q.	13	Var.	29·92	29·88	78	47	30	
				30·36	29·06	78	30	3·79	2·62

NOTES.—a. Thunder, with large hail, about noon.

b. Windy, showery.

c. Fair day, but with much *Cirrostratus*, and a bank of clouds in the W at sunset: very wet night.

d. Moderate rain the whole day.

e. Snow: hail at intervals: two or three *swallows* on the wing.

f. Hoar frost; steady northerly winds of late.

g. Much snow between 5 and 6 a. m.: hail at noon: steady rain an hour before sunset.

h. The ground white with snow by 2 a. m. of which there fell three or four inches, followed by rain.

i. Cloudy, windy. The land bordering on the Lea is again much inundated.

k. a. m. steady rain, giving sparks *negative.*

l. The cuckoo articulates. The nightingale has sung since the 25th.

m. After a clear drying day, appear the returning indications of wind and rain. A wet night and day ensued.

n. Large hail about 11 a. m.: about 2 p. m. thunder, in a *Nimbus* bearing SW: soon after four, rain in large drops, giving sparks *negative* as did the air itself afterwards; but there was still a dense shower in the SE in which the bow appeared.

o. A faint red blush on the evening twilight, followed by hoar frost.

p. A strong but not clear twilight; very calm nights of late, with much dew.

q. At sunset a bank of rocky *Cumulus* in the NW surmounted by *Cirrus,* pointing upward.

Additional Notes, &c.—Fourth Mo. 20. *Cumulostratus* clouds p. m. inosculating with a dense continuous cloud above them. 25. Overcast a. m., a breeze with small rain : *Cirrocumulus* p. m. with *Cirrus* below it pointing downward, and succeeded by *Cirrostratus.* 26. Rain more or less all day.

Fifth Mo. 3. A weak negative electricity in a clear air. 4. Maximum of temperature at 9 a. m. windy and overcast. 12. Barometer unsteady.

On the night of the 1st of this month about 10 o'clock, a very brilliant meteor was observed from several places in the neighbourhood of Whitehaven. It is described as having an apparent diameter of five or six inches, with a tail of three feet in length.—(PUB. LEDGER).

Bath, April 26.—Seldom at this season of the year has so heavy a fall of snow been seen in this climate. The storm (on the 21st) was incessant for nearly eighteen hours, and the ground was covered upon an average to the depth of sixteen inches. In this neighbourhood trees of very considerable size were bent double, and stript of their branches by the weight, and material injury has been done to the orchards.— (ATHENÆUM).

RESULTS.

Winds Variable.

Mean height of barometer .... 29·83 in.

thermometer .... 48·58°

Evaporation .... 3·79 in.

Rain, &c............................... 2·62 in.

Of which noted by day 0·41, by night 0·72.

𝔐eteorological 𝔒bservations

MADE CHIEFLY AT

PLAISTOW, AND OTHER STATIONS,

NEAR LONDON,

IN THE YEARS

1809, 1810, 1811.

TABLE XXXII.

1809.			Wind.	Pressure.		Temp.		Evap.	Rain, &c.
				Max.	Min.	Max.	Min.		
5th Mo.	New M.	May 14	E	29·92	29·81	77°	54°	23	
		15	E	29.81	29·71	68	50	8	8
		16	E	29·80	29·71	75	50	29	
		17	SE	29·80	29·76	77	57	39	
		18	SE	29·76	29·55	80	61	28	1
		19	Var.	29·74	29·51	75	53	14	25
		20	SW	29·90	29·74	64	47	12	—
		21						—	
	1st Q.	22	Var.	30·23	29·90	69	45	29	
		23	NE	30·23	30·14	69	44	27	
		24	NE	30·14	30·04	69	48	29	
		25	SE	30·04	29·89	64	50	11	
		26	SW	29·89	29·57	67	54	20	—
		27	S	29·61	29·57	70	46	22	
		28	SW	29·57	29·52	69	51	—	—
	Full M.	29	SW	29·75	29·48	66	41	38	25
		30							—
		31	S	29·80	29·45	66	49	40	—
6th Mo.		June 1	Var.	29·35	29·30	76	46	33	—
		2	SW	30·00	29·30	59	42	26	—
		3	Var.	30·00	29·64	66	50	19	7
		4	SW	29·64	29·27	67	48	18	6
	L. Q.	5	S	29·56	29·25	66	52	20	11
		6	SW	29·68	29·56	63	49	20	6
		7	SW	29·82	29·75	64	48	26	—
		8	SW	29·75	29·53	62	50	6	13
		9	W	29·56	29·54	61	46	10	4
		10	NW	29·71	29·56	64	50	12	6
		11	SW	30·05	29·71	59	48		
		12				68	53	24	
				30·23	29·25	80	41	5·83	1·12

NOTES.—Fifth Mo. 14. *Nimbi* in the W, and lightning from 10 to 11 p.m.: a small meteor. 15. a.m. thick air to the SE: a little rain: about 9 p.m. a *Nimbus* in the S, illuminated by frequent lightning: more distant lightning in the NW. 16, 17. *Cirrus* with *Cirrostratus*. 18. a.m. *Cirrocumulus*: p.m. overcast with *Cirrostratus*. 19. p.m. a most violent storm of thunder, hail, and rain, of which see the particulars at the end. 21. The cuckow very noisy at 11 at night. 31. Cloudy and windy.

Sixth Mo. 1. At night, barometer falling, wind SE, no dew: the sky overcast with clouds, among which is the *cymose*, or waved

Cirrostratus. 2. The foregoing prognostics of stormy weather have been fully verified. In the night, the wind rose and blew a tempest from SE: while it shifted to SW this morning, we had wet squalls of excessive violence: many trees have been blown down and the foliage and young fruit torn off in great quantities. 4. A rainbow at 4 a. m.: showers and wind. 5. Wet morning: stormy evening and night. 6. A brilliant bow at half past 5 p. m. squally: large *Cirri.* 7. Windy, showers: a bank of clouds in the NW at sunset. 8. Windy: a faint bow at 6 p. m. 9. Windy: a bow near sunset, and an extensive bank on the horizon. 10. Windy. 11. Calm at night.

RESULTS.

Winds Easterly in the fore part of the period, in the latter part Westerly.

Mean height of the barometer.........29·73 in.

thermometer........58·89°

Evaporation ....5·83 in.

Rain........ ....1·12 in.

Account of the Hail Storm on the 19*th of Fifth Month,* 1809.

The day had been sultry like some preceding ones, and overcast with clouds, which during the afternoon gave evident demonstrations of an approaching discharge of electricity. Large and deep *Cumulostrati* were ranged side by side, mingled with the *Cirrocumulus* and *Cirrostratus*, the whole having that peculiar almost indescribable character, which these charged conductors assume, when wrought up to the highest state of *tension.* About five in the afternoon, being at the laboratory, and perceiving a continued roll of thunder, with vivid lightning approaching from the south, and the appearance of a heavy shower in that quarter, I anticipated a storm of no common violence. We were proceeding to take measures for the safety of some glass utensils, when in an instant there opened upon us a volley of hail of such tremendous force, as in ten, or at most fifteen, minutes demolished most part of the skylights and south windows in the neighbourhood. These *icy bullets*, some of them a full inch in diameter, were discharged almost horizontally from a cloud to windward, and in such quantity as to be drifted in large masses under the walls. Whether borne by the impetuous blast that came with them, or carrying the air thus before them, I could not determine, but such was the velocity of their motion that in many instances a clear round hole was left in the glass they pierced: and one large pane (which I saw) had *two* such perforations distinctly formed. The water in the river, lashed

6

by the hail and raised by the wind, resembled a cauldron boiling violently, rather than waves with breakers. The electrical discharges were incessant, approaching with the cloud, and passing off with it: so that the whole resembled, in effect, the more mischievous artillery of human invention; inspiring, in spite of philosophical consideration, and delight at the grand and unusual phenomena, a sense of actual danger.

This sudden irruption over, it rained for a while moderately. The wind was at first E, then S during the hail, then W, then E, then W again. About seven, the clouds all at once put off their stormy character, and appeared as if going to sleep after this prodigious expenditure of power. The remainder of the evening was calm and pleasant.

A person who was on the road from London to Bow, during the storm, informed me that he experienced nothing but continued thunder and lightning, and *very heavy rain*, the latter appearing luminous on the ground on each side of him; which it often does in very heavy storms. It was evident from other circumstances that the hail was bounded in a western direction by the village of Bow: and it reached eastward from thence only about three miles. Its course appears to have been from S to N, over Blackheath, Greenwich, Blackwall, Bromley, Plaistow, West Ham, and Stratford, and so up the country between the rivers Lea and Roding, terminating probably on Epping Forest.

The damage done by the hail was very great: a London newspaper estimated it, from the accounts which reached the Editor, at 200,000 squares of glass broken in sashes, sky-lights, conservatories, hot-houses, &c. besides the injury done to the crops in fields and gardens. The foliage of large elms was cut off, and scattered on the ground to a furlong's distance to leeward; and fruit trees, besides being thus stript, received wounds in their bark which were visible long after. A West-Indiaman in passing Blackwall during the storm had her fore and main-top sails blown over the side, and one man drowned. We had no account of any individual being struck by the lightning.

Great as was the disturbance of the atmospheric electricity on the above occasion it was far exceeded in France, at the same season in 1808. *Cotte*, Journal de Physique, v. 68, p. 334, has this note under May 21. " Hail storms, which ravaged twenty parishes in the department of La Loire, and fifteen in that of Lot."

On this day, I observe, the weather changed to wet at London: on the other hand, the day of our hail-storm was a wet day, with some thunder at Paris.

A letter from Palermo states, that a fall of snow had taken place in the course of the last month at *Messina*, which for several days lay on the ground to the depth of six inches, *a circumstance* (says the writer) *never before known in this country.*—(Pub. Ledger).

Meteoric Explosions in Scotland.

A letter from Stewarton, dated July 18, says, on Tuesday the 11th current, at this place and for some miles round, a noise loud as thunder, but of much longer continuance, and very different in sound, was heard in the air.—The sky at this time was remarkably calm and serene.—The noise began with three loud distinct reports, accompanied with a kind of whizzing, resembling very much the firing of cannon. After that it was, to appearance, the rumbling of a vast heap of stones, which continued for a considerable time, and then turned to something very like the rattling of carriages: this continued so long that some people could hardly be persuaded that it was in the air—to those however who were out of their houses, the noise appeared as being almost perpendicular above their heads.—About the same time some persons are said to have seen red hot balls of fire rise, to appearance, from the earth, near Rosemount, on the road to Ayr, which after ascending entire a great height, fell in pieces into the sea, upon which a most tremendous noise immediately succeeded.—(Edinburgh Star).

These phenomena, compared with those which have in many instances accompanied the fall of meteoric stones, may be deemed sufficient evidence of some having fallen in the sea W of Scotland: their having appeared to " rise from the earth " is only a proof that the meteor came in a horizontal direction from the eastward.

M

TABLE XXXIII.

1809.			Wind.	Pressure. Max.	Min.	Temp. Max.	Min.	Evap.	Rain, &c.
6th Mo.	New M.	June 13		30·06	29·89	71°	52°	—	
		14		29·89	29·84	69	53	24	
		15	NW	29·93	29·84	65	43	9	
		16	SW	29·93	29·77	68	51	18	
		17	W	29·78	29·73	73	52	25	
		18	NW	29·85	29·78	65	44	15	
		19	NW	29·97	29·85	72	53	14	
	1st Q.	20	NW	30·17	29·97	78	51	14	
		21		30·28	30·17	78	50	—	
		22	SE	30·28	30·25	73	50	34	
		23	E	30·30	30·22	79	52	23	
		24	NE	30·39	30·28	75	55	24	
		25	N	30·39	30·37	70	44		
		26	NE	30·39	30·16	67	43		
	Full M.	27	N			72	54		
		28	N			66	52		
		29	NE	30·03	29·99	63	48		
		30	SE	29·94	29·90	67	50		
7th Mo.		July 1	S	29·94	29·78	72	49		
		2	Var.	29·78	29·73	71	50		
		3	NW	29·74	29·49	61	47		
	L. Q.	4	Var.	29·48	29·43	60	48		
		5	N	29·51	29·50	62	52		
		6	E	29·73	29·63	68	55		
		7	N	29·75	29·74	71	60		
		8	E	29·80	29·77	74	53		
		9	NW	29·80	29·78	68	52		
		10	NW	29·98	29·82	62	50		
		11	NW	30·03	30·02	70	51		
				30·39	29·43	79	43		

NOTES.—Sixth Mo. 16—18. Windy. 19, 20. A blush on the evening twilight. 22. Cloudy, a little rain p. m.: a *Stratus* at night. 23. The same. 24. Cloudy with wind.

Seventh Mo. 1. Heavy showers a. m. 2. Very wet p. m. 3. Showery. 4. Very wet with thunder p. m. 5, 6. Showers, followed by wind. 7. Evening very stormy: lightning almost incessant, thunder distant. 8. Fair. 9. Very wet. 10. Wet a. m., fair p. m.

RESULTS.

Mean height of barometer.............. 29·92 in.

thermometer .... 59·37°

In consequence of my absence from home, the accounts of Evaporation and Rain were left imperfect. It appears by the Register of the Royal Society, that the gauge at Somerset House collected 2·53 in. of rain during this period.

Whirlwind.—July 6.—The inhabitants of Cirencester were alarmed by the appearance of a *Tornado*. It was first observed about three miles to the S of the town, where it assumed the appearance of a large conical hay-rick covered with smoke. It moved rather slowly at first towards Cirencester, throwing down many trees in the parish of Siddington. Some persons had time to get upon the tower of Preston Church to observe its course. When it approached nearer it moved with a velocity almost incredible, and making towards the bason of the Canal, where it did considerable damage, skirted the town, and entered Lord Bathurst's Park from the Tetbury road. Here its fury seemed to be at its height : for timber trees from six to ten feet in girth were torn completely up by the roots, whilst others were stript of their branches, or literally cut asunder. After crossing the park it entered an orchard, where it threw down several trees and seemed to disperse as it could no longer be traced by the eye.—(Sun Paper).

TABLE XXXIV.

	1809.		Wind.	Pressure.		Temp.		Evap.	Rain, &c.
				Max.	Min.	Max.	Min.		
7th Mo.	New M. July	12	NW	30·06	30·01	75°	61°		
		13	NW	30·10	30·04	71	48		
		14	SW	30·14	30·06	73	55		
		15	NW	30·04	29·97	73	56		
		16	NW	29·93	29·81	75	41		
		17	NW	29·78	29·69	69	49		
		18	N	29·91	29·83	60	46		
		19	NW	29·99	29·99	69	53		
	1st Q.	20	NW	30·16	30·05	70	41		
		21	SE			72	50		
		22	NE	30·00	29·93	64	53		
		23	N	29·85	29·80	75	48		
		24	N	29·80	29·75	73	58		
		25	NE	29·73	29·66	81	59		
	Full M.	26	W	29·70	29·67	76	61		
		27	NW	29·67	29·62	75	61		
		28	W	29·63	29·57	70	52		
		29	SW	29·70	29·66	69	57		
		30	W	29·55	29·50	69	52		
		31	S	29·53	29·48	68	57		
8th Mo.	Aug.	1	W	29·66	29·60	71	55		
		2	W	29·68	29·51	70	56		
	L. Q.	3	S	29·35	29·31	69	49		
		4	SW	29·68	29·38	62	54		
		5	SW	29·70	29·49	65	57		
		6	S	29·34	29·24	70	55		
		7	NW	29·74	29·57	66	53		
		8	W	29·90	29·89	73	54		
		9	SE	29·88	29·83	75	53		
		10	E			82	63		
				30·16	29·24	82	41		

NOTES.—Seventh Mo. 14. Cloudy, windy. 15. A slight shower a. m. 16—19. Fair. 20. Some rain. 21—24. Fair: a *Stratus* at night the 23d. 25. A thunder storm began at one a. m. and continued till three: the lightning unusually vivid, the thunder distant. 26. Distant thunder again, with frequent lightning this evening. 28—30. Showers. 31. Fair.

Eighth Mo. 1—3. Showers. 4. Heavy rain p. m. 6. The same: rainbow twice. 7. Showers. 8, 9. Fair. 10. A storm of thunder, lightning, and heavy rain from 2 to 5 a. m.

2

RESULTS.

Mean height of barometer .... 29·75 in.

thermometer .... 61·95°

Rain omitted (together with the Evaporation) through my absence from home. The gauge of the Royal Society afforded 1·51 in. up to the 16th of eighth month, (when my account recommences) making 4·04 in. for the whole interval: and I found in the cistern of my own, on my return, 3·50 in. a portion (at least equal to the difference) having doubtless escaped.

Effects of Lightning.—Returning from Scotland through Newcastle-upon-Tyne, I was invited to examine the house of David Sutton, then undergoing considerable repairs from the damage done to it by a stroke of lightning, during a storm here on the third of eighth month. Out of several circumstances more usual in such cases, I selected the following as interesting:—The lightning first took (as it appears) a chimney, which it threw down, bringing a quantity of bricks and soot into a room in which a party of eight persons were at tea: *after the first explosion* a ball of fire made its appearance under the door opposite the chimney, where it remained long enough for the whole company to notice it. It then moved into the middle of the room, and separated into parts, which again exploded like the stars from a rocket. The electric fluid seems to have pervaded the whole house, and *every inch* of the bell-wires was oxidized and dispersed in beautiful mossy ramifications on the walls.

TABLE XXXV.

1809.			Wind.	Pressure.		Temp.		Evap.	Rain, &c.
				Max.	Min.	Max.	Min.		
8th Mo.	New M.	Aug. 11			29·63	77°	55°		
		12		29·63	29·62	66	54		
		13		29·74	29·68				
		14							
		15							
		16							
		17	SW	29·80	29·76	79	56	20	44
	1st Q.	18	SW	29·76	29·75	71	55	17	4
		19	SW	29·92	29·76	69	55	24	
		20	S	29·88	29·78	67	55	17	6
		21	SW	29·79	29·76	65	49	8	—
		22	SW	29·79	29·57	67	49	22	5
		23	SW	29·57	29·45	67	45	17	
		24	SW	29·45	29·40	66	46	19	17
	Full M.	25	SW	29·77	29·40	66	47	3	8
		26	SW	29·72	29·67	65	51	11	—
		27	W	29·96	29·72	65	53	8	—
		28	Var.	30·06	29·96	69	54	22	
		29	S	30·06	29·83	65	47	25	
		30	N	29·80	29·67	78	61	14	5
		31	SW			65	57	—	
9th Mo.	L. Q.	Sept. 1						—	—
		2	N	29·67	29·57	72	58	31	19
		3	E	29·57	29·57	72	57	13	—
		4	E						
		5	Var.	29·57	29·48	74	57	19	32
		6	SE	29·48	29·22	70	57	14	12
		7	Var.	29·28	29·20	67	54	4	4
		8	NW	29·53	29·28	65	50	7	
				30·06	29·20	79	45	3·15	1·31

NOTES.—Eighth Mo. 17. A fine day: at sunset a dense bank of clouds in the E, low in the horizon, among which lightning played incessantly. 18. A wet morning. 20—22. Much wind. 24. Showers p.m. with a strong variable electricity: the bow appeared twice. 25. Thunder showers in various directions near us: the rod in high charge, mostly *negative:* rainbow. 26. The rod gave sparks, *positive,* without rain at the time, the sky being overcast with *Cirrus* and *Cirrostratus.* 27. A strong *positive* electricity in an atmosphere very clear below, with *Cumulus* and *Cirrocumulus.* 28, 29. The air *positive,* and serene. 30. About 5 p.m. a shower of large warm drops, strongly *negative*, heavier rain to the NE: about seven, at temperature 67°, a heavier shower, *positive* : about nine, a third charge in the rod: the bell rang much, and the pith balls with their threads showed luminous. Rain followed, with a changeable electricity, and much thunder and lightning in the W.

Ninth Mo. 5. The wind shifted during a heavy rain from E to W, the shower was once tried and found *positive*. 7. The wind veered to N by night.

RESULTS.

The Winds chiefly Westerly.

Mean height of barometer .... 29·66 in.
thermometer .... 61·15°
Evaporation (in 23 days) .... 3·15 in.
Rain (in the same) .... 1·31 in.

There was generally much dew, the frequent rains notwithstanding.

TABLE XXXVI.

	1809.		Wind.	Pressure. Max.	Min.	Temp. Max.	Min.	Evap.	Rain, &c.
9th Mo.	New M.	Sept. 9	W	29·63	29·53	68°	54°	13	
		10	W	29·65	29·63	63	57	8	2
		11	NW	29·76	29·65	65	53	8	
		12	SW	29·82	29·76	63	46	8	2
		13	SE	29·76	29·64	62	54	9	26
		14	NE	29·95	29·76	64	54	6	15
		15	SW	30·13	30·11	66	44	9	
	1st Q.	16	W	30·11	29·95	64	57	11	—
		17	S	29·95	29·85	63	50	7	—
		18	W	29·67	29·55	64	52	14	
		19	Var.	29·72	29·43	63	42	8	6
		20	W	29·53	29·27	67	54	21	8
		21	SW	29·81	29·57	64	52	8	17
		22	SW	29·67	29·59	69	58	19	
	Full M.	23	SW	29·80	29·67	65	50	21	5
		24	SW	29·86	29·66	58	53	11	13
		25	NW	29·95	29·80	54	38	3	27
		26	W	29·95	29·45	58	43	12	1
		27	W	28·78	29·33	65	38	10	13
		28	NW	29·94	29·87	55	35	8	8
		29	SW	29·98	29·90	55	39	8	1
		30	Var.	30·12	29·80	58	47	2	—
10th Mo.	L. Q.	Oct. 1	N			60	45	—	—
		2	SW	30·32	30·30	61	56	9	—
		3	Var.	30·22	30·17	63	55	4	
		4	E	30·17	30·05	64	40	13	
		5	E	30·05	30·04	63	48	14	
		6	E	30·04	30·02	64	45	14	
		7	NE	30·09	30·02	61	35	14	
		8	E	30·09	30·05	58	42	17	
				30·32	29·27	69	35	3·09	1·44

Notes.—Ninth Mo. 11. Between 5 and 6 p. m. a shower from clouds formed overhead: the rain showed no electricity, but a moderate *positive* charge ensued as it ceased. 14. A hard shower, p. m. mixed with hail: at night heavy rain, with lightning in the S. 18. A blush on the evening twilight: windy. 19, 20. The same. 21. Lunar halo. 22. Stormy. 24. The moon rose clear but with a red recess (as it were) round it: the night was stormy. 25. Rain most of the day. 26—29. Hoar frosts: windy. 28. A wet squall mixed with hail p. m.: much redness, as if from elevated *Cirri* at sunset.

Tenth Mo. 4. A steady breeze, E. At sunset *Cirrostratus* in the W, beautifully tinted, and in the E, much dewy haze. 5. Abundance of dew: *Gossamer: Cirrus, Cirrocumulus.* At sunset a dewy haze highly coloured in the E, succeeded by much red in the W. 7. Hoar frost, misty morning: much dew at sunset. 8. Misty: windy: *Cirrocumulus.*

RESULTS.

Winds chiefly Westerly till near the close of the period.

Mean height of the barometer......... 29·84 in.

of the thermometer .... 55·20°

Evaporation .... 3·09 in.

Rain .... 1·44 in.

Mount Vesuvius.—A letter from Naples, dated the 9th ult. gives the following account of the state of this mountain:—" After a few days of tranquillity, the eruptions of Vesuvius recommenced. On the 4th of September, a new crater opened to the south-east, much larger than that which was formed in 1807. This mouth has, since that time, constantly vomited a vast torrent of lava, which takes its direction towards the town of Torre, formerly destroyed by an eruption of the volcano. This torrent divides in two branches, and forms an island, at the extremity of which the lava unites, and produces a lake of fire, in the district of Atrio del Cavallo. No damage has yet taken place; and there is nothing to be feared, unless the torrent should change its direction, or become more impetuous. In the night of the 5th, as in the month of August 1782, Vesuvius vomited an immense quantity of ashes, stones, and smoke, forming upon the mountain a second mountain, the summit of which appeared to rise to the skies; but these substances receiving an impulsion absolutely perpendicular, have fallen into the crater, or in its neighbourhood, but have not done any injury either to the inhabitants or the houses, nearest the eruption."— (PUB. LEDGER).

Plymouth Dock, Sept. 23.—It has blown quite a hurricane for the last two nights, and about two this morning its violence considerably increased; nor is it now (5 p.m.) much abated, there is still a great surf in the Sound.—Both the last and the preceding nights, we had also an unusually heavy rain. We have scarcely had a dry day these three months.

The General Beresford, Capt. Appleby, from Honduras to Bristol, was totally lost the 21st September, in the Gulph of Florida, during a heavy gale of wind. Capt. A. writes that seven other vessels were lost there at the same time.—(PUB. LEDGER).

TABLE XXXVII.

	1809.		Wind.	Pressure		Temp.		Evap.	Rain, &c.
				Max.	Min.	Max.	Min.		
10th Mo.	New M.	Oct. 9	NE	30·05	30·02	52°	35°	20	
		10	NE	30·01	30·00	50	40	27	
		11	NE	30·10	29·99	52	37	—	
		12	NE	30·19	30·05	51	31	38	
		13	E	30·27	30·19	51	29	—	
		14	NW	30·33	30·27	55	27	—	
		15	NW	30·27	30·10	52	32	18	
	1st Q.	16	SW	30·10	30·07	56	48	0	—
		17	SW	30.07	30·05	60	52	10	7
		18	W	30·13	30·06	58	54	—	
		19	W	30·15	30·12	58	52	15	—
		20	SW	30·12	30·10	54	50	2	
		21	S	30·08	30·06	55	50	5	
		22	Var.	30·06	29·98	54	49	4	
	Full M.	23	SE	29·98	29·89	55	46	—	
		24	SW	30·10	29·89		47	—	
		25	Var.	30·27	30·10	64	48	31	
		26	NE	30·27	30·25	67	43	4	2
		27	NW	30.26	30·22	63	46	2	2
		28	S	30·26	30·24	54	45	0	3
		29						0	—
		30	NE	30·24	30·07	53	45	1	3
	L. Q.	31	N	30·05	30·02	52	39	1	1
11th Mo.		Nov. 1	N	30·20	30·11	53	41	7	
		2	N	30·11	30·10	50	37	8	
		3	N	30·10	29·89	47	41	8	7
		4	NE	29·84	29·76	46	37	1	25
		5	NE	29·90	29·84	44	36	1	12
		6	NE	30·05	29·84	46	36	2	2
				30·33	29·76	67	27	2·05	0·65

NOTES.—Tenth Mo. 9—12. Fine. 13. Foggy morning: hoar frost. 14. Hoar frost. 15. Cloudy; very misty night. 16. Cloudy. 17. Overcast by *Cirrostratus*, which was coloured by the setting sun: after which, rain towards morning. 18—22. Cloudy. 23. A *Stratus* at night. 24. Clear; much wind at night. Misty mornings have been frequent for two weeks past. 27. Wind N a. m., W evening, with much mist.

Eleventh Mo. 1. Misty morning: very fine day. 2. Fine clear morning. 3—6. Rain.

RESULTS.

Prevailing Wind Northerly.

Mean height of the barometer..........30·08 in.
thermometer........47·74°
Evaporation........................ 2·05 in.
Rain............................. 0·65 in.

The rain of the 17th, which fell in the midst of a month of fine weather, was probably the effect of a southerly current. The barometer had not fallen three-tenths, and it was high at the time: but the mean temperature on the preceding day had risen 10°, after two or three frosts, and the evaporation was quite suspended. The products of the rain gauge from the 26th to the 31st inclusive were *wholly derived from copious mists and dews*. The latter part of the tenth month appears by the papers to have been stormy in the West Indies: a heavy gale of wind at Martinique, at least, is particularly mentioned.

On the 25th of tenth month my friend Thos. Forster, found by means of a small balloon sent up from Clapton about 2 p.m., that there prevailed on that afternoon no less than *four* different currents in the atmosphere, at different heights, in the following order, beginning from the lowest: ESE, N, SW, SSE. The wind below was variable and very gentle, and there were a few clouds in the sky, of the modification *Cirrus*. He found by similar means, on the following day, *three* currents, *viz.* ENE, a little above, SE, and lastly, SSW.

Lisbon, Oct. 27.—On the 25th of this month, at fifty minutes past nine o'clock at night, the shock of an earthquake was felt. The people of this city, who have not forgotten the fatal catastrophe of the 1st of Nov. 1755, were greatly alarmed, and with reason. It lasted but an instant, and did no damage. Some persons say, that they felt another about two the next morning; but this is not generally believed, as fear will sometimes produce that effect on the imagination.—(Papers).

TABLE XXXVIII.

1809.			Wind.	Pressure. Max.	Min.	Temp. Max.	Min.	Evap.	Rain, &c.
11th Mo.	New M.	Nov. 7	NW	30·42	30·05	52°	36°	3	
		8	NE	30·43	30·41	47	41	2	
		9	N	30·41	30·32	49	45	—	
		10	SE	30·32	30·12	50	42	25	
		11		30·12				—	
		12	SW		29·74	46	41	7	3
		13	NW	29·77	29·72	47	32	1	
	1st Q.	14	N	29·69	29·64	46	35	6	17
		15	N	29·72	29·68	40	28	—	
		16	NW	29·75	29·50	39	27	17	—
		17	W	29·72	29·44	46	27	—	—
		18	N	30·27	29·72	38	29	—	—
		19	NW	30·47	30·27	38	22	—	
		20	NW	30·30	30·24	38	26	—	
		21	W	30·30	30·12	42	30	—	
	Full M.	22	W	30·12	30·04	51	37	19	1
		23	SW	30·04	29·37	49	41	—	35
		24	NW	29·76	29·37	44	34	7	1
		25	SW	29·76	29·10	45	36	5	1
		26	NW	29·63	29·10	45	26	—	
		27	NW	29·95	29·63	41	27	—	—
		28	W	29·95	29·88	39	33	4	1
		29	SW	29·88	29·85	40	31	—	
	L. Q.	30	SW	29·30	29·25	47	32	4	32
12th Mo.		Dec. 1	W	29·60	29·30	44	36	7	4
		2	W	29·92	29·60	43	31	—	
		3	SW	29·60	29·50	50	37	10	3
		4	NW	29·90	29·43	49	35	5	3
		5	SW	30·11	30·00	46	34	6	
		6	SW	30·00	29·82	54	47	6	
				30·47	29·10	54	22	1·34	1·01

NOTES.—Eleventh Mo. 8—13. Misty mornings. 14. The same: rain at night. 17. Snow early, preceded by much redness at sunset: wind very boisterous during the day, with showers. 18. p. m. The ground covered by several snow showers. 19. Hoar frost. 22. Misty morning: wind. 23. A stormy night. 24. Hoar frost. 25. Wind in the night. 26. Morning overcast: day pretty fine. 27. A very white frost on the ground. 28. Misty day. 29, 30. Misty mornings, the latter a stormy night.

*(*Notes, &c. *continued).*

Twelfth Mo. 1, 2. Much wind by night. 3. *Cirrrostratus* a. m. with a turbid sky. 5. Hoar frost: windy night. 6. Cloudy, windy: the maximum temperature at 9 a.m.: a brisk gale all night.

RESULTS.

Prevailing Winds Westerly.

Mean height of the barometer..........29·86 in.
       thermometer .... 39·32°
Evaporation .... 1·34 in.
Rain .... 1·01 in.

Dover, Nov. 27.—It has blown very hard, and several ships have gone past into the Downs with damage.—(Pub. Ledger).

TABLE XXXIX.

1809.			Wind.	Pressure. Max.	Min.	Temp. Max.	Min.	Evap.	Rain, &c.
12th Mo.	New M.	Dec. 7	SW	30·25	29·82	54°	35°	8	4
		8	SW	29·82	29·65	48	37	4	4
		9	SW	29·65	29·15	53	42	8	3
		10	SW	29·30	29·10	49	37	9	15
		11	SW	29·45	28·94	46	36	7	10
		12	SW	29·20	29·05	48	36	12	1
	1st Q.	13	SW	29·40	29·20	43	30	5	—
		14	SE	29·20	28·70	44	33	6	19
		15	S	28·83	28·70	41	32	4	1
		16	SE	28·80	28·68	40	31	3	4
		17	SW	28·38	28·25!	44	34	6	45
		18	W	29·22	28·25	42	39	4	—
		19	NW	29·73	29·22	44	34	4	—
		20	SW	29·74	29·60	43	32	3	11
	Full M.	21	NW	29·95	29·74	44	32	5	
		22	W	29·95	29·90	44	35	1	
		23	Var.	29·93	29·84	37	28	—	—
		24	SW	30·05	30·00	36	28	3	3
		25	Var.	30·05	29·77	43	35	2	8
		26	Var.	29·83	29·74	43	38	1	7
		27	W	30·10	29·83	41	34	2	1
		28	SW	30·10	29·75	45	33	3	—
				30·25	28·25	54	28	1·00	1·36

NOTES.—Twelfth Mo. 8. Misty: very slight hoar frost: windy night. 9. Misty a. m., stormy night: the barometer had descended lower and was found rising. 10. The day pretty calm, the night windy. 11. A lunar halo, large and faint: at night the wind rose again and blew very hard till morning, with rain. 12. A squall at 11 a. m. with rain and some hail. 13. At 8 a. m. several *Nimbi* to the S; hoar frost: about 1 p. m. a squall with rain and hail. 14. Hoar frost, misty: at evening a large lunar halo: the fore part of the night a violent storm from the SE. 15. a. m. calm with *Nimbus*. 16. Hoar frost, calm and cloudy, with *Cirrostratus* and *Nimbus;* a large lunar halo just perceptible: rather windy night. 17. The wind now and then strong, but on the whole moderate by day: the night stormy and very wet. 18. a. m. very dark with clouds. The barometer very unsteady, having thrice changed its direction since last time of noting. 19. Windy. 20. a. m. calm and cloudy: wind moderate afterwards, with rain. 21. Hoar frost, with frozen drops of considerable size ad-

2

hering to the grass. *Cirri* p. m. in fine groups: clear night. 22. Hoar frost. 23. Misty. 24. Snow, a. m. 25, 26. Misty, overcast. 28. Cloudy, after a windy night. 29. Much wind by night: the rain came from the NW.

RESULTS (of twenty-two days).

Prevailing Winds Southerly.

Mean height of the barometer .... 29·44 in.
thermometer........39·20°
Evaporation....................... 1·00 in.
Rain.............................. 1·36 in.

Of which there fell by night 1·15 in. by day 0·09 in.

This great and long continued depression of the barometer came to a *crisis* on the 18th of twelfth month at seven in the morning: from which time a steady equable rise took place for twenty-four hours: and was prolonged at a less rate, for three or four days after. The depression was very extensive, being probably not limited by the northern, and certainly not by the southern extreme latitude of our islands. Thus, at Manchester, the barometer descended to 28 in. *(Hanson)*; and at Paris it was down on the 17th to about 28·5 in. English, *(Journ. de Physique)*, the wind there being SSE, and the weather stormy and wet. In fact, during a series of tempestuous *nights*, (it was chiefly by night that the barometer fell) the whole Northern Atlantic atmosphere appears to have been in general and rapid motion to the northward. Yet the rapid rise of the barometer was not (in this neighbourhood) connected with any violent movement in the contrary direction.

State of the Barometer within 36 years —By the Register of Mr. John Mills, of Bury, it appears, that on the 12th of Jan. 1773, it was at 28·25.— On the 11th of March, 1773, at 28.—On the 20th of Jan. 1791, at 28; and on the 17th and 18th of Dec. 1809, at 28.—(PAPERS).

Deal, Dec. 12.—We have experienced a very heavy gale of wind from the SSW the greater part of last night and this morning, and it still continues to blow very hard.—(PUB. LEDGER).

Plymouth Dock, Dec. 13.—It has blown a strong gale at SSW for several days past, and still continues, accompanied with heavy deluges of rain.—(PUB. LEDGER).

Greenock, Dec. 15.—The Mars, Captain Harvey, arrived here yesterday from Quebec, and has experienced, since the 3d instant, the most severe weather. To preserve the ship a considerable part of the cargo was thrown overboard.—(PUB. LEDGER).

Montrose, Dec. 16.—The whole coast is strewed with wreck.—(PUB. LEDGER).

NOTES, &c. *(continued).*

Plymouth, Dec. 16.—Wind SW. The violence of the wind on Thursday night has seldom been exceeded, even in this place, which is exposed to the furious blasts of the Atlantic. It suddenly veered round to the SE, and threw such a tremendous swell into the Bay, that the Temeraire was for some time in great danger of foundering.—(PUB. LEDGER).

Greenock, Dec. 18.—The Argus, Fortune, from Le Have, in Nova Scotia, arrived at the Tail of the Bank on Saturday morning last, after a passage of 32 days, during a great part of which she experienced very heavy gales of wind.—(PUB. LEDGER).

Torbay, Dec. 18.—It has blown quite a hurricane from the S and W the whole of the last week.—(PUB. LEDGER).

Plymouth, Dec. 18.—We are still visited by the same severe blowing weather latterly noticed. The wind has now shifted to the north, and seems inclinable to veer round to the eastward.—(PUB. LEDGER).

Deal, Dec. 18.—It blew very hard last night, and early this morning, from the S, and SSW.—(PUB. LEDGER).

Whitehaven, Dec. 19.—We have had nearly a week of the most stormy and tempestuous weather that can be imagined, particularly on Friday morning, when there was a most violent gale of wind from the southward, accompanied with hail and sleet, which was followed by a very heavy rain. On Friday evening, there was a very awful storm of thunder and lightning, such as, perhaps, was never witnessed here at this season of the year; the latter was uncommonly vivid, and the peals of thunder were exceedingly loud, and in rapid succession. It continued, with little intermission, the whole night. Several chimnies have been blown down, roofs damaged, &c. We learn, that a great deal of snow has fallen in the interior of the country. It is said to be lying in drifts of nine feet deep in some places on the east side of this county, and the adjoining part of Northumberland.—(PUB. LEDGER).

The Meteor called Falling Stars in connexion with an Earthquake.

Cape of Good Hope, Dec. 8, 1809.—On the 5th instant we were still in a state of alarm, on account of the earthquake the night before at half-past ten: we had three successive shocks of a tremendous nature, accompanied by a most awful sound. The whole lasted about three minutes, with an interval of hardly two seconds between each shock. During all the time there was a serene sky: the stars shone bright, but we observed *an uncommon number of stars falling.*—(TIMES).

THE PRESENT PERIOD terminates, as the Reader will probably have remarked, at the end of three weeks, instead of the lunar month as usual. The reason is, that I here determined on the adoption of a new period, beginning with the Moon's last quarter, which I found to be more convenient for investigating the variations of the Barometer and Thermometer, as affected by the two principal changes in the position of that planet with regard to the Sun and the Earth: and of which I may have occasion to treat hereafter. *This period I have used in my Tables ever since.*

Having ceased to publish my Observations when the ATHENÆUM was discontinued at Midsummer, 1809, I did not resume the printing of them for three years. In this interval, from causes which it is not important to mention, they were more than once interrupted; and as it will be found in the conclusion of this work, that a continuity of Observations in the same neighbourhood is of greater importance than I then supposed, I have endeavoured to supply the deficiencies in the following manner. I take the variations of the Barometer for a part of the years 1810 and 1811, with remarks on the clouds and state of the sky, from the Register of my friend *Thos. Forster, junior,* of Clapton. His notes will be found to contain much curious matter. With respect to his Barometer I find it to range about *a tenth of an inch higher* than my own, instead of as much lower, which it would have done, from the higher level of Clapton, had the two been adjusted: notwithstanding which I prefer its results to those of the Barometer then used at the Laboratory. The Observations of my partner *John Gibson,* at the latter station, in great measure supply the materials for the remainder of the deficient Tables. I shall distinguish the Notes from each other by affixing the initials of their respective authors.

TABLE XL.

1809.			Wind.	Pressure.		Temp.		Evap.	Rain, &c.
				Max.	Min.	Max.	Min.		
12th Mo.	L. Q.	Dec. 29	NW	29·70	29·50	49°	41°	9	6
		30	W	29·86	29·68	50	41	8	
1810.		31	SW	30·09	29·98	51	45	7	
1st Mo.		Jan. 1	SW	30·22	30·09	51	45	5	
		2	S	30·22	30·10	45	40	3	
		3	SW	30·25	30·10	50	43	3	
		4	S	30·34	30·25	50	44	4	3
	New M.	5	SE	30·30	30·25	47	44	5	
		6	SW	30·31	30·22	49	42	4	
		7	S	30·22	30·00	43	37	3	
		8	SE	30·00	29·85	46	39	1	—
		9	S	29·95	29·85	48	42	—	—
		10	SE	29·98	29·90	43	41	6	
		11	SE	29·90	29·87	46	42	5	—
		12	E	29·94	29·87	44	32	14	—
	1st Q.	13	E	30·03	30·01	33	27	—	
		14	NE	30·01	29·89	30	24	—	—
		15	N	29·89	29·78	28	16	—	—
		16	NW	30·02	29·78	28	10	—	—
		17	SE	30·26	30·02	31	14	—	—
		18	Var.	30·28	30·25	36	20	—	—
		19	E	30·25	30·19	31	14	—	—
	Full M.	20	NE	30·19	30·05	29	21	—	—
		21	E	30·05	29·97	36	30	—	—
		22	E	30·09	29·95	33	32	35	—
		23	N	30·22	30·09	37	33	1	3
		24	Var.	30·31	30·20	38	33	3	
		25	NE	30·36	30·31	33	31	—	
		26	E	30·31	30·26	36	31	—	
		27	NW	30·26	30·22	31	30	5	
				30·36	29·50	51	10	1·21	0·12

NOTES.—Twelfth Mo. 29. Much wind in the night. 30. The same. 31. Windy, overcast.

1810. First Mo. 1, 2. Windy, cloudy. 4. *Cirrostratus, Cirrocumulus.* 5. Fine grey morning: p. m. with the wind SE. *Cirri* pointing upward and westward: after which much loose *Cumulus,* 6 a. m. wind SW: pretty clear. 7. Grey morning. 8. a. m. quite overcast: a little drizzling from SE. 9. Very cloudy a. m. the wind rising at S, with some rain. 11. Fair, though cloudy. 12. Overcast, drizzling: a breeze at E, which increased at night to a strong wind; this continued till the next evening, the evaporation being more than double in consequence. 14. Snow, a. m. from NE. In the evening I observed wild ducks migrating to the W, the temperature 22°, with a breeze at N. The wind then passed to W, and blew strong. 16. Wind a. m. SW. Snow at intervals most of the day, cristallized in stars. 17. Much rime on the trees a. m., a little snow in the night. 18. Hail balls a. m., snow at intervals. 19. At night misty, with much rime and some hail balls. 20. A very clear day with us, though mist and smoke prevailed to the S: the trees, being quite shrouded in rime, are extremely beautiful. After sunset the temperature went down to 21°, then rose again by 10 p. m. to 26°, the wind springing up at NE, with cloudiness. 21, 22. Snow in stars. 23. Misty a. m., a little rain and sleet p. m. 24. Very cloudy: small rain. 25, 26, 27. Misty.

RESULTS.

Winds Westerly in the forepart, afterwards Northerly and Easterly.

Barometer: Greatest height . . . . 30·36 in.

Least. 29·50 in.

Mean of the barometer . . . . 30·07 in.

Thermometer: Greatest height . . . . 51°

Least. . . . . 10°

Mean of the thermometer . . . . . . 36·43°

Evaporation . . . . 1·21 in.

Rain . . . . 0·12 in.

February 19, 1810.—The greatest cold last winter at Moscow was in the night of January 11, when mercury exposed to the open air in a cup was frozen so hard, that it could be cut with shears, and even filed.—(PUB. LEDGER).

This effect requires a temperature considerably below − 40° of Fahrenheit: indeed the thermometer is said to have been seen at − 44°, ere the quicksilver made its final sudden retreat into the bulb by congealing. The date by our style (if not already corrected) would be the 23d of the month, and probably 1809.

TABLE XLI.

	1810.		Wind.	Pressure.		Temp.		Evap.	Rain, &c.
				Max.	Min.	Max.	Min.		
1st Mo.	L. Q.	Jan. 28	NW	30·32	30·22	31°	30°	—	
		29	E	30·48	30·32	34	31	—	
		30	Var.	30·48	30·39	32	30	—	
		31	SW	30·39	30·16	47	30	7	5
2d Mo.		Feb. 1	SW	30·16	30·08	50	47	—	—
		2	Var.	30·08	29·77	47	42	4	18
		3	Var.	29·87	29·68	46	42	1	3
	New M.	4	NW	30·07	29·87	45	30	—	
		5	SW	30·07	30·03	48	30	10	
		6	SW	30·03	29·95	47	41	5	—
		7	S	29·96	29·93	49	44	4	—
		8	SE	29·93	29·82	48	42	5	2
		9	SW	29·76	29·73	47	42	3	—
		10	NW	29·85	29·76	54	42	6	2
	1st Q.	11	E	29·76	29·40	43	39	—	17
		12	SW	29·40	29·04	43	36	8	23
		13	SW	29·28	28·98	45	33	1	11
		14	NE	29·67	29·28	43	33	18	—
		15	NW	30·08	29·67	39	23	—	
		16	N	30·12	30·08	41	25	—	—
		17	N	30·22	30·18	38	22	—	
		18	SW	29·98	29·84	34	26	—	—
	Full M.	19	N	30·24	29·98	35	23	—	—
		20	NW	30·49	30·24	33	11	—	—
		21	Var.	30·50	30·30	32	14	—	
		22	SW	30·30	29·50	36	29	—	—
		23	SW	29·54	29·40	45	36	—	
		24	SW	29·48	29·45	50	43	—	10
		25	W	29·91	29·48	50	35	40?	1
				30·50	28·98	50	11	1·07	0·92

NOTES.—First Mo. 28. In the evening some lightning in the NE: very calm: sleet. 29. Grey morning: calm: barometer fluctuates, scarce any sun for a week past. 31. The wind changing to SW with rain, a period was put the continued stagnation of temperature about the freezing point; which, with cloud and calm, has obtained ever since the full moon.

Second Mo. 3. A breeze at N a. m. 7. A little rain a. m. 10. The wind shifting suddenly to NW, brought the smoke of the city over us in the unusual form of a dense *elevated cloud*, with a clear space below: a little rain, and *Cumulostratus* followed. 11. Wind pretty strong:

cloudy and drizzling. 12. Very misty. The barometer has fallen 0·36 in. with a steady wind at E. The night was windy at SW, with some snow afterwards. 13. Squally. 15. A little snow: a lunar corona. 16. Hoar frost: some snow in the night: the moon showed of a *golden* colour. 17. Very fine day: large *Cumuli:* the moon *pale* at night, with a purplish colour round it. 18. Overcast a.m., much snow p.m., a slight thaw. 19. Clear morning, snowy evening. About nine, at the breaking of the clouds, the sky presented a fine spectacle. For some space around the full moon, it was of the usual lively grey: next to this appeared a pink, or pale rose colour, deepening as it receded towards the horizon. Against this unusual ground the silver lights, common to the broken *Cumuli*, were degraded to a cream colour, not less beautiful: and as any of these passed under the moon, they presented iridescent *spectra*, in which the red and blue were uncommonly vivid and distinct. These appearances were doubtless owing to the presence of an unusual quantity of frozen haze, out of which the snow was all the while forming. 20. About 11 a.m. a pinky haze appeared in a snow storm under the sun, as before under the moon. 27. A fine day. 28. Windy night.

RESULTS.

Winds Variable, but for the most part Westerly.

Barometer: Greatest height . . . . 30·50 in.

Least. . . . . 28·98 in.

Mean of the period . . . . 29·91 in.

Thermometer: Greatest height . . . . 50°

Least. . . . . 11°

Mean of the period . . . . 37·63°

Evaporation . . . . 1·07 in.

Rain . . . . 0·92 in.

TABLE XLII.

1810.		Wind.	Pressure. Max.	Min.	Temp. Max.	Min.	Evap.	Rain, &c.
2d Mo.								
L. Q.	Feb. 26		29·88	29·78	49°	42°	—	—
	27		30·01	29·72	54	47	17	—
3d Mo.	28	SW	30·01	29·86	56	39	11	—
	March 1	SW	29·86	29·69	54	51	7	—
	2	W	29·69	29·57	54	44	1	5
	3	SW	29·68	29·54	54	32	3	10
	4	Var.	29·54	29·24	50	40	1	22
New M.	5	Var.	29·24	28·92	46	31	4	36
	6	SW	28·92	28·81			—	—
	7	SW	28·96	28·81	48	36	8	14
	8	S	29·11	28·96	54	46	8	5
	9	SW	29·36	29·11	58	50	1	19
	10	SW	29·90	29·36	60	46	—	
	11	SW	29·90	29·68	57	50	24	48
	12	Var.	30·00	29·68	56	48	—	—
1st Q.	13	NE	30·00	29·95	48	38	13	—
	14	E	30·02	29·86	42	32	20	
	15	E	29·86	29·57	40	34	—	
	16	E	29·66	29·57	40	35	—	
	17	NE	29·89	29·66	43	30	44	
	18	E	29·95	29·89	45	24	—	
	19	SE	30·04	29·95	47	28	—	
	20	N	29·95	29·80	50	31	—	
Full M.	21	N	29·93	29·70	50	31	28	
	22	NE	30·17	29·95	45	27	—	
	23	NE	29·95	29·76	52	26	—	
	24	E	29·80	29·76	50	35	44	
	25	E	29·92	29·84	41	33	53?	
	26	E	29·92	29·77	43	34	19	
	27	SW	29·66	29·61	55	42	9	10?
			30·17	28·81	60	24	3·15	1·69

NOTES.—Third Mo. 1. Misty, drizzling. 2. Cloudy. 3. Much drizzling rain p. m. 4. Wet night. 5. Snowy forenoon. 7. The Lea has risen over its banks. 9. A heavy squall about 1 p.m., with rain: a stormy night: much water out. 11, 13, 14. Windy: cloudy. On the latter evening a remarkable lunar halo. A circle of *white* surrounded the moon at a considerable distance, which was well defined on the inner side, so as to make the included space appear dark, while on the outer it faded insensibly into the colour of the sky. 15. A little snow a. m. 17—20. Hoar frosts. In the day time, during this interval the *Cumulus* appeared, passing to *Cumulostratus*, and evaporating at evening. 21. Much wind at times in the night. 22, 23. Hoar frosts. 24, 25. The deposition of dew suspended. 26. A moderate gale at E, as yesterday. 27. Cloudy a. m., a breeze at SE: wet p. m.: night windy, SW. The rain is put down by estimate.

RESULTS.

Winds Westerly in the forepart with rain, afterwards Easterly with a dry atmosphere.

Barometer: Greatest height .... 30·17 in.

Least .... 28·81 in.

Mean of the period .... 29·67 in.

Thermometer: Greatest height .... 60°

Least .... 24°

Mean of the period .... 42·08°

Evaporation .... 3·15 in.

Rain .... 1·69 in.

A Friend who went from our neighbourhood to reside this winter at Cowley Bridge, near *Exeter*, has favoured me with the following Observations made there with a Six's Thermometer, during the Second month. The place is two miles and a half above the city, on the bank of the Ex.

1810.		Max.	Min.	1810.		Max.	Min.
2d Mo. Feb.	1	54°	46°	2d Mo. Feb. 15		34°	18°
	2	51	44		16	35	29
	3	46	40		17	34	18
	4	45	27		18	39	31
	5	45	29		19	36	33
	6	52	42		20	36	17
	7	51	35		21	34	25
	8	51	38		22	45	30
	9	55	38		23	50	44
	10	55	38		24	54	47
	11				25	52	34
	12	44	43		26	53	45
	13	44	38		27	57	42
	14	45	32		28	58	45

The mean temperature of this month near Exeter, appears by the above observations to have been 40°, while with us it was 39°—the extremes, 58° and 17°, with us 56° and 11°.

Portsmouth, March 25.—Wind Easterly. It has blown very hard all day, and still continues.

Deal, March 25.—Wind NE, blows hard.—(PUB. LEDGER).

TABLE XLIII.

1810.			Wind.	Pressure.		Temp.		Evap.	Rain, &c.
				Max.	Min.	Max.	Min.		
3d Mo.	L. Q.	March 28	W	29·90	29·66	52°	36°	22	
		29	Var.	29·92	29·88	55	35	10	
		30	SW	29·88	29·75	51	41	6	7
		31	S	29·75	29·42	52	41	—	4
4th Mo.		April 1	E	29·69	29·42	52	42	12	17
		2	SE	29·73	29·60	52	40	6	2
		3	SW	29·40	29·35	58	41	7	7
	New M.	4	W	29·70	29·40	50	34	9	
		5	SW	29·70	29·36	54	42	12	
		6	SE	29·36	29·30	48	41	7	2
		7	SE	29·45	29·30	52	41	6	
		8	E	29·47	29·45	56	42	15	10
		9	E	29·50	29·40	53	43	8	35
		10	NE	29·65	29·50	45	34	7	14
	1st Q.	11	NE	29·90	29·65	43	33	—	—
		12	NE	29·92	29·88	42	30	—	
		13	Var.	29·92	29·86	42	33	26	—
		14	SW	29·86	29·82	48	35	8	
		15	SW	29·82	29·57	52	33	17	
		16	Var.	29·50	29·37	52	36	12	10
		17	SW	29·60	29·54	56	41	—	
		18	S	29·72	29·54	60	43	33	
	Full M.	19	S	29·90	29·72	61	47		
		20	NE	30·12	30·07	63	44		
		21	W	30·12	30·12	61	43		
		22	W	30·12	30·12	66	45		
		23	E	30·18	30·12	71	46		
		24	E	30·18	30·13	74	37		
		25	NE			65	41		
				30·18	29·30	74	30	2·23	1·08

NOTES.—Third Mo. 28. A little rain. 29. A shower mixed with hail about noon.

Fourth Mo. 3, 4. Cloudy: much wind. 5. A fine day: 6 a.m. much wind and cloud from the S. 7. The trees dripping, as from a mist this morning. 8. Windy. *Cumulostratus* was formed by inosculation, and passed to Nimbus. 9. Windy, cloudy: steady rain p.m. 10. A stormy wet day. 11. Snow in small quantity. Hail mixed with rain at intervals. 13. A little snow and some hail balls. 18. A single swallow: bats on the wing: stormy indications p.m. Some very faint coruscations in the W and NW last night. 19—25. Fine days.

RESULTS.

Wind Variable, with a large proportion of Easterly.

Barometer: Greatest height................30·18 in.

Least.......................29·30 in.

Mean of the period............29.89 in.

Thermometer: Greatest height................ 74°

Least........................ 30°

Mean of the period............ 47°

Evaporation (in 22 days) .... 2·23 in.

Rain .... 1·08 in.

TABLE XLIV.

1810.		Wind.	Pressure.		Temp.		Evap.	Rain, &c.
			Max.	Min.	Max.	Min.		
4th Mo. L. Q.	April 26	E	30·08	30·00	64°	42°	26	
	27	E	30·09	30·05	60	35	42	
	28	E	30·09	30·05	67	34	30	
	29	NE	30·05	29·96	72	37	33	
	30	NE	29·96	29·85	75	37	36	
5th Mo.	May 1	NW	29·85	29·83	67	42	30	2
	2	NE	29·83	29·71	57	39	7	
New M.	3	NW	29·73	29·70	58	35	8	
	4	Var.	29·80	29·70	53	36		
	5	NE	29·85	29·80	53	37		
	6	NE	29·80	29·78	52	32		
	7	NE	29·78	29·45	51	38		
	8	SW	29·67	29·41	53	46		
	9	S	29·81	29·75	63	40		
	10	W	30·02	29·90	63	47		
1st Q.	11	E	30·07	30·02	66	56		
	12	NE	30·02	29·92	58	41		
	13	NE	29·92	29·80	61	52		
	14	NE	29·80	29·48	63	47		
	15	Var.	29·48	29·30	62	48		
	16	NW	29·41	29·38	55	50		
	17	S	29·41	29·39	63	51		
	18	E	29·65	29·41	49	35		
Full M.	19	SE	29·96	29·65	50	35		
	20	SE	29·95	29·80	63	34		
	21	SW	29·80	29·58	61	45		
	22	Var.	29·88	29·80	63	42		
	23	NW	30·06	29·88	60	36		
	24	NW	30·15	30·06	63	37		
			30·15	29·30	75	32		

NOTES.—Fourth Mo. 26—30. Clear days with dew and hoar frost: a strong breeze on the 27th, whence so great an evaporation.

Fifth Mo. 1. Wind N at intervals: no dew. 2. Cloudy morning: in the evening *Cirrocumulus*, which changed to *Cirrostratus*, 3. At 5 a. m. the same modification, and about 8, a refreshing rain. 5, 6. Fair. 7. Dripping. 8. Showers. 9. Cloudy. 10. Little rain. 11, 12. Fair. 13. Dripping. 14. Windy. 15. Rain in the night. 16. Rain a.m. 17, 18. Wet. 19, 20. Fine. 21. Wind and some rain. 22—24. Fine.

RESULTS.

Winds chiefly Easterly.

Barometer: Greatest height................30·15 in.

Least........................29·30 in.

Mean of the period............29·80 in.

Thermometer: Greatest height...............   75°

Least........................   32°

Mean of the period............50·53°

Having been during part of this period at the coast on account of indisposition, the Observations on the Barometer and Thermometer were continued by my son: a few omissions are supplied from those made at Stratford. The Evaporation and Rain were not registered further.

A *Whitehaven* Paper of the 11th says: the weather for some days past has been the coldest ever remembered at this season of the year; there has been a sharp frost every night—the Isle of Man and the Scotch hills are covered with snow: some vessels had six inches of snow upon their decks.

At *Manchester* on the morning of the 5th, the temperature was 27°, and on the two following days there were slight falls of snow, with a high wind at N. On the 8th, frequent heavy showers of hail, wind NE.—*Hanson.*

At *Malton,* the wind in May was N, NE, or E, for twenty days, and the mean heat of the month 47°.—*Stockton.*

The Papers make mention of hail storms in the west of France, on the 18th and 19th of this month, which were very destructive to the corn and vines; more especially in a district about *Nerac,* where " people who were in the fields were obliged to fly, covered with contusions, to whatever shelter they could find, and numbers of birds and poultry were found dead:" in another place, " some of the hail stones were as large as a goose's egg; and every pane of glass exposed to the storm was broken."

TABLE XLV.

	1810.		Wind.	Pressure. Max.	Min.	Temp. Max.	Min.	Evap.	Rain, &c.
5th Mo.	L. Q.	May 25	N	30·18	30·10	70°	36°		
		26	NE	30·08	30·00	73	37		
		27	NE	30·00	29·98	68	46		
		28	N	30·29	30·10	63	32		
		29	E	30·41	30·40	69	34		
		30	E	30·41	30·36	69	40		
		31	E	30·37	30·35	74	42		
6th Mo.		June 1	E	30·40	30·30	74	40		
	New M.	2	E	30·29	30·27	79	42		
		3	NE	30·31	30·29	72	39		
		4	Var.	30·30	30·25	76	44		
		5	NE	30·30	30·29	58	45		
		6	Var.	30·28	30·28	72	53		
		7	Var.	30·18	30·15	74	49		
		8	Var.	30·14	30·07	78	55		
		9	Var.	30·01	29·95	79	57		
	1st Q.	10	Var.	30·00	29·95	72	45		
		11	NW	30·00	29·95	66	52		
		12	W	30·00	29·92	67	51		
		13	W	30·00	29·92	66	43		
		14	NW	30·21	30·13	63	44		
		15	NW	30·29	30·19	70	39		
	Full M.	16	NE	30·00	29·90	66	37		
		17	N	30·00	30·00	70	40		
		18	W	30·00	30·00	73	55		
		19	W	30·00	29·98	71	58		
		20	SW	30·20	30·00	73	55		
		21	W	30·20	30·20	76	61		
				30·41	29·90	79	32		

NOTES J. G.—*May* 27. A shower in the evening. 28. A thunder storm at night, the wind SE, lightning nearly incessant and vivid: some heavy showers. 29. Considerable rain p. m.

June 4. Wind NE about nine: in the course of the day it shifted round against the sun, fixing in the NE in the evening. 10. A slight shower a. m., distant thunder p. m. 13. Some refreshing showers in the course of the day. 18. Lunar halo. 20. Showers p. m.

NOTES T. F.—*June* 2. *Cirrus, Cirrostratus,* and fleecy *Cumulus* observed. The *Cirrostratus* in particular was highly coloured. 3. Small meteors, usually called falling stars, seen this evening.

4. *Cirrostratus.* 5. Black clouds which threaten rain. 7. Distances very hazy. 13. *Cirrostrati* early this morning. 17. *Cirri* early. About 5 p. m. I observed the modification of *Cirrostratus* approximating to *Cirrocumulus,* and disposed in long arcs extending from N to S, passing on gently with the wind. 18. Towards evening, an apparent confusion of the electrical state of the two sheets of clouds took place, which threatened rain.

The weather has been so dry for the last three weeks, that all my hygrometers have ceased to be of any use.

RESULTS.

Winds Easterly in the fore part, with dry weather; Westerly with some showers, towards the end.

Barometer : Greatest height . . . . 30·41 in.
Least. . . . . 29·90 in.
Mean. . . . . 30·14 in.
Thermometer : Greatest height . . . .   79°
Least . . . .   32°
Mean . . . . 54·20°

The case of a house struck by lightning at East Thriston, (see the note) suggests *two precautions* in a time of thunder storms—not to suffer loaded fire-arms to be about, as the least spark of electricity reaching the powder will cause it to explode—and, on having reason to suppose that any part of a building has been struck, immediately to see that no combustible matter has been kindled by the iron, which may have conducted the electric fluid, and in so doing have been heated or even melted.

June 4.—In a tremendous storm of thunder and lightning, the electrical fluid entered the house of Mr. Cowens, farmer, at East Thriston, shivered to pieces a press-bedstead in the kitchen, ran along and melted the bell-wires attached to three rooms ; in the parlour a closet-door was split. At this moment Mr. Cowens came in, and his dog, which was close behind him, was killed ; *a loaded gun in the passage exploded soon after.* When the alarm had a little subsided, it was discovered that the rooms on the second floor were on fire, the lightning *having set fire to the curtains and bed-hangings*; the flames were, however, soon happily extinguished. In the staircase window, which fronts the east, there is only one pane of glass left whole.—(PUB. LEDGER).

TABLE XLVI.

		1810.		Wind.	Pressure. Max.	Min.	Temp. Max.	Min.	Evap.	Rain, &c.
6th Mo.	L. Q.	June	22	Var.	30·20	30·20	74°	48°		
			23	SE	30·40	30·20	75	44		
			24	SE	30·20	30·20	81	50		
			25	NE	30·20	30·11	83	57		
			26	N	30·11	30·04	60	49		
			27	NW	30·04	30·00	67	47		
			28	E	30·06	30·05	74	55		
			29	SE	30·10	30·04	75	50		
			30	Var.	30·18	30·14	75	50		
7th Mo.	New M.	July	1	Var.	30·18	30·14	76	59	14	20
			2	W	29·85	29·84	76	56	30	
			3	S	29·79	29·44	70	53	11	24
			4	SW	29·64	29·44	62	51	09	22
			5	Var.	29·91	29·81	69	58	10	
			6	W	30·05	29·95	72	49	14	
			7	SW	30·05	30·02	79	52	23	
	1st Q.		8	Var.	29·90	29·88	73	52	09	17
			9	W	29·99	29·86	72	59	24	05
			10	SW	29·80	29·79	75	48	19	—
			11	Var.	29·69	29·55	81	59	—	—
			12	W	29·69	29·61	74	59	36	42
			13	S	29·69	29·65	74	55	29	
	Full M.		14	SW	29·91	29·75	67	50	13	34
			15	N	30·10	29·96	68	49	12	06
			16	W	30·13	29·97	65	52	10	20
			17	Var.	29·84	29·70	64	53	11	24
			18	NE	29·90	29·85	67	52	18	
			19	N	29·90	29·85	67	46	10	
			20	N	30·10	29·96	64	50	16	
					30·40	29·44	83	44	3·18	2·14

Notes T. F.—*June* 28. Early in the morning *Cumuli* were observed floating at different altitudes: about 11 p. m. a very hard thunder storm came on.

July 1. Rain and lightning continued through the night. 7. *Cirrostratus* in dark spots to NW about sunset. 8. The same succeeded by storms. 12. The clouds appear mountainous and electric, with drops of rain. 16. Fleecy *Cumuli* beneath *Cirri*. 18. Fine towering *Cumuli*, and rather windy. 19. Spots of cloud before the moon. The air remains dry, notwithstanding the rain.

Notes J. G.—*July* 1. About seven o'clock in the evening, the

1

atmosphere began to thicken in the W, at which time there were two very distinct currents, the lower E and SE—the upper W, which soon prevailed. In about an hour after, a vivid flash of lightning—the thunder very distant, followed by heavy rain. Wind tempestuous, lightning very frequent in the night. 6. A *Stratus* on the marshes at night. 11. About half-past 2 p. m. wind E, the western horizon began to darken, soon after which there was a hurricane, the wind blowing in all directions, but most furiously from the W, which finally prevailed: this was immediately succeeded by a storm of thunder, lightning, and heavy rain, which continued about an hour: the evening afterwards was fine. 14. About 2 p. m. a very considerable thunder storm from the W, which continued about half an hour, accompanied with heavy rain, and large hail at intervals: the whole afternoon unsettled, with thunder at a distance, the lightning at first very vivid: at half-past four an uncommonly loud clap of thunder, resembling the report of a cannon: evening fine. 16. Some thunder in the afternoon: a *Water-spout* at Ramsgate, (see note).

RESULTS.

Winds Variable.

Barometer: Greatest height . . . . 30·40 in.

Least. . . . . 29·44 in.

Mean of the period . . . . 29·94 in.

Thermometer: Greatest height . . . . 83°

Least. . . . . 44°

Mean of the period . . . . 62°

Evaporation (in 20 days) . . . . 3·18 in.

Rain (in the same) . . . . 2·14 in.

Thunder storms appear to have prevailed extensively. That of the first of the seventh month did considerable damage at Sheffield: that of the 14th was attended with a number of accidents (detailed in the Papers) in Middlesex, Kent, &c. On the 15th, (if the date be not incorrect) there was a hail storm at Windsor, in which hailstones as large as common marbles ravaged the gardens, and beat the fruit off the trees.

Ramsgate, July 16.—During a squall of thunder and rain, a stream of water, seven or eight feet in diameter, issued from a very heavy black cloud in the east; which, after taking a horizontal direction towards the south, for about half a mile, suddenly fell into the sea, a very little distance from the shore, with a rushing noise, agitating the water all around in a most extraordinary manner, and rebounding again to the height of several yards.—(PUB. LEDGER).

TABLE XLVII.

1810.		Wind.	Pressure.		Temp.		Evap.	Rain, &c.
			Max.	Min.	Max.	Min.		
7th Mo. L. Q. July 21		NW	30·10	29·96	60°	48°	14	—
	22	N	30·20	30·12	67	44	14	—
	23	W	30·21	30·21	71	53	17	—
	24	SW	30·20	30·10	78	45	21	
	25	E	29·94	29·86	77	59	20	—
	26	SE	29·86	29·71	71	52	15	41
	27	SW	29·68	29·52	68	53	09	13
New M.	28	SW	29·65	29·61	66	52	13	69
	29	W	30·00	29·94	71	56	16	
	30	W	29·80	29·78	67	52	14	
	31	SW	29·86	29·83	70	41	13	31
8th Mo. Aug. 1		NW	29·97	29·85	68	45	12	—
	2	NW	30·01	29·92	73	57	09	07
	3	SW	29·90	29·80	75	56	17	25
	4	SE	29·65	29·58	71	54	14	20
	5	SW	29·74	29·70	70	50	12	20
	6	W	29·82	29·79	72	57	—	
1st Q.	7	NW	29·70	29·64	72	59	28	41
	8	NW	29·92	29·60	64	53	—	—
	9	NW	30·04	29·99	65	52	30	33
	10	SW	29·82	29·73	70	53	—	—
	11	W	29·73	29·67	68	53	53	18
	12	NW	29·90	29·70	68	57	13	14
	13	NW	29·88	29·52	66	55	—	
Full M.	14	W	29·85	29·84	67	57	42	65
	15	NW			68	49	—	—
	16	NW			53	47	14	40
	17	NW			63	40	—	
	18	W			70	49	29	
	19	W			67	47	—	—
	20	NW			69	49	17	
			30·21	29·52	78	40	4·56	4·37

NOTES T. F.—*July* 21. Moon appears bright; and its edges well defined. 23. *Cirri* and *Cirrostrati.* 24. Small meteors, commonly called falling stars, observed. 25. *Cirri* and spongy *Cumuli.* Distant thunder in the evening. 26. *Cirrostrati* prevail—very dark night. 27. High wind in squalls.

August 6. *Cirrostratus* prevails. 9. Fine rows of *Cirrostratus* above *Cumuli.* The moon appears hazy and obscure. 10. *Cirri* and *Cirrostrati* preceded high wind in squalls. 11. Windy. Swifts, *Hirundines apodes*, still common. 12. Windy and small rain. 13. The modification of cloud called *Cirrus* prevalent; squalls of wind and flying *Cumuli.* Some heavy black clouds came over about 6 p. m. which threatened rain. 14. Clouded hazy day. At night the moon appeared hazy, and indicated rain, which ensued during the night.

NOTES J. G.—*July* 25. Some showers p. m. 27. Wind extremely high all the afternoon. 28. About 1 p. m. a thunder storm with heavy rain and hail: several other thunder showers in the course of the afternoon. A house was struck at Ilford, (see the note). 31. A very heavy shower with thunder about 5 p. m.

August 3. A rainy night. 5. Thunder p. m. 13. Wind extremely high from the NW. 15. Rainy morning. 16. Wind high and bleak, with small rain most of the day. 18. Foggy morning. 19. Slight showers p. m.

RESULTS.

Winds Westerly.

Barometer:	Greatest height	30·21 in.
	Least...........	29·52 in.
	Mean (of 25 days)	29·848 in.
Thermometer:	Greatest height	78°
	Least........	40°
	Mean of the period	59·98°
Evaporation		4·56 in.
Rain		4·37 in.

Thunder storms appear to have prevailed very extensively on and about the fourth of the eighth month in the north of England, and in Scotland. An anecdote in the notes, which appears to belong to this date exemplifies the terror with which even wild animals are inspired by near discharges of electricity.

Saturday se'nnight, about one o'clock, during a heavy storm of thunder, lightning, and rain, the electric fluid struck the roof of the Coach and Horses public-house, Ilford, Essex. In its course it took the bell-wires, which it melted into the size of small shot; it then took the door-posts of the parlour, which, with the force, were shivered to splinters nearly the whole height; it then rolled a considerable way into the parlour, where, being evaporated, it totally disappeared, leaving behind it a most sulphurous smell, and on the floor a place scorched so much as to be entirely black. Although several persons were in the parlour and rooms adjacent, they received not the least injury.

At Carnforth, about six miles from Lancaster, during a dreadful thunder storm, as a person of the name of Taylor was sitting alone in his house, he heard something scratch at the kitchen-door; thinking it was the cat, he opened the door; a large foulmart (or polecat) rushed past him, and sat down by the fire. The man, astonished, sat down in his chair and looked earnestly at the strange visitor, which also stared hard at him; at length, after a loud clap of thunder, the foulmart leaped upon the man's lap, who, putting it into a wire cage, exhibited it to his neighbours.

TABLE XLVIII.

	1810.		Wind.	Pressure.		Temp.		Evap.	Rain, &c.
				Max.	Min.	Max.	Min.		
8th Mo.	L. Q.	Aug. 21	W			71°	44°		
		22	NW			78	49	38	
		23	NW			80	48		
		24	Var.			83	49	27	
		25	NE			83	58	21	
		26	W	30·13	30·03	75	54	11	
		27	N	30·17	30·14	73	52		
		28	NW	30·21	30·20	73	49	25	
		29	E	30·19	30·12	72	55		
	New M.	30	NW	30·08	30·01	74	59	22	8
		31	E	29·96	29·95	82	62	—	
9th Mo.		Sept. 1	SE	29·96	29·93	83	61	44	
		2	E	29·99	29·95	85	63	31	
		3	SW	29·90	29·78	68	53	—	—
		4	NW	29·95	29·75	61	48	28	44
		5	W	30·05	30·05	64	54	—	
	1st Q.	6	W	30·27	30·02	67	40	32	
		7	E	30·40	30·31	68	38	—	
		8	N	30·24	30·16	71	39	26	
		9	NE	30·06	30·04	73	51	—	
		10	W	30·00	29·96	68	48	28	—
		11	SW	29·95	29·70	61	48	—	12
		12	N	30·00	29·72	59	41	10	—
	Full M.	13	NW	30·15	30·08	62	48	—	
		14	NW	30·38	30·16	69	41	23	
		15	NE	30·40	30·37	62	50	—	
		16	NE	30·30	30·20	64	57	18	09
		17	E	30·14	30·10	72	49	—	
		18	NW	30·10	30·10	66	53	14	
		19	N	30·16	30·13	61	50	—	
				30·40	29·70	85	38	3·98	0·73

Notes T. F.—*August 26.* Slight squalls of wind about 10 h. m.
27. Summer lightning during night. 28. *Cirrostratus* disposed in long
strata, extending from N. to S. 29. Light fleecy *Cumuli*; fog at intervals, very partial. 30. Sky thinly covered with *Cirrostratus* early.
Summer lightning and showers at night. 31. Very loud peal of thunder, and hard rain about 3 a. m. *Cirri* and *Cirrocumuli*; at night
very vivid lightning succeeded by hard thunder showers.

September 1. *Cirri* and *Cirrocumuli*: temperature increasing. 2. *Cirri*,
&c. Wind (as is usual in hot weather) rises soon after noon, and
falls towards night. 3. Tufts of *Cirrus* early. Rain set in about
5 p. m. and continued through the night, accompanied by lightning.

6. Windy in the day. 10. *Cirrostratus* and *Cirrocumulus*; the former prevailing. Rain succeeds. 12. Very beautiful sunset. I observed two bars or streaks of *Cirrostratus* of a rich crimson colour, extending from SW to NE on a ground of almost golden hue, at the apparent altitude of about 20°. 13. *Cirrostratus* disposed in beds of small spots, succeeded by small rain, and warmer air. 14. *Cirrostratus* and *Cirrocumulus*. 19, 20. Calm, heavy, dull weather. Horizon very foggy, and sky overcast.

Notes J. G.—*August* 21. A *Stratus* at night on the marshes. 22. Foggy morning. 29. The same. 30. Distant thunder and lightning about 11 p. m. 31. About half-past 2 a. m. a tremendous peal of thunder resembling repeated discharges of artillery: it was preceded by extremely vivid lightning; the latter prevailed through the night, and played particularly for a considerable time upon the surface of a barge-load of *coke*, which lay near the laboratory: about 10 p. m. a distant storm again perceptible.

RESULTS.

Winds for the most part Northerly and Easterly.

Barometer: Greatest height .... 30·40 in.
Least...................... 29·70 in.
Mean of the period .... 30·08 in.
The whole for 25 days.
Thermometer: Greatest height............... 85°
Least...................... 38°
Mean of the period........... 60·65°
Evaporation .... 3·98 in.
Rain .... 0·73 in.

TABLE XLIX.

1810.			Wind.	Pressure		Temp.		Evap.	Rain, &c.
				Max.	Min.	Max.	Min		
9th Mo.	L. Q.	Sept. 20	NE	30·04	30·03	63°	55°	—	
		21	NE	30 03	29·97	71	55	—	
		22	SE	29·97	29·86	71	51	31	
		23	Var.	30·03	29·86	63	54	—	
		24	NE	30·09	30·03	68	57	—	
		25	NE	30·05	30·03	73	53	—	
		26	NE	30·03	29·95	71	45	—	
		27	NE	29·87	29·85	71	47	59	
	New M.	28	NE	30·00	29·87	69	44	—	
		29	E	30·00	29·93	67	52	—	
		30	E					—	
10th Mo.		Oct. 1	E	30·23	29·93	71	43	31	
		2	E	30·23	30·22	56	41	—	
		3	NE	30·24	30·13	66	41	—	
		4	NE	30·13		63		22	
	1st Q.	5	NE		29·85		41	—	
		6	Var.	29·92	29·86	62	44	—	
		7	SW	29·93	29·90	69	45	19	
		8	Var.	29·90	29·88	69	50	—	
		9	NE	29·88	29·80	62	54	—	—
		10	NE	29·88	29·80	59	48	—	
		11	NE	29·88	29·83	61	47	32	
	Full M.	12	N	30·20		58		—	
		13	N		29·80		32	—	
		14	NE	30·20	30·15	60	39	—	
		15	E	30·15	29·88	58	41	36	
		16	SE	29·88	29·65	57	43	—	49
		17	S	29·35	29·30	65	55	—	—
		18	SW	29·72	29·35	63	46	—	—
		19	SW	29·72	29·68	59	56	41	29
				30·24	29·30	71	32	2·71	0·78

NOTES.—Ninth Mo. 20. Grey sky: slight indications of *Cumulus* uniting with *Cirrostratus*. 21. A *Stratus* this night and the last. 22. Mostly clear a. m. then various clouds ending in *Nimbus*. 23. *Nimbus* to the S: a shower. 24. A very lively pink tinge in the W at sunset, and a fainter red in the E: the wind sprung up fresh, and the night was cloudy, with a faint flash or two of distant lightning. 25. A fine day. 26, 27. Clear: *Stratus* at night. 28. The same: about 7 p. m. a bright blue meteor descended obliquely in the E. 29. Cloudy: some appearance of an active electricity. 30. Rain early, after which fair.

Tenth Mo. 7. The nocturnal *Stratus* has prevailed (with gossamer) since the beginning of the month. 8. In the course of this day, the wind, which was first N with mist, went slowly by S to E, when distant *Nimbi* appeared: and after sunset a swift *scud* came from that quarter. 9. About 7 a. m. a slight shower or two. 10, 11, 12. Brisk winds: *Cirri*: lunar halo. 13. Hoar frost. 14, 15. Windy: *Cirri.* 16. a. m. wind E, *Cirrocumulus* passing to *Cirrostratus*, in bars from N to S, nearly crossing the visible hemisphere: in the latter quarter a haze tinged purplish by the sun. The wind shifted for a while to S, and there was rain at intervals p. m. 17. Rain. 18, 19. Much wind.

NOTES T. F.—*September* 22. A great disturbance of the electric state of the atmosphere was conspicuous. A fog covered the ground at sunrise; about noon I observed *Cirri* spread about at a great altitude: these were succeeded by *Cirrostrati, Cirrocumuli,* and *Cumuli* of various appearances; some large and lowering, others loose dark coloured fleeces, floating in a lower region. Towards evening the wind rose, and the barometer fell; but the night turned out calm and clear, and summer lightning prevailed. 23. Several modifications of cloud. Clear night, and summer lightning. 25. Overcast at sunrise; very clear day afterwards; falling stars. 26. Clear day, and rather windy in the middle; calm clear night. Small meteors. 27. *Cirrostratus* during the day, disposed in beds of small aggregates, extending in arcs across the zenith. Clear night; small meteors frequent. 29. Foggy at sunrise. After it cleared off, *Cirrostratus,* dispersed about in the atmosphere; in some places in thin films, in others in rows of small spots. *Cirrocumulus* also appeared. Loose flocks of dark reddish *Cumulus* in a lower region. At sunset a very highly coloured *Cirrostratus,* on an almost golden sky, gave the W horizon a very beautiful appearance. Rain came on during the night. 30. The western sky appeared deep red after sunset.

October 1. At night the stars' light suddenly diminished, and a lucid burr (not a halo) was observed round Jupiter. 2. Electric state of the atmosphere very much disturbed; various modifications of cloud. A breeze rose from E at 10 a. m. Clouds highly coloured at sunset. 3. Clear day; only *Cumuli* passed over with the wind. 6. *Cirrocumuli*; heat increasing. 12. Swallows and martins last seen. 16. A total change in the weather took place to-day. An intervening current from the south, and the prevalence of *Cirrus* and *Cirrostratus*, indicated rain, which came on during the evening. 17—20. During this period, the weather was warm, accompanied by rain and high wind at intervals. *Cirrocumulus* and *Cirrostratus* prevailed between the showers.

RESULTS.

Prevailing Winds NE and E nearly to the close of the period.

Barometer: Greatest height ....30·24 in.
Least........................29·30 in.
Mean of the period............29·91 in.
Thermometer: Greatest height................ 71°
Least...................... 32°
Mean of the period............ 56°
Evaporation .... 2·71 in.
Rain .... 0·78 in

TABLE L.

	1810.		Wind.	Pressure. Max.	Min.	Temp. Max.	Min.	Evap.	Rain, &c.
10th Mo.	L. Q.	Oct. 20	SW	29·68	29·60	64°	49°	—	91
		21	SW	29·60	29·30	62	56	—	30
		22	SW	29·59	29·30	60	48	53	
		23	NW	29·73	29·50	57	43	—	
		24	N	30·16	29·73	58	36	—	
		25	N	30·35	30·16	52	33	32	
		26	NE	30·35	30·21	52	44	—	
		27	Var.	30·21	29·68	—	—	—	—
	New M.	28	Var.	29·68	29·54	54	31	14	92
		29	NW	29·85	29·60	45	30	—	—
		30	NW	30·01	29·85	45	27	—	
		31	W	29·85	29·71	46	35	19	27
11th Mo.		Nov. 1	NE	29·74	29·68	53	37	—	—
		2	NE	29·95	29·93	48	46	—	—
		3	NE			47	43	—	30
	1st Q.	4	NE	29·93	29·68	46	31	18	—
		5	SW	29·68	29·17	45	35	—	—
		6	Var.	29·17	28·98	47	33	—	—
		7	Var.			—	—	—	39
		8	Var.	29·43	28·98	49	32	12	
		9	NE	29·43	28·60	53	38	—	—
		10	NE	29·18	28·50	50	43	—	10
	Full M.	11				46	43	—	
		12				48	31	—	06
		13		30·15	28·50	48	30	—	
		14	SE	30·15	29·54	54	36	—	
		15	SW	29·54	29·27	58	53	—	1·25
		16	SW	29·30	29·20	58	49	—	
		17	SW	29·46	29·20	54	45	10	35
				30·35	28·50	64	27	1·58	4·85

NOTES.—Tenth Mo. 20. Windy: much drizzling rain. 21. Windy: small rain: stormy night. 22. Much wind: cloudy p. m., stormy night. 23. Windy, cloudy. 24. Hoar frost: *Cumulus*, with the superior clouds, ending in *Nimbus:* some appearance of lightning, with a small meteor. 25. Hoar frost: *Cumulus* ending in *Nimbus:* meteors. 26. No dew: cloudy. 28. Rain from before sunrise to near sunset. 29. *Cirrus:* snow in the night: brisk wind. 31. Rain at intervals.

Eleventh Mo. 1—8. Cloudy, windy: wet at intervals. 9. A shower mixed with hail about noon. Lofty piles of *Cumulus* and *Cumulostratus*, p. m., the moon first very pale, then faintly obscured by elevated *Cirri* and haze, with a faint halo of the largest diameter. At 9 p. m. the barometer, which had risen to 29·43 began to fall with a strong gale from the eastward and abundance of rain. 10. At noon the rain abated: the wind going by S to SW, but presently returning to NE. 11. a. m. Cloudy: no dew: NW. 12, 13. Hoar frost: cloudy: windy: misty at night. 14. Very wet night. 15. Maximum of temperature at 9, with the wind SW: stormy wet night. 16. Windy: some heavy showers early.

NOTES T. F.—*October* 21. The weather has been much damper ever since the 16th, so as to render the *glass stands* of a peculiar sort of electrical instrument of no use, by making them serve as *conductors*. 22. High wind all day. 29. Stars hazy; very cold air. *November* 9. Wind rose during the night, and became very high. 10. High wind all day. A small coloured corona or burr observed round the moon about 10 p. m. 14. Before sunrise I observed the clouds towards the east very highly coloured with red and deep crimson. 15. Much rain fell during the night.

RESULTS.

Winds Northerly in the middle, SW at the beginning and end of the period.

Barometer: Greatest height . . . . 30·35 in.
     Least . . . . 28·50 in.
     Mean of the period . . . . 29·59 in.
Thermometer: Greatest height . . . . 64°
     Least . . . . 27°
     Mean of the period . . . . 45·48°
  Evaporation . . . . 1·85 in.
  Rain . . . . 4·85 in.

METEOR.—An article in the German Papers, dated from Waly on the Meuse, the 22d ult. says:—On the 19th of this month, between five and six o'clock in the evening, a luminous meteor appeared to the south, about the distance of a quarter of a league from the small commune of Brezeau; persons who attentively examined it assert, that it was nearly a quarter of an hour in collecting, floating over the place where it was first seen, and that when all its parts had united, it appeared all at once as a very considerable globe of fire, taking a northerly direction; it spread terror among the inhabitants of the village, who believed their houses would be burnt, and they themselves perish. This globe was accompanied by a frightful noise, which was heard at the distance of more than a league and a half, and sometimes resembled the rolling of a rapid chariot, at others, the noise of rain violently driven by the wind. It was followed by a very thick fog, and carried up from the ground every thing that it met in its passage. In crossing a river it absorbed the water, which soon afterwards fell in rain. It wandered for some time near the village. One thing certain is, that the roof of a house was thrown down, which is the only trace it has left. It was accompanied and followed by an abundant rain, much lightning, and loud claps of thunder. Continuing in the same direction, it suddenly turned into a column of fire, which, with the fog, rose towards the heavens. This made many persons believe the fog was smoke. It remained about a quarter of an hour in this state, a quarter of a league to the north of the village, and at a short distance from the forest of Beaulieu. This column now sunk a little, and at last suddenly disappeared, leaving a thick fog, which had no smell. This phenomenon lasted three quarters of an hour, and travelled over the space of half a league.

TABLE LI.

	1810.		Wind.	Pressure. Max.	Min.	Temp. Max.	Min.	Evap.	Rain, &c.
11th Mo.	L. Q.	Nov. 19	SW	29·70	29·61	53°	42°	—	
		20	NW	29·71	29·58	52	43	—	34
		21	SE	29·58	29·48	57	45	—	—
		22	Var.	29·85	29·58	52	43	21	1·16
		23	SE	29·86	29·83	54	45	—	—
		24	E	29·79	29·68	53	39	—	38
		25	SE	29·64	29·56	48	36	—	—
	New M.	26	Var.	29·39	29·19	47	37	—	63
		27	E	29·22	29·09	46	39	—	—
		28	W	29·08	28·94	47	35	—	—
		29	SE	29·13	29·04	43	34	14	36
		30	W	29·42	29·19	40	29	—	
12th Mo.		Dec. 1	S	29·70	29·47	37	30	08	
		2	SW	29·89	30·02	35	25	—	
	1st Q.	3	SE	29·95	29·95	43	31	—	—
		4	W	29·96	29·92	49	38	—	—
		5	W	29·95	29·89	50	45	—	—
		6	S	29·63	29·42	52	41	14	42
		7	SW	29·42	29·36	46	34	—	
		8	W	29·79	29·54	39	26	—	—
		9	NW	29·94	29·89	36	26	—	
	Full M.	10	SE	29·60	29·38	38	34	—	74
		11	NW	29·90	29·56	36	28	—	
		12	NW	29·76	29·60	49	40	—	
		13	W	29·93	29·80	51	49	32	—
		14	W	29·78	29·66	52	38	—	98
		15	NW	30·06	29·88	45	36	—	
		16	Var.	30·25		42	30	—	
		17	SW	30·10	29·94	49	42	—	—
		18	Var.	29·65	29·45	50	38	19	53
				30·25	28·94	57	25	1·08	5·54

Notes T. F.—*November* 16. Very windy showery night. 17. Flash of lightning about half-past 6 p. m. 20. The maximum of thermometer at 11 p. m. 21. Thunder clouds about. 22. Showers of hail and rain; lightning at night. 23. Flash of lightning at night. 30. Moon well defined; but yellowish. Flashes of lightning observed.

December 2. *Cirri, Cirrostrati,* and *Cirrocumuli,* early; succeeded by change of weather. 4. Very damp. 5. *Cirrostratus* and *Cirrocumulus.* Windy night. 9. Sky overspread with *Cirrocumulus,* p. m. 10. Upper current NNE to-night. 11. A burr round the moon, about half-past 10 p. m. a little coloured with yellow, red, and green, at its extremities. 13. *Cirrostratus* and *Cirrocumulus* in the afternoon: rain came on at night, accompanied by high wind, and increasing temperature. 14. Very windy showery day; but clear night. 15. Early a. m. *Cirrostratus* was spread about the sky, and threatened rain. It however cleared: and at night light tufts of *Cirrus*, approximating to *Cirrostratus*, scattered about, presented a very curious sky by moonlight.

NOTES J. G.—*November* 22. Rainy morning—a heavy shower of hail about 1 p. m.—afternoon extremely stormy, with hail. 23. An *ignis fatuus* observed in the marshes in the evening, which continued for a considerable time. 26th. Several *ignes fatui* seen in different parts of the marshes at night—some of them extremely bright. 27, 28. The *ignis fatuus* again seen in the evening—the following day the marshes were flooded from an overflow of the river, since which time they have not been observed.

December 3. White frost. 4. Very foggy morning. 5. Foggy morning. 6. Rainy morning. 9. White frost. 10. Very rainy morning—some snow mixed with rain. 14. Night very stormy, with heavy rain. 16. Evening very foggy. 18. Stormy day.

RESULTS.

Winds Variable: Easterly rather predominated in the former, Westerly in the latter part of the period.

Barometer: Greatest height............... 30·25 in.
     Least........................ 28·94 in.
     Mean of the period .... 29·66 in.
Thermometer: Greatest height.............. 57°
      Least....................... 25°
      Mean of the period .... 41·15°
  Evaporation....................... 1·08 in.
  Rain .... 5·54 in.

TABLE LII.

1810.			Wind.	Pressure.		Temp.		Evap.	Rain, &c.
				Max.	Min.	Max.	Min.		
12th Mo.	L. Q.	Dec. 19	NW	29·84	29·50	42°	34°	—	
		20	SW	29·81	29·38	50	39	—	—
		21	W	29·38	29·33	44	41	28	28
		22	W	29·68		52	41	—	—
		23	SW			52	40	—	80
		24	NE			51	40	—	76
		25	W			49	40	31	32
	New M.	26	NW	29·85	29·80	52	43	—	
		27	NW		29·54	46	38	—	—
		28	NW	30·30	30·20	37	31	—	19
		29	NW	30·50	30·36	34	27	—	
		30	NW	30·50	30·50	34	25	—	
1811.		31	N	30·51	30·35	32	27	40	
1st Mo.	1st Q.	Jan. 1	NE	30·28	30·15	28	24		—
		2	NE	30·05	29·85	28	23		—
		3	Var.	29·76	29·71	28	18		—
		4	E	29·95	29·91	31	24		
		5	NE	29·89	29·83	28	25		
		6	NE	29·94	29·84	28	23		
		7	NE	29·94	29·90	29	24		
		8	NE	29·86	29·82	29	20		
	Full M.	9	NE	29·94	29·86	30	20		—
		10	E	29·97	29·94	42	20		—
		11	S	29·82	29·74	44	35	—	36
		12	S	29·68	29·49	48	35	—	—
		13	SW	29·70	29·66	46	39	—	36
		14	Var.	29·70	29·66	50	43	12	—
		15	W	29·60	29·62	45	32	—	12
		16	W	30·06	29·88	46	53	—	—
				30·51	29·33	52	18		3·19

Notes T. F.—*December* 16. *Cirrostratus* disposed in beds, p. m.
17. *Cirrocumulus* in the intervals of the fog a. m.; *Cirrostratus* p. m.
18. A burr round Jupiter. 20. Hard squalls of wind during the night.
21. Very high wind, particularly at night. 25. High wind accompanied by flashes of lightning all night. 26. The marshes along the Lea flooded.

1811. *January* 4. Very high wind. 5. Dark fleeces of *Cumulus* seen floating beneath *Cirri,* which were in a calm region above.
12. The doors of the house swelled with damp. 16. *Cirrostratus* and *Cirrocumulus.*

Notes J. G.—*December* 21. Rainy morning. 22. Wind very high all night. 24. Very windy night, with heavy rain. 25. Wind high all day, with rain; frequent lightning in the evening from SE. 26. Wind very boisterous early in the morning—day fine: the rain of the last three or four days being impeded in its passage to the Thames by the spring tides, overflowed the banks, and filled the marshes. 27. Rainy morning—wind blowing strong from the NW. 28. Some snow in the afternoon. 29. A little snow in the afternoon. 30. Clear, frosty morning—some snow in the evening. 31. Ground covered with snow in the morning.

1811. *January* 1. Snow to the depth of about two inches fell in the course of the day. 2. Snow at intervals during the day. 3. Snowy morning—wind S. Thermometer 24° at 9 a. m. 4. Wind very bleak and high during the day; continued all the night of the 4th to blow furiously from the E and NE. 5. Wind very high all day. 6. Day very fine; wind abated; moon extremely bright in the evening. 7. Wind again very boisterous in the morning; evening cloudy. 9. Snowy morning; some appearances of a thaw in the afternoon; evening cloudy. 10. Hoar frost, very thick fog; thermometer 20° at 9 a. m.: the trees had a most beautiful appearance from the frost, till the afternoon, when the wind shifted to the SE and S, and a thaw commenced, accelerated by a gentle rain in the evening. 11. Rainy morning; thermometer 42° at 9 a. m. The evaporation since the beginning of the month has been very considerable, most of the snow having disappeared during the continuance of the frost; an accident to the gauge prevented its being measured.

RESULTS.

Winds for the most part Northerly.

Barometer : Greatest height................30·51 in.

Least......................29·33 in.

Mean (some of the lower observa-

tions wanting) ....29·88 in.

Thermometer : Greatest height............... 52°

Least...................... 18°

Mean of the period............35·86°

Evaporation to the end of the year ..... 0·99 in.

The remainder of the period imperfect.

Rain .... 3·19 in.

Comparison of the Winds and Weather, during the foregoing two periods, with the same in the Island of Iceland.

A Register of the Weather kept at Reikiavik in Iceland, and published in Sir Geo. Mackenzie's Travels, offers the following interesting points of comparison with the weather in Britain, during the fore part of this winter:—

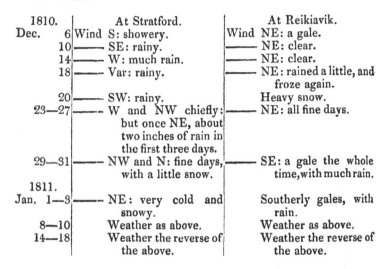

1810.		At Stratford.	At Reikiavik.
Dec. 6	Wind	S: showery.	Wind NE: a gale.
10	——	SE: rainy.	—— NE: clear.
14	——	W: much rain.	—— NE: clear.
18	——	Var: rainy.	—— NE: rained a little, and froze again.
20	——	SW: rainy.	Heavy snow.
23—27	——	W and NW chiefly: but once NE, about two inches of rain in the first three days.	—— NE: all fine days.
29—31	——	NW and N: fine days, with a little snow.	—— SE: a gale the whole time, with much rain.
1811. Jan. 1—3	——	NE: very cold and snowy.	Southerly gales, with rain.
8—10		Weather as above.	Weather as above.
14—18		Weather the reverse of the above.	Weather the reverse of the above.

From a third Register, kept at the intermediate distance of 180 miles from Stratford, and on the opposite side of our Island, viz. that of Thos. Hanson, at Manchester, it may now be shown that the contrast of Stratford with Reikiavik, was not the result of local causes.

At *Manchester*, Dec. 6. Wind S: rainy. 10. E: snow; (here the intermediate station has an intermediate current.) 14. S: boisterous with rain. 18. S: fine. 20. S: rainy. 23—28. SW chiefly: about an inch of rain. 29—31. N: fair. 1811. Jan. 1—8. N, NE, and E, with snow. 9, 10. SE: clear and variable. 14—18. Westerly gales, with about an inch of rain.

Thus it appears, that on many days in the course of the first six weeks of the winter of 1810—11, the greater part of Britain was subject to a current proceeding in an opposite direction to that which prevailed in Iceland: and this obtained whether the Winds were Northerly or Southerly. That it should snow and freeze in Iceland, while we have mild weather about London, seems perfectly natural:

but it has probably never been suspected, that *our* periods of frost are contemporaneous with *their* intervals of rain and thaw.

This subject is certainly deserving of a more particular investigation, but the paucity of observations with instruments in Iceland presents a great obstacle: yet a good comparison may be formed on the general reports of such as visit that island in trading vessels.

EARTHQUAKES IN ITALY.

" At Parma, in the night between the 24th and 25th Dec. (1810,) there was a very severe shock of an earthquake. Several persons, who came from the midnight mass, were thrown down in the streets. In Genois-street, all the chimnies tumbled down, and several houses sustained considerable damage."

" The earthquake seems to have extended quite across Italy.—An article, dated Verona, December 25, says:—' We yesterday experienced here the shock of an earthquake, the heaviest I ever witnessed. It lasted ten seconds, and its direction was from N to S.—Some few persons were hurt, but no lives were lost.'—Another from Genoa, of the 26th, says:—' At one o'clock in the afternoon, we felt here a violent shock of an earthquake, in the direction from E to W. It lasted from eight to ten seconds, set the bells a ringing, and produced much confusion amongst the furniture of the inhabitants. Three old houses tumbled down. No farther mischief was done either here or in the neighbourhood.' "

WINTER THUNDER STORMS.

" The phenomenon of a thunder storm on Christmas day, (1810,) was not confined to this country, but was experienced in several places in Germany, and followed by so dreadful a gale of wind, that many houses were blown down, and the heavy laden waggons on the public roads overturned."

WINTER IN SIBERIA.

" Letters from various districts in Siberia, dated in October last, mention that the winter has commenced unusually early in that country. The cold set in in the middle of September, and the snow fell in such quantities, as to prevent the harvest, which was very abundant, from being entirely got in."—(PAPERS.)

TABLE LIII.

	1811.		Wind.	Pressure. Max.	Pressure. Min.	Temp. Max.	Temp. Min.	Evap.	Rain, &c.
1st Mo.	L. Q.	Jan. 17	NW	29·82	29·75	51°	39°	30	
		18	W	29·96	29·80	40	32	—	
		19	NW	30·48	30·40	41	27	—	
		20	SE	30·41	30·15	40	30	—	
		21	SW	30·15	30·10	38	31	—	
		22	NW	30·20	30·20	36	31	12	
		23	SE	30·31	30·30	40	33	—	
	New M.	24	N	30·50	30·47	39	31	—	
		25	Var.	30·54	30·51	37	27	—	
		26	W	30·36	30·08	42	31	08	6
		27	NW	29·64	29·60	40	25	—	
		28	NW	29·65	29·63	31	19	—	
		29	NW	29·76	29·63	27	14	—	
		30	N	29·78	29·55	34	21	—	
	1st Q.	31	E	29·18	29·08	45	34	13	31
2d Mo.		Feb. 1	S	29·43	29·29	46	35	—	—
		2	E	29·68	29·55	46	38	—	
		3	S	29·90	29·60	49	29	—	
		4	SE	30·13	30·10	48	30	—	
		5	E	30·05	29·72	46	35	—	
		6	S	29·65	29·52	51	45	—	
		7	SW	29·82	29·60	52	39	45	18
	Full M.	8	SW	29·74	29·58	49	40	—	—
		9	N	29·78	29·62	47	41	—	—
		10	S	29·85	29·70	52	48	—	—
		11	SW	29·55	29·37	52	45	34	47
		12	SW	29·34	29·30	49	35	—	—
		13	SW	29·34	29·25	40	35	16	44
		14	NW	29·68	29·21	42	35	—	
		15	E	29·80	29·15	42	36	—	—
				30·54	29·08	52	14	1·58	1·46

Notes T. F.—*January 20.* *Cirri* and *Cirrostrati* observed. 26. Cold increasing, although the wind was SW; a white frost. 27. I observed an arc of *Cirrostratus* to extend across the zenith in the direction of the wind. Snow fell during the night.

February 2. About 9 p.m. I observed a lunar halo. I took the diameter of its area with a quadrant, which was about 40°. 3. Showery morning; towards evening I observed red coloured *Cirrostrati* in an apparently calm region; while fleecy *Cumuli* floated beneath them in the wind. The *Cirrostrati* refracted a fine red tint, while the *Cumuli*, passing under, and making the same angle with the sun, appeared blackish. 4. White frost, succeeded by thaw. About 8 p. m. a lunar

NOTES, &c. *(continued)*.

halo of about 40° diameter appeared for a few minutes during the passage of a *Cirrostratus* before the moon. 5. Sky variously spotted, streaked and freckled with *Cirrostratus* in the morning, and with *Cirrocumulus* at night. 6. Temperature much increased. In the evening I observed a double lunar corona; that is, a small one within a larger one. I have observed that coronæ as well as halones are generally prognostics of approaching rain, &c. 7. *Cirrus, Cirrostratus,* and *Cirrocumulus* preceded showers of rain and hail. 8. Sky highly coloured at sunrise; at night I observed, by the motion of the clouds, that there were two currents of air. 10. Frogs observed about. Thrush sings. 13. Hard shower of hail about noon.

NOTES J. G.—*January* 20. White frost. 28. A little snow in the morning. 29, 30. White frost. Thermometer 13° at Plaistow.—Wind very boisterous in the evening; a considerable quantity of snow (about three inches) fell during the night, wind E, blowing very strong. 31. Rainy morning: barometer falling, evening stormy.

RESULTS.

Winds Variable.

Barometer: Greatest height .... 30·54 in.
Least...................... 29·08 in.
Mean of the period .... 29·80 in.
Thermometer: Greatest height .... 52°
Least. .... 14°
Mean of the period .... 38·05°
Evaporation....................... 1·58 in.
Rain .... 1·46 in.

METEOR.—" On the night of the 26th of January, a large globe of fire, of a very deep red, was seen to pass through the air, at Wasserbourg, a small town of Bavaria, ten leagues from Munich; it descended gently, and fell into the Inn without explosion." Supposing the appearance in question to have been a large elevated *Meteor*, this apparent fall into the *river* was probably a deception of sight.—L. H.

SUBMARINE VOLCANO.

St. Michael's, Azores, Feb. 17.—" On the 26th, 27th, and 28th of January this island again experienced several very severe shocks of an earthquake. On the 31st, a tremendous explosion of smoke and flames issued from the water, at the distance of half a league, or two English miles from the shore, to the W of our island. The scene was awful beyond all description, and from the bowels of the inflammatory substance, upwards of 80 fathoms deep in the ocean, issued smoke, fire, cinders, ashes, and stones, of an immense size. Quantities of different kinds of fish floated on the surface of the sea towards the shore. This dreadful eruption of fire has perhaps been the saving of the island and its inhabitants."—(PAPERS.)

TABLE LIV.

	1811.			Wind.	Pressure.		Temp.		Evap.	Rain, &c.
					Max.	Min.	Max.	Min.		
2d Mo.	L. Q.	Feb.	16	NW	29·95	29·36	40°	25°	—	18
			17	NW	30·20	30·18	38	29	47	
			18	SE	30·14	30·05	45	31	—	
			19	E	29·99	29·82	41	30	—	
			20	E	29·74	29·70	40	34	—	
			21	E	29·51	29·27	51	39	15	
			22	S	29·18	29·09	54	35	—	
	New M.		23	SE	29·25	29·25	54	41	16	24
			24	SE	29·18	29·04	51	37	—	—
			25	SW	29·40	29·14	49	38	—	18
			26	W	29·30	29·25	55	42	—	
			27	W	29·65	29·50	49	40	42	—
			28	S	29·78	29·50	52	40	09	08
3d Mo.		March	1	W	29·78	29·58	47	38	—	—
	1st Q.		2	W	29·80	29·80	52	46	26	05
			3	W	30·00	30·00	54	42	—	
			4	W	30·02	30·00	55	46	—	
			5	S	29·80	29·35	55	42	—	—
			6	W	29·63	29·46	52	48	46	33
			7	W	29·39	29·30	57	48	—	—
			8	S	29·78	29·33	58	34	—	52
			9	NW	30·46	30·20	45	34	35	
	Full M.		10	W	30·52	30·47	53	40	—	
			11	Var.	30·48	30·46	56	41	—	
			12	SE	30·46	30·42	53	38	—	
			13	NE	30·40	30·34	52	39	27	
			14	NE	30·38	30·34	49	36	—	
			15	E	30·39	30·35	48	29	—	
			16	E	30·31	30·25	54	27	30	
					30·52	29·04	58	25	2·93	1·58

Notes T. F.—*February* 17. This afternoon appeared *Cirrostratus* of various figures, *Cirrocumulus* and fleeces of *Cumulus*. 18. Various modifications of cloud again to day. 24. Gentle showers with clear intervals. 25. The moon appeared of a deep brazen colour. 28. A lunar corona succeeded by showers.

March 1. *Cirrostratus* as usual prevailed between the showers. 3. *Cirrostratus* and *Cumulus* prevail. A lunar corona round the moon all the evening, and sometimes a halo. 4, 5. A few *Cirri* and *Cirrostrati*; windy. 6. I observed *Cirrus*, approximating to *Cirrostratus*, disposed in faint whitish transverse bars, and forming a kind of reticular plexus, or net work in the zenith: a windy night. 8. The wind got to the NE, and was high at night. 10. Fleeces of *Cumulus* richly coloured by the rising sun. 13. Clear morning: as the day advanced I observed *Cirri* of various shapes ramifying in all directions, *Cirrostrati*, and *Cirrocumuli*. In a lower region *Cumuli* floated along in the wind. The general appearance of the clouds to-day indicated a great disturbance of the electric state of the atmosphere: very similar kind of weather prevailed during last September, and a curious circumstance which then took place, happened again on the return to day of the same kind of weather; namely, the irregular pulsation of the electric bells of De Luc's column. 14. Only *Cumuli* passed over with the wind. The electric bells of De Luc's column pulsate very irregularly. 15. Clear night, falling stars seen.

NOTES J. G.—*March* 1. About a quarter past 2 p. m. a *Nimbus* passed over, discharging rain; after which a rainbow appeared, the colours of the complementary bow extremely distinct. 2. Wind very high all night. 3. A large halo round the moon in the evening. 6. A shower of hail about noon—evening rainy—wind high. 7. Wind high in the morning—afternoon very showery—evening and night stormy. 8. Rainy morning. 12. Foggy morning.

RESULTS.

Winds Variable.

Barometer: Greatest height................30·52 in.
Least........................29·04 in.
Mean of the period............29·82 in.
Thermometer: Greatest height................ 58°
Least.. .... 25°
Mean of the period............43·93°
Evaporation........................ 2·93 in.
Rain............................. 1·58 in.

TABLE LV.

		1811.	Wind.	Pressure. Max.	Min.	Temp. Max.	Min.	Evap.	Rain, &c.
3d Mo.	L. Q.	March 17	Var.	30·17	30·17	61°	26°	11	
		18	Var.	30·15	30·15	62	35	—	
		19	W	30·23	30·20	55	42	—	
		20	W	30·15	30·07	57	49	—	
		21	SW	30·05	30·05	60	45	31	6
		22	NE	30·38	30·13	50	29	—	
		23	NW	30·50	30·48	56	30	—	
	New M.	24	E	30·40	30·28	54	36	20	
		25	E	30·21	30·11	53	32	—	
		26	NE	30·24	30·10	56	31	18	
		27	NE	30·45	30·23	57	26	—	
		28	Var.	30·60	30·48	62	36	—	
		29	NE	30·61	30·55	61	31	—	
		30	NE			62	37	38	
	1st Q.	31	NE	30·22	30·16	50	43	03	
4th Mo.		April 1	E	30·10	30·00	55	34	—	
		2	Var.	29·98	29·90	64	43	—	
		3	W	30·04	29·98	62	39	25	
		4	N	30·15	30·09	63	41	—	
		5	NE	30·18		57	28	—	
		6	N			61	33	—	
		7	NW	29·55	29·52	40	31	30	6
	Full M.	8	NE	29·62	29·59	46	28	—	
		9	N	29·79	29·67	46	26	—	
		10	Var.	29·90	29·85	54	37	—	—
		11	Var.	30·22	30·03	51	32	38	
		12	Var.	30·23	30·12	55	46	—	35
		13	SW	30·15	30·00	65	54	—	—
		14	W	30·20		65	51	33	15
		15	NW			61	52	—	—
				30·61	29·52	65	26	2·47	0·62

Notes T. F.—*March* 18. This evening *Cirri*, becoming *Cirrostrati*, observed. De Luc's electric bells quite silent. 19. *Cirrocumulus* and gentle showers. Electric bells ring weak, but regularly. 20—22. During this period mild winds and damper air prevailed. Electric bells pulsated pretty regularly: on the evening of the 22d fleecy evanescent *Cumuli* indicated clear weather. The electric bells became silent at night. 23. Cloudy; fine purple and yellow coloured sunset. Bells silent. 24. Very clear; only faint streaks of the *Linear Cirrus*. 25. Early I observed *Cirrus*, ramifying about in all directions, and

5

becoming *Cirrostratus* and *Cirrocumulus:* fleecy *Cumuli* floated in the wind beneath them. Bells silent. 27. *Cirrus* prevailed this evening, and became *Cirrostratus*, coloured by the setting sun. Bells begin to ring again. 28, 29. *Cirrus* and *Cirrostratus*. Bells ringing irregularly, or at intervals.

April 1. A meteor seen to SW about 9 p. m. 2, 3. *Cirrostratus* and *Cirrocumulus* alternately prevail. 4. This afternoon fleecy, rocky, and mountainous *Cumuli;* in a higher region *Cirrostratus* and *Cirrocumulus* in different places, the latter most abundant during the day, but the former ultimately prevailed, and at night exhibited a lunar halo, of the usual diameter, *i. e.* between 40° and 50°. From the 24th March the electric bells rang irregularly till about the 14th April, when they rang regularly and loud till the 18th, when they ceased, and have not rung since.

NOTES J. G.—*March* 16—18. White frost. 20. Evening much clouded—wind high. 21. A little rain early in the morning. 23. White frost. 30. Morning foggy. 31. Very cloudy day.

April 6. White frost and foggy. 7. Some gentle rain and some snow in the morning, day very cloudy and cold. 14—16. Some gentle refreshing showers.

RESULTS.

Winds variable, but chiefly Northerly.

Barometer: Greatest height ....30·61 in.
Least...................... 29·52 in.
Mean of the period........... 30·12 in.
Thermometer: Greatest height .... 65°
Least.................... .. 26°
Mean of the period .... 46·75°
Evaporation..................... .... 2·47 in.
Rain.............................. 0·62 in.

FRANCE. *Hail Storm.*—The departments of Agin, and the Upper Marne, in France, were, at the beginning of April, visited by a dreadful hail storm, *which killed many persons,* as well as cattle, destroyed the vines, and did much mischief besides. Many of the hail stones were five inches long, and two inches diameter, and weighed six ounces. The storm was succeeded by a frost, which lasted two days.

IRELAND. *Meteor.*—April 5, at ten o'clock at night, a large and most brilliant meteor passed in a northerly direction over the eastern extremity of Belfast. It was of a most beautiful and vivid white, and appeared to expand itself considerably as it passed, with an undulatory motion.—(PAPERS).

TABLE LVI.

		1811.	Wind.	Pressure. Max.	Min.	Temp. Max.	Min.	Evap	Rain, &c.
4th Mo.	L. Q.	April 16	Var.	30·00		62°	41°	—	17
		17	NW			61	45	18	
		18	SE	29·34	29·22	62	34	—	—
		19	E	29·29	29·28	62	48	—	9
		20	S		29·34	65	52	35	12
		21	S	29·68	29·59	68	48	—	
	New M.	22	E	29·68	29·60	69	52	—	—
		23	SE	29·76	29·62	77	48	47	
		24	NW	29·87	29·80	75	52	—	
		25	Var.	29·85	29·80	66	49	—	
		26	NW	29·78	29·66	69	40	36	
		27	NE	29·64	29·62	66	39	—	
		28	Var.	29·65	29·55	68	51	—	
		29	SW	29·71	29·67	58	46	—	—
	1st Q.	30	SW	29·81	29·78	61	48	52	5
5th Mo.		May 1	S	29·68	29·65	64	54	—	11
		2	SW	29·98	29·66	61	46	23	—
		3	SW	30·05	29·91	64	54	—	22
		4	SW	29·94	29·91	67	55	—	—
		5	SW	29·95	29·73	62	40	39	18
		6	E	30·08	29·93	58	46	—	31
		7	Var.	29·83	29·72	56	46	—	10
	Full M.	8	Var.	29·85	29·72	63	51	—	—
		9	S	29·58	29·52	62	46	22	40
		10	Var.	29·78	29·75	69	51	—	—
		11	SW	29·85	29·84	74	50	—	
		12	Var.	29·78	29·66	82	56	40	4
		13	E	29·59	29·52	83	60	—	
		14	S	29·74	29·51	72	43	30	
				30·05	29·22	83	34	3·42	1·79

NOTES T. F.—*April* 17. Only *Cumulus* observed. 18. *Nimbi* pouring hail and rain. 19. *Cirrus* appeared early, followed by *Cirrostratus*, *Cirrocumulus*, and *Cumulus*, and eventually by *Nimbus* and showers. 21. *Cirrus* extending its fibres along with the current of air. In a lower region *Cumuli* float along in different planes, the lower ones black and lowering. The cuckow and swallow first seen. A lucid meteor observed about 3 p. m. 22—25. *Cirrus*, *Cirrostratus*, *Cirrocumulus*, *Cumulus*, and *Cumulostratus*, of various figures continually prevail, with summer lightning and dry air. 26, 27. Same kind of clouds with showers. 28. The multiform appearance of the *Cirrostratus* exhibited a beautiful sky this afternoon; in some places it was finely *undulated*, then became *reticular*, and lastly *confused vapour*. *Cirrocumulus* and *Cumulus* also seen: showers late in the evening. 29, 30. *Nimbi* (with cirrose fibres extending from them) pouring down showers.

May 1. Continued showers through the day. 2, 3. Showers with clear intervals. 4. Only *Cumuli* to-day. 6–8. Showery at times. 9. Rainy. 11. *Cirrostratus* coloured by setting sun. 12–14. *Cirrus*, &c. Sky deep blue in the eastern horizon. On the 13th only *Cumuli;* 14th, *Cirri* and *Cumuli*. 15. Only *Cumuli* in forenoon. Towards evening *Cirrus*, *Cirrostratus*, and *Cirrocumulus*, in different altitudes, by approaching and collasping, formed very dense *Nimbi:* and exhibited very various tints and unusual appearances, which ended in rain. The electric bells of De Luc's column were silent till the 7th May, when they began to ring, and have continued ringing, more or less regularly, till the present time.

NOTES J. G.—*April* 18. Showers and sunshine—in the twilight of the evening a brilliant meteor descended in the E, of the apparent magnitude of Venus—there were heavy clouds in the NE horizon at sunset, with much haze above them. 19. *Cirrus, Cumulus,* and *Cirrostratus* clouds, with a strong easterly breeze. 20. The first swallow made its appearance this forenoon. 21. Some lightning in the evening. 22. Some distant thunder heard this morning—much lightning in the NW at night. 23. Some rain about 5 a.m.—the cuckow heard this morning. 24. Foggy morning. 29. Wind high from the SW in the morning, with showers—afternoon, wind increased, with rain mixed with hail—some thunder at a distance.

May 1. Gentle showers during the day. 2. Showery day. 3. Very cloudy, with showers. 5. Wind very high from the SW all day, with showers—generally clouded. 6. Morning very fine, wind gone down, changed in the course of the last night to the eastward; afternoon very cloudy, with showers;—night rainy, wind high. 9. Rainy morning; continued to rain till towards evening, which was fine—some distant thunder in the afternoon. 10. Showery and fine; towards evening a rainbow, and some distant thunder. 13. Some lightning in the evening.

RESULTS.

Winds variable, but mostly Southerly.

Barometer: Greatest height .... 30·05 in.
Least........................ 29·22 in.
Mean of the period .... 29·70 in.
Thermometer: Greatest height............... 83°
Least........................ 34°
Mean of the period .... 57·19°
Evaporation .... 3·42 in.
Rain .... 1·79 in.

TABLE LVII.

	1811.		Wind.	Pressure, Max.	Min.	Temp. Max.	Min.	Evap.	Rain, &c.
5th Mo.	L. Q.	May 15	SE	29·82	29·81	72°	55°	14	—
		16	E	29·90	29·82	70	39	—	
		17	NE	29·92	29·90	77	48	—	
		18	NE	29·94	29·89	76	57	—	
		19	NE	30·00	29·98	61	51	35	51
		20	NE	29·94	29·82	74	51	—	30
		21	SE	29·75	29·72	70	55	—	—
	New M.	22	E	29·75	29·69	76	54	28	
		23	NW	29·95	29·81	72	50	—	
		24	SE	29·95	29·93	73	57	—	
		25	W	30·10	30·01	81	54	52	
		26	E	30·05	30·04	84	56	—	
		27	Var.	30·00	29·78	80	61	—	—
		28	SW	29·80	29·68	70	60	—	—
		29	SW	29·97	29·72	62	48	66	14
	1st Q.	30	W	30·00	29·81	77	56	—	14
		31	SE	29·62	29·48	75	52	26	7
6th Mo.		June 1	Var.	29·72		79	53	—	
		2	NE	29·54	29·49	67	51	25	49
		3	W	29·98	29·70	64	50	—	
		4	SW	29·90	29·85	71	59	—	—
		5	S	29·79	29·68	67	53	53	27
	Full M.	6	SW	29·89	29·81	75	53	—	
		7	S	30·05	29·99	80	53	—	
		8	Var.	29·96	29·85	88	53	56	
		9	W	30·24	30·22	74	46	—	
		10	E	30·15	29·98	77	54	—	
		11	SW	29·98	29·98	72	52	65	
		12	SW	30·08	30·05	73	47	—	—
				30·24	29·48	88	39	4·20	1·92

NOTES T. F.—*May* 16, 17. *Cirrus* and *Cirrostratus* followed by *Nimbi;* but no rain fell hereabout. Distant thunder heard. 18. Close day: in the afternoon various modifications of cloud appeared; in some places they showed a tendency to *Cirrocumulus*, in others *Nimbification* seemed rapidly going on; about 5 p.m. the sky, seen behind a large *Cumulostratus* under the setting sun, was of a deep brownish lake colour: as evening approached mountainous clouds rose majestically in the horizon, while others above were fringed with bright gold: rain succeeded in the night. 20. A uniform mass of cloud obscured the sky at sunrise: as the day advanced it broke, and divided itself into several distinct modifications. *Cirrocumulus* of various figures, in some places looking like wind-rows of hay, in others consisting of small round *Nubeculæ*, appeared; as well as *Cirrus* spread out in continuous sheets approaching to *Cirrostratus*, while flocks of *Cumulus* floated along in the wind below. In the wind about noon I observed a *Cirrus* cloud of a very remarkable figure; it consisted of many light tufts of a sort of *horse-shoe figure*, rising one above another. In the evening the distinct modifications were lost in a general haziness of a reddish colour; in some places blackish spots appeared, which were the *Nuclei* on which *Nimbi* formed, and thunder storms continued through the night. 21, 22. Thunder storms with fair intervals. 23. *Cumuli* alone early: in evening, streaks of *Cirrus* above them; some clouds showed a tendency to *Cirrocumulus*. 24. *Cirrocumulus* followed by increased heat, and evening lightning. 25. *Cirrostratus* strewed at different altitudes, also *Cirrocumulus* and *Cumulus*: in the evening *Cirrostratus* becomes dense, and approaches to *Nimbus*. 26. Various clouds, evening lightning. 27. *Cirrus* ramifying about, becomes *Cirrostratus*, which obscures the sky. 30. *Cirri* and *Cumuli*, followed by undulated and plane *Cirrostratus*. 31. Stormy day; upper currents blow in various directions.

June 4. Various clouds through the day: in the evening extensive beds of *Cirrocumulus*. 5. Showers. 6. *Cirrocumulus, Cumulus*, &c. In evening a thin sheet of *Cirrostratus* exhibited a faint ill-defined simple lunar halo. 7. In the evening a simple lunar corona. 12. Various clouds, with light showers: while at Plaistow with Mr. Howard, I observed large *Cumuli* pass under ramifying *Cirrus*, which rapidly (in consequence) became small flimsy *Cirrocumuli*, and presented a beautiful piece of sky.

NOTES J. G.—*May* 15th. Very considerable appearances of a storm in the W in the evening. 16. Some distant thunder in the afternoon—appearance of a heavy storm in the W at the same time. 18. Some gentle rain in the morning. 19. Gentle rain most of the day. 20. Lightning very frequent in the evening—thunder at a distance—about 10 p.m. a very vivid flash; immediately after which, a brilliant meteor descended in the SE. 21. About 4 a.m. a heavy thunder storm, the thunder remarkably loud—morning rainy. 22. Evening rather stormy; very frequent lightning, and thunder at a distance. 27. Evening rather clouded, with some lightning. 28. Very cloudy morning, wind high from the SW 29. Wind very high all day—frequent lightning in the evening. 31. Some rain in the morning, with thunder.

June 2. Very rainy day. 8. Frequent thunder and lightning in the evening, with high wind.

RESULTS.

Winds Variable.

Barometer: Greatest height . . . . 30·24 in.
Least. . . . . 29·48 in.
Mean of the period . . . . 29·88 in.

Thermometer: Greatest height . . . . 88°
Least. . . . . 39°
Mean of the period . . . . 63·19°

Evaporation . . . . 4·20 in.
Rain . . . . 1·92 in.

EXTRAORDINARY METEOR.

On the first day of this period, about half-past 8 p. m. a very singular luminous meteor was seen to the NNW of *Geneva*, passing over France : the same appears also to have been seen from *Paris*. It had at first the shape of the letter S, and was very slow in its apparent motion, being visible for seven or eight minutes, till concealed by a cloud : it was accompanied by a kind of whizzing noise. Professor Pictet, on comparing the different observations, concluded that it was elevated 24½ leagues above the earth's surface.—(See NICHOLSON'S JOURNAL, vol. xxx. pa. 216.)

TORNADO.

On the 12th of May, about five o'clock in the afternoon, a destructive phenomenon appeared at Bonsall, in the Peak of Derbyshire. A singular motion was observed in a cloud of a serpentine form, which moved in a circular direction, from S, by W to N, extending itself to the ground. It began near Hopton, and continued its course about five or six miles in length, and about four or five hundred yards in breadth, tearing up plantations, levelling barns, walls, and miners' cots. It tore up large ash trees, carrying them from 20 to 30 yards; and twisted the tops from the trunks, conveying them from 50 to 100 yards distance. Cows were lifted from one field to another, and injured by the fall ; miners' buddle-tubs, wash-vats, and other materials, carried to a considerable distance, and forced into the ground. This was attended with a most tremendous hail storm, stones, and lumps of ice were measured from nine to 12 inches circumference, breaking windows, injuring cattle, &c.—(P. LEDGER.)

AGITATION OF THE SEA AT PLYMOUTH.

An alarming and most uncommon flux and reflux of the sea took place May 31, commencing about 3 a. m. and not finally terminating till ten. The sea fell instantaneously about four feet, and immediately rose about eight feet : universal consternation pervaded the whole of the port. The vessels in Catwater were thrown about in the greatest confusion; many dragged their anchors, some drifted, and several lost their bowsprits and yards. About a quarter before seven the sea rose to the height of 11 feet, and again receded. At half-past nine the tide (half-flood) suddenly stopped ; and, in a moment, ebbed six inches and a half; at ten it ebbed again in the same most extraordinary manner, and then flowed as usual to high water. Two gales from SSW and E., preceded this astonishing phenomenon; but at the time of its occurrence the wind was light at SSW.

June 8.—About four o'clock, the tide again flowed and ebbed several feet in as many minutes, which continued at intervals for the space of four or five hours, during which the immense swell, commonly called a *boar*, drove into the harbours of Sutton Pool and Catwater, at the rate of four knots an hour, subjecting the vessels at anchor there to great danger. The wind was variable, but mostly south-west. During the operation of the *boar*, it thundered and lightned excessively —(P. LEDGER).

STORMS OF HAIL, THUNDER, &c.

From the various accounts of these in the Papers at this season, the following are selected, for the purpose of comparison with phenomena under the corresponding dates in our Register :—

We learn by a letter from his Majesty's ship Indefatigable, that the ships which sailed under her convoy for the East Indies, met with a most violent storm of hail, rain, thunder, and lightning, on the 20th of April, in lat. 46° 46', long. 11° 39' west. Four fire-balls passed along the Indefatigable, but they happily did no other damage than setting fire to the foretop-mast, which did not burn long. The Warley was also set on fire; the Perseverance had her maintop-mast struck; and the Warren Hastings her foretop-mast.

April 23.—A violent hail storm at night at Wynnstay, the seat of Sir W. W. Wynn. Some of the hail stones measured $2\frac{1}{2}$ inches in circumference. The whole new range of hot houses, which were only that day finished, had 1123 panes of glass broken.

His Majesty's brig Piercer, being off Ushant, on the night of the 28th of April, encountered a most severe gale of wind at SW, &c. (Details of a tremendous sea follow).

April 29.—One of the severest hail storms ever remembered came on at Welton, and lasted for above a quarter of an hour. The damage done to the gardens was very great, all the fruit-trees being stripped of their blossoms, and the glass of the hot-houses perforated as if bullets had been shot through them.

May 19.—At Sheffield, there was a dreadful storm of thunder and lightning, accompanied with hail. The stones, which measured from one to five inches in circumference, were pieces of ice encrusted with frozen snow. The damage is beyond precedent. At Beauchief, a whirlwind tore up seven trees by the roots, broke several in the middle; many buildings were unroofed, and haystacks thrown down; *nearly all the water was carried out of Mr. Stead's milldam, and dispersed in the air.*

May 20.—A severe thunder storm came on at Ingatestone, accompanied with a deluge of rain. In ten minutes the water ran about three feet deep in the streets. The same night at Potter-street, on the road to Newmarket, hail stones nearly as large as pigeons' eggs fell, accompanied with the most terrific thunder and lightning.

HEREFORD, *May 27.*—We were visited by a dreadful storm of thunder and lightning, accompanied by torrents of rain, very destructive eastward of this city. It commenced about three o'clock in the afternoon, and continued with little intermission till past eight. (This storm, and the consequent inundation of several thousand acres of land, destroyed a number of lives and much property).

June 8.—A severe storm of rain, hail, and lightning took place in Birmingham and the neighbourhood. The hail, or rather pieces of ice, which fell, are described of prodigious size, and considerable damage has been done to the windows.

WORCESTER, *June 8.*—This day another most tremendous storm of thunder, lightning, and rain took place about 11 a.m. equal to that of the 27th ult. except the hail.

A storm was also experienced (June 28), at Bury and its neighbourhood. The lower part of the houses were filled with water, which lay in the street five feet deep. The hail stones, 5-8ths of an inch in diameter, broke near 5000 panes of glass.—About 26 head of cattle were killed last week by the lightning, at Risby and Walsham, in Norfolk.

TABLE LVIII.

	1811.		Wind.	Pressure. Max.	Min.	Temp. Max.	Min.	Evap.	Rain, &c.
6th Mo.	L. Q.	June 13	W	30·17	30·15	72°	46°	—	
		14	SW	30·10	29·88	73	49	61	
		15	SW	29·96	29·88	73	47	—	
		16	W	30·10	30·04	72	50	31	—
		17	NW	30·22		72	45	—	
		18	SE	30·40	30·34	79	46	27	
		19	NW	30·26	30 14	75	55	—	
	New M.	20	NW	29·95	29·78	66	46	30	
		21	N	29·90	29·78	62	43	—	—
		22	N	29·94	29·91	60	44	36	5
		23	N	29·87		60	51	—	
		24	N	29·80	29·75	75	46	20	
		25	E	29·95	29·84	79	53	—	
		26	Var.	29·98	29·95	76	57	32	41
		27	NE	29·95	29·89	74	61	—	9
	1st Q.	28	N	29·96	29·91	67	60	—	50
		29	N	29·91	29·89	68	59	—	
		30	N	29·89	29·88	65	58	17	
7th Mo.		July 1	NE	29·88	29·86	66	61	—	
		2	Var.	29·96	29·90	76	58	—	18
		3	NE	30·13	30·00	59	48	16	—
		4	NW	30·15	30·14	58	43	—	
		5	SW	30·17	30·13	67	47	—	
	Full M.	6	N	30·10		71	54	—	—
		7	N	29·98	29·97	67	54	36	7
		8	N	30·00	29·98	73	52	—	
		9	SE	30·10	30·06	75	57	—	
		10	Var.	30·14	30·10	80	52	—	
		11	NW	30·19	30·15	78	58	69	
				30·40	29·75	80	43	3·75	1·30

NOTES T. F.—*June* 13. *Cumuli* early; in evening *Cirrus*, *Cirrostratus*, and *Cirrocumulus*. 14. Early the sky was filled with light *Cirrose* fibres, *Cumuli* floated beneath them; afterwards *Cirrocumulus*, *Cumulostratus*, &c. appeared. 16. Early appeared the *Cirrus*, followed by general obscurity and a gentle shower; afterwards *Cirrocumulus* was observed overhead through the general mistiness; about 3 p. m. the sky again became obscure, and it rained. In the evening the sky exhibited very interesting and beautiful phenomena. Long thin sheets of the *Cirrostratus* appeared in the N and NW, some of them acquired the appearance of the architectural cyma, they perpetually changed their figures, and some which extended over dense *Cumulostratus* became (apparently in consequence) *Cirrocumulus,* which cloud ultimately prevailed; at 11 p. m. I saw an extensive bed of it. 17—20. Fair weather. *Cirri* continually observed above, while *Cumuli* float beneath them. On the 20th the morning was cloudy with few drops of rain, and a clear evening. 21. Sky well covered with clouds all the morning; afterwards it cleared, and *Cumuli* appeared. About 5 p.m. a smart shower came on, after which *Cumulostrati* continued to pass over, while *Cirrostratus* and *Cirrocumulus* appeared to extend over them in a higher region. Spongoid evanescent *Cumuli* also appeared. 25. Wind strong in the day. The evening was calm, and the sky above the setting sun of a rich yellow colour; a large bed of *Cumulostratus* passing over exhibited very beautiful deep red tints. About half-past nine a small meteor appeared in the SE, it was simply a stationary accension, and lasted scarcely a second. 27—*July* 3. Sky obscured with clouds almost all this time, with occasional rain, damp, and mostly warm air, and a north wind; it sometimes cleared for a short time, when several strata of clouds of various kinds appeared. 5. *Cirrus* seen early, followed by beautifully arranged beds of undulated *Cirrostratus* and *Cirrocumulus,* and lastly by *Cumulostratus.* 6, 7. Various clouds as heretofore. 8. A change in the electricity of the atmosphere was observable to day: a cloudy sky was followed by *Cirri* ramifying chiefly toward the east; in the evening the modifications appeared in their natural order, *Cirrus* the highest, and *Cirrocumulus, Cirrostratus,* and *Cumulus,* in successively lower regions.

NOTES J. G.—*June* 27. Rainy morning. 28. Very rainy day: some lightning in the evening.

RESULTS.

Prevailing Winds Northerly.

Barometer:	Greatest height	30·40 in.
	Least........................	29·75 in.
	Mean of the period	30·02 in.
Thermometer:	Greatest height...............	80°
	Least........................	43°
	Mean of the period	60°
Evaporation		3·75 in.
Rain		1·30 in.

SUBMARINE VOLCANO.

(From a coloured plate of the phenomenon, published by Boydell, 1812).

" This eruption broke out in the sea on the 13th of June, 1811, about a mile from St. Michael's, (Azores). On the 17th, Capt. Tillard, commander of H. M. S. Sabrina, in company with Mr. Reid the Consul, and two other gentlemen, proceeded to the cliff nearest to it, which was between 300 and 400 feet above the level of the sea. Its appearance when quiescent was that of an immense body of white smoke, revolving almost horizontally on the water; when suddenly would shoot up a succession of columns of the blackest cinders, ashes, and stones, in form like a spire rising to windward, at an angle of from 70° to 80° from the horizon, and to a height of between 700 and 800 feet above the sea. The columns at their greatest height broke into branches, resembling magnificent pines: and as they fell, mixing with the festoons of white feathery smoke, assumed at one time the appearance of plumes of black and white ostrich feathers, at another that of the light wavy branches of a weeping willow. These bursts *were accompanied by the most vivid lightning,* and by a noise like the continued firing of cannon and musquetry intermixed : and as the cloud of smoke rolled off to leeward, *it drew up a number of waterspouts,* which formed a beautiful and striking addition to the scene. While the party were viewing the eruption, *a crater* began visibly to be formed above water, though the volcano was then only four days old, and the depth of the sea was 30 fathoms ! "

The sequel of the account states that in little more than a fortnight a complete island was formed, nearly a mile in circumference, and about 300 feet high: Captain Tillard took a drawing and plan of the phenomenon. From the *crater,* which continued open, flowed *a stream of boiling water* six yards over. This island gradually disappeared towards the close of the year, being converted by the waves into an extensive shoal: but in two months afterwards smoke again issued from the sea near the place.

THUNDER STORMS.

Waterford.—During the earlier part of Monday, July 1, the atmosphere was sultry almost in the extreme. Dark clouds were seen towards the NE, and thunder was heard at times from the same quarter. Till two o'clock, the wind blew from NW, at which hour it suddenly changed and blew from the E, and this city was almost immediately visited by one of the most tremendous storms that have, in all probability, ever been experienced in Ireland. For the space of an hour, the thunder and lightning succeeded each other in awful and nearly uninterrupted rapidity, accompanied by torrents of rain, together with hail stones, of a size greater than any inhabitant of the city remembers to have ever before seen. Some of them, were two, and others three inches in circumference. The streets were every where overflowed; after the storm had subsided, the windows all along the quay, and wherever the houses were unprotected, exhibited the appearance of having been exposed to the assault of fire-arms, few of the panes of glass having escaped. The whole scene exhibited one of those sublime but terrible conflicts of the elements, which the English and Foreign Journalists have had frequent occasions recently to describe, but which are of rare occurrence in this island.

July 1.—As the sloop Diligence, bound from Truro to Plymouth, was coming round Penlee Point, a flash of lightning struck the vessel, by which the crew and passengers were laid senseless on the deck. In this state the greater part of them remained some minutes, but none of them received any serious injury. The mast of the vessel was split in every direction, and the top-mast was shivered to pieces, as if cut up by an axe. *Fortunately the lightning did not come in contact with the cargo, which consisted of gunpowder !*

July 2.—In the neighbourhood of Wonash, near Guildford, after one of the most sudden strokes of thunder, a deluge of rain took place, so instantaneously as to carry away or destroy every thing which impeded its progress—The damage is estimated at not less than 1,500*l.* The same day a thunder storm, accompanied with torrents of rain, came on at Oxford and its vicinity. A barn at Finstock was set on fire by the electric fluid; the flames communicated to another barn, and both were destroyed, with a large quantity of wheat contained therein.

North Wales has been recently visited by several awful storms, and very serious injury has been sustained by many persons. On Wednesday, July 10, the town of Llanidloes was drenched with torrents of rain, accompanied by dangerous lightning; the windows of several houses were shattered at Trefeglwys, and a great number of sheep have been destroyed on the neighbouring hills.

A heavy storm of rain suddenly came on at Salisbury, July 10, attended with a phenomenon *supposed to be a water-spout*, which hung in a spiral form from the clouds, with the end waving to and fro like the tail of a kite. At first it appeared about a mile to the north of the city, and was drawn up into the clouds about the same distance to the SE. It passed over rather near to the earth, and *a rushing noise* was heard at the same time. Though the rain fell so very heavy there, it extended only two miles.—(PAPERS).

Effects of Lightning on growing Potatoes.

During a thunder storm in the summer of 1811, in my absence, the lightning was seen to strike the ground in a potatoe field, about a furlong to the SW of my house at Plaistow. The spot where it fell (which was rather lower and moister than the surrounding parts) presented the following appearances. The immediate impetus of the lightning was clearly seen in a *smooth round excavation*, from which about a bushel of earth had been *dispersed:* from this proceeded a semicircular *furrow* of about five yards in length, losing itself imperceptibly by ramification. The growing potatoes, for the space of about a square rood, chiefly to the E of the cavity and furrow (and in the probable direction of the stroke) had suffered considerably, but without the least sign of scorching, or other immediate injury to the tops. The damage was *confined to the stems,* which for a few inches in height were partly burnt or *riven*, partly turned *pulpy.* What is extraordinary is, that the far greater part of the plants *survived* the stroke, continuing to grow by means of the small remaining connection of the top with the root. The pulpy ones died (precisely as by *frost*) mixed in every part with the rest, and many of them 20 yards distant from the cavity and furrow above mentioned.

Meteorological Observations

MADE AT

PLAISTOW, NEAR LONDON,

IN THE YEARS

1811 and 1812.

(First published Monthly in Nicholson's Philosophical Journal).

TABLE LIX.

	1811.		Wind.	Pressure. Max.	Min.	Temp. Max.	Min.	Evap.	Rain &c.
7th Mo.	L. Q.	July 12	NW	30·03	29·91	76°	59°	—	
		13	W	29·91	29·79	75	60	24	42
		14	SW	29·83	29·76	65	59	—	—
		15	S	29·83	29·80	71	59	—	—
		16	SW	29·85	29·83	70	54	43	—
		17	S	29·85	29·75	72	54	—	12
		18	SW			71	60	—	57
		19	SE	29·94	29·75	73	54	—	—
	New M.	20	W	29·94	29·90	64	53	47	79
		21	Var.	29·88	29·82	60	53	—	1·61
		22	W	30·01	29·88	70	50	—	
		23	NW	30·11	30·01	69	50	—	
		24	NW	30·15	30·11	72	54	—	
		25	NW	30·14	30·12	73	55	57	
		26	SW	30·12	30·09	72	51	—	
		27	N	30·09	29·91	74	55	—	
	1st Q.	28	SE	29·91	29·85	78	54	32	
		29	NE	30·11	29·85	75	54	—	
		30	N	30·11	30·08	67	49	—	
		31	NE	30·08	—	68	52	42	
8th Mo.		Aug. 1	NE	—	29·90	76	53	—	
		2	S	29·90	29·69	73	51	—	
		3	SW	29·67	29·58	76	51	43	11
	Full M.	4	NW	29·73	29·60	68	54	—	27
		5	S	29·65	29·62	66	52	29	
		6	S	29·59	29·48	63	50	—	26
		7	NW	29·60	29·50	67	54	—	15
		8	SW	29·49	29·35	62	50	32	74
		9	NW	29·60	29·48	64	45	—	33
		10	NW	29·86	29·60	61	44	29	—
				30·15	29·35	78	44	3·78	5·37

Notes.—Seventh Mo. 15. Small rain about 2 p. m. 19. A thunder shower early: fine day. 20, 21. Forty-eight hours rain. 22. Temperature 60°, (the maximum for 24 hours) at 8 a. m. 26. Orange coloured *Cirri* at sunset. 27. Thunder clouds: a few drops p. m.: much dew. 28. *Cirrocumulus* cloud, very beautiful, interchanging with *Cirrostratus*, succeeded by large *Cumuli*. In the evening some appearance of a thunder storm far in the NW. 29. Evening, parallel bars of *Cirrostratus*, stretching E and W: a blush on the twilight. 30. Windy, cloudy.

Eighth Mo. 2. Large elevated *Cirri*. 3. *Cirrocumulus*, followed

5

by *Cirrostratus:* evening overcast: rain by night. 4. Windy, at SW, by night. *Cumulostrati,* in various quarters, at sunset. 7. Opaque twilight, with *Cumulostratus.* 8. Very wet, a.m.; at noon a thunder shower; at 6 p.m. a heavy squall from NW, with rain and hail; the *Nimbus,* as it receded, presenting a perfect and brilliant bow: windy night. 9. Large *Cumuli* rose, and at noon inosculated with the clouds in a superior stratum: a thunder shower ensued before 2 p.m., after which appeared the distinct strata again: about 6 p.m. a second thunder shower, long very dense in the SE, where the bow was conspicuous above an hour. This day was nearly calm. 10. Rain fell again about noon, upon the union of two strata of cloud.

NOTES T. F.—*Clapton, July* 16. Showers, with fair intervals a.m., evening clear, the sky abounding with *Cirri,* ramifying about, and generally pointing to the E: their kind of (internal) motion may be compared to that of a piece of cheese full of mites. 17, 18. Warm with various clouds.

August 10. Showery day: very clear night: small meteors were very abundant, with this remarkable circumstance that their trains lasted longer, after the extinction of the lucid head, than usual.

RESULTS.

Prevailing Winds Westerly.

Barometer: Greatest height................30·15 in.
Least......................29·35 in.
Mean of the period...........29.835 in.
Thermometer: Greatest height...............78°
Least......................44°
Mean of the period...........61°
Evaporation........................3·78 in.
Rain...............................5·37 in.
Character of the period changeable, with much rain.

July 19.—A most tremendous storm of thunder, lightning, rain, and hail, visited the parish of Guilsheld, Montgomeryshire; a large oak tree, containing about 50 feet of timber, growing near Varenwell, was rent asunder from top to bottom, and the bark thrown to a distance of upwards of 20 yards.

July 21.—A heavy fall of rain did much damage at Stamford. In the meadows about Barrowden, Wakerley, Harringworth, Thorpe, Caldecot, and Bringhurst, immense quantities of hay were carried away. Several thousand loads, it is supposed, have been destroyed, in addition to which the injury sustained by the land has been very great. The water rose thirteen feet in less than four hours in Wood Newton parish.

TABLE LX.

1811.			Wind.	Pressure. Max.	Pressure. Min.	Temp. Max.	Temp. Min.	Evap.	Rain, &c.
8th Mo.	L. Q.	Aug. 11	NW	30·16	29·86	61°	42°	—	
		12	NW	30·16	30·10	64	50	—	
		13	SW	30·24	30·10	73	52	33	—
		14	NW	30·25	30·09	66	47	—	
		15	SW	30·25	29·97	68	51	—	—
		16	SW	30·14	29·97	68	57	37	—
		17	NW	30·13	30·03	70	45	—	
		18	E	30·03	29·76	72	55	—	2
	New M.	19	Var.	29·72	29·65	68	54	30	35
		20	W	30·05	29·72	64	57	—	5
		21	SW	30·08	30·04	68	56	—	
		22	W	30·04	29·92	71	52	41	—
		23	S	29·92	29·73	68	55	—	4
		24	E	29·73	29·52	70	55	—	—
		25	SW	29·70	29·50	65	48	16	39
		26	SW	29·78	29·70	67	56	—	
	1st Q	27	W	30·07	29·74	68	44	—	1
		28	Var.	30·11	30·03	66	51	42	
		29	SW	30·17	29·96	69	46	—	2
		30	NW	30·20	30·13	69	47	24	
		31	SW	30·02	29·97	71	53	—	
9th Mo.		Sept. 1	NW	30·22	30·02	68	45	—	
	Full M.	2	N	30·29	30·26	65	45	—	
		3	NE	30·29	30·24	64	53	34	
		4	NE	30·24	30·18	62	53	—	
		5	E	30·18	30 15	71	52	—	
		6	E	30·17	30·13	73	44	35	
		7	NE	30·19	30·13	72	43	—	
		8	E	30·20	30·17	74	47	22	
				30·29	29·50	74	42	3·14	0·88

Notes.—Eighth Mo. 11. *Cumulostratus*, dense about noon, but which soon after dispersing, a brilliant sunset ensued. 12. a. m. cloudy: wind SW. 13. A few drops at intervals: rain in the S by inosculation. 14. a. m. *Cumulus*, with haze gradually increasing above: p. m. clouds below disperse: a fine elevated veil of *Cirrus*, coloured at sunset. 15. Elevated clouds, with traces of *Cumulus*: some large drops about noon: at sunset, the western sky richly coloured with red and yellow, on *Cirrocumulus* passing to *Cirrostratus*: windy night. 16. a. m. Windy: p. m. small rain: clear evening, with coloured *Cirrus* and *Cirrocumulus*. 18. Evening, large *Cirri*, pointing upwards. 19. a. m. thunder showers,

chiefly to SSW and N: a strong variable charge in the insulated conductor: evening fair and windy, with *Cumulostratus*. 20, 21. Windy:
much dew. 22. Light rain a. m.: showers p. m. 23. Misty morning:
Cumulus, with *Cirrostratus* from the S: about *one*, these inosculated,
and showers prevailed p. m. 24. Misty morning: *Cumulostratus:* a
few drops of rain: evening, *Cirrostratus.* 25. Misty, and raining at
half-past 8 a. m.: wind SE: evening, *Cumulostratus* evaporating, beneath a veil of *Cirrus*, which at the moment of sunset, was of a light
silver grey, and during twilight, passed through yellow, orange, red,
and purple, to dull grey; and lastly became again somewhat red:
much dew, with a very moist air. 26. A small lunar halo, on clouds
moving in a northerly current. 27. Windy, a m.: small rain, evening:
much dew. 28. Windy. 30. a. m. *Cirrus,* with points dependent and
crossing, and *Cumulus* forming beneath: at 9 p. m. *Cirrocumulus,* with
much dew: the barometer unsteady. 31. Fine day: *Cumulus, Cirrus,
Cirrocumulus:* a diffused blush on the twilight, which begins to be
very luminous.

NOTES T. F.—*Clapton, August* 21. All the modifications appeared
in the day; at times it was quite clouded: in the evening the upper
currents, by two Montgolfier balloons, appeared to be S and NW.
23. A balloon, which did not ascend much, went first with a SW
then a S wind. 28. Early appeared *Cirri* in a lofty region, while *Cumuli* floated below; cloudy by night. 29. Clear warm morning with
various clouds and a strong current from the south, cloudiness and rain
followed; a small air-balloon that I launched at one o'clock went with
a SW wind. Mr. Sadler in his balloon went with a W and SW wind
alternately: the evening was cloudy, and wind various both below and
above, as appeared by many fire-balloons. 30. Large *Cirri* ramified
about in the morning, and *Cumuli* flew along in an under current; at
night the wind next the earth was N, above it E, and still higher
long beds of *Cirrocumulus* passed over from SW. 31. Flimsy *Cirrocumulus* from SW.

RESULTS.

Wind Westerly, with little exception, to the time of Full Moon, when
it came round by N to the Eastward.

Barometer: Greatest height................30·29 in.
Least......................29·50 in.
Mean of the period ....30·025 in.
Thermometer: Greatest height............... 74°
Least...................... 42°
Mean of the period ....59·20°
Evaporation....................... 3·14 in.
Rain ....0·88 in.

TABLE LXI.

	1811.	Wind.	Pressure. Max.	Pressure. Min.	Temp. Max.	Temp. Min.	Evap.	Rain, &c.
9th Mo.	L. Q. Sept. 9	E	30·17	30·15	75°	51°	—	
	10	E	30·15	30·05	77	46	—	
	11	Var.	30·15	30·00	80	55	45	
	12	E	30·19	30·15	73	53	—	
	13	E	30·11	30·01	71	40	—	
	14	E	30·02	29·98	74	45	56	
	15	NE	30·07	30·02	70	55	—	
	16	E	30·05	29·98	66	44	—	
	New M. 17	E	30·05	29·95	70	47	—	
	18	E	29·95	29·87	71	44	95	
	19	SE	29·87	29·50	74	47	—	
	20	SE	29·53	29·50	74	51	—	—
	21	S	29·80	29·53	66	54	32	8
	22	Var.	29·80	29·60	65	52	—	
	23	SW	29·60	29·43	61	43	—	14
	24	NW	29·62	28·92	64	49	30	14
	1st Q. 25	W	29·26	28·86	60	48	10	46
	26	W	29·21	29·17	61	39	—	10
	27	SW	29·33	29·20	51	40	18	45
	28	NW	29·54	29·33	62	44	—	—
	29	NW	29·71	29·54	64	51	—	20
	30	SW	29·72	29·55	62	50	22	16
10th Mo.	Oct. 1	W	29·76	29·47	63	46	—	4
	Full M. 2	SW	29·87	29·85	64	39	19	
	3	E	29·85	29·46	61	52	—	26
	4	S	29·57	29·46	69	59	—	2
	5	SW	29·75	29·57	67	54	35	15
	6	SW	29·95	29·87	63	52	—	18
	7	SW	29·92	29·85	69	54	21	1
	8	SW	30·00	29·90	67	53	10	
			30·19	28·86	80	39	3·93	2·39

NOTES.—Ninth Mo. 9. Before sunset, after a serene day, *Cirrus* clouds, pointing downwards from the W. 11. *Cirrocumulus*, some dry haze: wind westerly by night, scarcely sensible. 14. *Cirri* and haze in the evening: twilight of a bright orange colour. 15. Much wind: clear. 16. a. m. overcast: p. m. clear: twilight duller, with *Cirrostratus*. 17. Much wind: very clear sky. 18. As yesterday: evening twilight luminous, orange surmounted with rose colour, the latter somewhat in converging streaks. 19. Morning twilight obscure, with dense *Cirri:* much dew: wind a. m. NE: thunder clouds at different heights, some of which moved from the SE: there were clouds throughout the night, with lightning. 20. Wind a. m. NE : thunder clouds again, which grouped, and passed, about 2 p. m. to the W, with a few drops: *Nimbi*, with a faint bow, in the distance: evening cloudy, with two strata: wind SE : much lightning in the SW. 21. a. m. cloudy: rain, with distant thunder at one and 2 p. m.: *Nimbi* and *Cumulostratus:* faint bow. 22. a. m. overcast: wind veered to NW, apparently by E : *Cirri*, in lines from NE to SW. 23. a. m. wind fresh from SW, with rain ; p. m. fair, with various modifications of cloud, which were finely coloured at sunset in the east. 24. a. m. clear : much dew: fair day, but with clouds indicating rain : twilight milky, with a blush of red : the moon disappeared early, behind *Cirrostratus* clouds, and it rained heavily in the night. 25. Cloudy and windy, with rain. 26. a. m. *Cirrus* with *Cumulus:* p. m. showers. 27. Windy : wet. 28. a. m. misty: p. m. showers, *Cirrostratus*, and a blush on the twilight. 29. Evening, lightning : wet night. 30. Lunar halo.

Tenth Mo. 1. a. m. wind SE, showery. 2. A little before sunrise I observed a *Stratus* in the marshes to the SE, very nearly resembling a sheet of water ; one which was seen from this village, in similar circumstances, about two weeks since, was actually taken by several persons for an extensive inundation : in the afternoon, large elevated *Cirri* and *Cirrostrati*, rapidly passing at sunset from red to grey, indicated a renewal of the wet weather. 3. Misty morning, with *Cirrostratus* above : very wet, p. m. 4. Much wind : cloudy night. 5. Squally. 6. a. m. cloudy, much wind : evening calm; large *Cirri* and *Cirrostrati*, with a blush on the twilight : a bright blue meteor in the NW: wet night. 7. Cloudy, with a gale of wind. 8. Fair.

RESULTS.

Barometer :	Greatest height	30 19 in.
	Least	28·86 in.
	Mean of the period	29·736 in.
Thermometer :	Greatest height	80°
	Least	39°
	Mean of the period	57·85°
Evaporation		3·93 in.
Rain		2·39 in.

From the Full Moon of the last period to the New Moon of the present, easterly breezes with clear days, and the *Stratus* by night. Evaporation went on increasing as the wind became stronger : dew fell in plenty, and the small meteors, called shooting stars, were abundant. The latter half of the present period brought the accustomed compensation, in rain from the westward: the approach of this was perceptible for several days beforehand; and the ground being dry, it was attended at the beginning with some discharges of electricity from the clouds.

HEAT ON THE CONTINENT.

In Silesia, Poland, and Russian Lithuania, the extreme heat, unaccompanied by the least rain, for near two months, has entirely destroyed the hopes of the husbandmen. The fields of wheat, oats, and barley, have been burnt up. The meadows are withered, and the brooks have disappeared; the trees are despoiled of their leaves, and the earth rendered so hard, that they cannot work it.

The excessive heats of July have occasioned the conflagration of several forests in the Tyrol, with the consequent destruction of 64 villages and hamlets, situated in their vicinity, and of the loss of near 10,000 head of cattle. The principal forest, which has thus been destroyed, is about seven miles in extent, and three in breadth, and distance three leagues from Inspruck. It caught fire on the 26th of July, and had not ceased burning on the 4th of August.—(PAPERS).

September 19:—A violent tempest of thunder and lightning was experienced at Stourport. The hail broke an immense number of windows. It came from the east. About the same hour, the storm was very violent at Worcester: the principal body of it seemed attracted towards Malvern Hills: it was preceded by unusual gloom, and a thick cloud of dust; the lightning was of a very vivid description; but we have not yet heard of any serious damage.

HARWICH, *September 25.*—Last night the wind blew a violent storm from the S and SSE, attended with rain.

PLYMOUTH, *September 25.*—It has blown a tremendous gale of wind the whole of this day from NNW.

HURRICANE AT SEA.

A letter, dated the 8th October, from an Officer belonging to the *Tartarus*, gives the following account of the late tempest on the American coast, in which that vessel and her consorts suffered so severely :—" On the morning of the 26th ult. we left Halifax to cruize off the American coast, under the orders of the *Æolus*, Lord James Townshend, and accompanied by the Africa, 64, and Spartan, 38; having fine weather, with every prospect of making a successful cruize; and no ships were ever in a better state of equipment to encounter the storms and danger of the perilous ocean. On the morning of the 30th of September, when pretty near our cruizing ground off New York, in lat. 40° 50′ north, long. 65° west, a heavy gale of wind came on at SE, and blew with tremendous fury. The Æolus, our Commodore, taking the advice of our old friend and companion *the never failing barometer*, made all snug; you may be assured that we followed the example, though every sail in the ship had been braced long before, our top-gallant-masts got on deck, our jibboom and spritsail-yard taken in, and every thing removed out of the tops.

The hurricane continued for four hours, with a mountainous sea, during which the quicksilver fell gradually in the barometer until scarcely a particle of it was to be seen above the wood; (probably to near the lower extreme of the scale, or 28 inches, L. H.) but the scenery of the sky is is impossible to describe. *No horizon appeared, but only a something resembling an immense wall, within ten yards of the ship*—at this moment we lost sight of our Commodore, who had only a short time before wore on the other tack : we were in the act of following his motions, when being before the wind, and just coming to, we were laid on our beam ends; our mizen and main-top-masts were blown away, although there was not any sail on them, and *without any person's hearing the crash ;* in an instant five feet water rushed into the hold, through the ports of the main deck (all the hatchways were battened down), and the water had reached the coamings of the quarter-deck, when orders were given to throw overboard the larboard quarter-deck guns and the sheet-anchor, or we must inevitably have foundered. In this state we remained nearly half an hour, expecting every moment to be our last, as the stillness of the ship convinced us that she was settling down. We were about to cut away our masts, when fortunately feeling the ship sally to windward, from that moment we entertained some hopes. During all this time the pumps were at work, yet we gained but little, for it was the rudder-coat which had burst in and occasioned the water to rise upon us, and this it still continued to do without our being able to prevent it. The bulk-head of the coal-hole having been washed down by the quantity of water in it, the coals were carried into the well, and three pumps were rendered useless. In the midst of this distress, a faithful servant stationed at the barometer to watch its change, called up through the sky-light (the frame of which had been washed overboard) with great joy, that *the quicksilver had risen an inch,* and still continued to rise. This inspired all hearts with fresh spirit, but no men could behave better, or do more than our crew did. After we had relieved the ship from a great quantity of water, we found the main-mast tottering, and every minute expected that and the bowsprit to go, but our greatest care was the preservation of the lower masts, to save the ship from foundering after the gale, for want of something to bring her to.—All our booms and boats were washed overboard, but we succeeded during the height of the storm, in lowering the main yard on deck, by which means we saved the main-mast from falling, as the rigging had sundered through its seizings, and nothing else could have preserved it. The gale had by this time considerably abated, and about six o'clock in the evening it cleared off, though still blowing hard. We again saw the Æolus about a mile and a half from us, and, with respect to her masts, she had suffered more than the Tartarus. I would, if possible, give you a description of the noise occasioned by the hurricane, but I am unequal to the task: if you can conceive, however, all the savage animals of the brute creation assembled to affright mankind by their roaring, you will have some faint idea of the deafening variety of sounds in the tempest we experienced! The day before yesterday all four ships met together off the harbour, and, under jury-masts—all went in together complete wrecks."

TABLE LXII.

1811.			Wind.	Pressure.		Temp.		Evap.	Rain, &c.
				Max.	Min.	Max.	Min.		
10th Mo.	L. Q.	Oct. 9	SW	30·00	29·99	67°	54°	—	
		10	SW	29·99	29·77	63	57	—	
		11	SW	29·77	29·60	65	51	—	
		12	S	29·80	29·60	62	48	40	8
		13	SW	29·90	29·86	62	49	—	3
		14	S	29·86	29·81	63	51	—	1
		15	S	29·76	29·75	73	53	33	
		16	S	30·03	29·76	70	55	—	
	New M.	17	Var.	30·10	30·03	71	47	—	
		18	SW	30·16	30·10	68	50	—	11
		19	Var.	30·21	30·18	65	49	20	
		20	SW	30·05	29·96	64	55	—	
		21	SW	29·96	29·50	65	56	20	
		22	S	29·52	29·46	64	50	—	15
		23	SW	29·50	29·48	60	49	—	
		24	Var.	29·48	29·35	57	42	20	8
	Last Q.	25	S	29·35	28·65	53	38	—	18
		26	Var.	28·80	28·65	54	41	15	32
		27	SE	28·84	28·81	56	43	—	11
		28	Var.	28·84	28·80	56	41	15	44
		29	SW	29·05	29·00	55	43	2	18
		30	Var.	29·55	29·00	58	43	—	14
	Full M.	31	W	29·77	29·68	59	48	15	14
11th Mo.		Nov. 1	SW	29·68	29·62	62	57	—	11
		2	SW	29·58	29·50	62	53	34	14
		3	SW	29·70	29·60	58	48	—	8
		4	W	29·98	29·80	60	42	17	
		5	SW	29·89	29·83	56	43	—	25
		6	SW	29·83	29·52	53	45	13	50
				30·21	28·65	73	38	2·44	3·05

NOTES.—Tenth Mo. 12. Windy: wet evening. 13. Much wind. 14. A shower before 9 a. m., at which time occurred the maximum of temperature. 15. Much dew on the grass: serene day: twilight milky, with converging streaks of red. 16. a. m., Much dew: a mist on the river: the smoke of the city was remarkably depressed, and sounds from thence unusually strong: some thunder clouds appeared and passed to E. 17. *Cumulus* clouds surmounted with *Cirrostratus*, and *Cirri* above. 18. A very wet mist a. m., wind NW: at 2 p.m. cloudy; very moist air, the *dew point* (or temperature at which a body colder than the air condenses water from it) being 63°: and about sunset, at temperature 63°, I found dew just beginning to be deposited on the grass: it rained hard about five next morning. 19. a. m. Misty, small rain : p. m. clear: evening, *Cirri* very elevated, and long coloured red ; a *Stratus* forming. 20. Misty : then overcast : the wind, which had been E, veering by S : abundance of *gossamer.* A quicken tree (*sorbus aucuparia*) exhibits a new set of leaves and blossoms along with the ripe berries. 21. Grey morning, with little dew and a strong breeze. 22. Dew scarce perceptible : wind veers to S, a breeze : p. m. very cloudy, with showers: much wind at night. 24. At mid day a drizzling rain, during which the vane turned to E. 25. Clear, fine day : wind veered to S : at sunset *Nimbi* and *Cirrostrati* in SW: heavy shower by 11 p. m. 26. Showery: a fine rainbow at 10 a. m. 27. *Nimbi* a. m. in different quarters, mixed with *Cumulus* and *Cirrostratus*, beneath large plumose *Cirrus* clouds. 28. Clear a. m., much dew, *Nimbi* forming amidst various clouds : vane at NE: p. m. a shower in the S, during which appeared, for a short time, a numerous flight of swallows : they had been last observed on the 15th : the wind returned by S to NW, with much cloud and rain. 30. At 9 a. m. the rain intermitting, the highest and most considerable mass of clouds was moving from W, an intermediate portion from S, and the wind below fresh at E : in this state of things sounds came very freely from the westward, and by eleven the wind was SW: at 3 p. m. distinct *Nimbi* and a bright bow : showery at night, with a lunar halo. 31. a. m. Clear : the sun and moon appeared red on the horizon ; at night, the wind being S, sounds came loud from the W.

Eleventh Mo. 1. Much cloud a. m.: wind fresh at SW. 2. As yesterday: stormy at night. 3. A rainbow at 8 a. m. 4. *Nimbi* a. m. to windward : at sunset, the dense clouds in the E finely coloured : rainbow : wind W. 5. Stormy a. m.: wet p. m. 6. Cloudy, showery: evening, abundance of *Cirrostratus:* a wet night.

RESULTS.

Barometer: Greatest height . . . . 30·21 in.

     Least.. . . . .28·65 in.

     Mean of the period . . . . .29·614 in.

Thermometer: Greatest height . . . . 73°

     Least. . . . . 38°

     Mean of the period . . . . .54·86°

  Evaporation . . . . 2·44 in.

  Rain . . . . 3·05 in.

Wind with little exception SW and S : very stormy on the W coast the last week of tenth month. The fore part of the period changeable; the latter wet, without the usual intervening frosty nights.

TABLE LXIII.

	1811.	Wind.	Pressure. Max.	Min.	Temp. Max.	Min.	Evap.	Rain, &c.
11th Mo.	Nov. 7	NE	29·58	29·43	52°	41°	—	7
L. Q.	8	S	29·65	29·35	52	42	6	19
	9	Var.	29·67	29·34	54	47	—	34
	10	SW	29·34	29·22	55	39	10	21
	11	NW	29·91	29·22	52	36	—	3
	12	NW	29·60	29·50	54	38	18	22
	13	W	29·73	29·69	52	35	—	
	14	SW	29·65	29·57	55	40	12	7
	15	W	29·57	29·49	48	36	11	8
New M.	16	NW	29·66	29·49	50	41	—	
	17	SW	30·17	29·66	49	44	—	
	18	N	30·25	30·17	50	45	—	1
	19	N	30·39	30·25	53	31	12	
	20	NW	30·39	30·32	48	28	—	
	21	SW	30·32	30·25	46	29	6	
	22	E	30·22	30·18	45	25	—	
1st Q.	23	NW	30·24	30·22	47	28	6	
	24	SW	30·35	30·24	46	33	—	
	25	NW	30·36	30·35	50	38	5	
	26	W	30·41	30·35	47	41	—	
	27	SW	30·40	30·35	44	40	6	
	28	W	30·35	30·27	48	39	—	
	29	SW	30·27	30·25	47	42	—	
Full M.	30	SW	30·25	30·10	50	41	8	
12th Mo.	Dec. 1	SW	30·10	29·50	52	48	13	
	2	NW	29·86	29·50	52	34	8	15
	3	SW	29·86	29·32	50	40	13	4
	4	W	29·45	29·21	46	31	—	
	5	N	29·96	29·74	32	22	—	
	6	SW	29·74	29·56	50	28	15	
			30·41	29·21	55	22	1·49	1·41

NOTES.—Eleventh Mo. 7. A calm pleasant day. 8. Cloudy, drizzling. 9. Wind, a. m. NW: a dripping mist: then clear and calm: *Cirrostratus*, evening: rain before nine the next morning. 10. *Nimbi* at sunset, with red haze on a brilliant twilight. 11. At sunset the clouds coloured in the E: a *Nimbus* in the W: windy night. 13. A clear sunset beneath dense clouds. 14. Windy, SW a. m. 15. Clear, windy: *Nimbi* at sunset to S. 18. Dripping mist. 19. Fair: *Cumuli* p. m., which evaporating at sunset, a beautiful red twilight ensued, with *Cirrostratus*. 20. Hoar frost and ice, the first this season: clear day, with *Cirrus* clouds: at sunset, the purplish haze of the dew was conspicuous, and the twilight of a rich crimson, with converging darker streaks upon it, probably the shadows of prominent objects on the earth. 21. Hoar frost: a *Stratus* in the evening. 22. Cloudy through the day in the superior atmosphere: twilight milky and luminous, with a blush of red. 23. Much rime on the grass, &c.: the sun emerged suddenly from the surface of a dense frozen mist, *Cirri* stretching from E to W, *Cirrostrati* and *Cumuli* beneath: the evening quite overcast. 24. Various modifications of cloud ending in *Cumulostratus*. 25. Morning twilight red. 26. Calm. lightly clouded. 27. Overcast: a few drops p.m. 29. At sunset, a *Stratus*, with a veil of superior clouds on the western sky richly coloured, the reflection from which gave considerable colour to the *Stratus* itself: wind above NW. 30. Cloudy. The weather has been calm since the 15th instant.

Twelfth Mo. 1. This morning the wind rose, bringing much cloud, with a few drops of rain: the night was stormy, and the evaporation was increased near sixfold: hence the formation of so great a mass of cloud, the superior atmosphere not being in a state to take up the water. 2. Rain commenced soon after 8 a.m.; about this time too the thermometer, which had been rising, began to fall; the barometer, which had been descending, to rise; and the wind, which had been SW, to go to the N. 3. Wind, a.m. fresh at SW: the sky overcast chiefly with *Cirrostratus*: stormy night: shower about 1 a.m., after which the wind abated. 4. Clear, windy a.m.: various clouds p.m. 5. Snow early this morning: wind N: evening twilight orange coloured, but with fainter horizontal streaks of cloud above it, which were also discernible at the ensuing sunrise, with *Cirrostratus* beneath: windy.

RESULTS.

Barometer: Greatest height .... 30·41 in.
Least .... 29·21 in.
Mean of the period .... 29·898 in.
Thermometer: Greatest height .... 55°
Least .... 22°
Mean of the period .... 42·95°
Evaporation .... 1·49 in.
Rain .... 1·41 in.

Wind chiefly SW and NW. The fore part of the period wet, the middle fair and tending to frost, the conclusion windy and changeable. There has been a strong tendency to the red refraction during twilight.

TABLE LXIV.

	1811.		Wind.	Pressure. Max.	Min.	Temp. Max.	Min.	Evap.	Rain, &c.
12th Mo.	L. Q.	Dec. 7	SW	29·56	29·35	54°	52°	—	—
		8	S	29·40	29·23	53	40	10	42
		9	SW	29·08	28·90	48	36	4	50
		10	Var.	29·67	29·08	41	32	—	2
		11	NW	29·96	29·67	45	32	—	—
		12	SW	29·96	29·86	49	33	—	1
		13	W	29·85	29·77	54	35	22	30
		14	SW	30·00	29·85	42	30	—	
	New M.	15	SW	30·00	29·50	47	35	—	18
		16	SW	29·58	29·39	42	36	—	1
		17	SW	29·86	29·58	40	31	—	
		18	SW	29·86	29·75	52	33	25	3
		19	SW	29·70	29·68	52	46	—	13
		20	W	29·68	29·60	53	49	—	16
		21	NW	30·15	29·60	49	27	—	3
	1st Q.	22	W	30·18	30·06	45	32	—	
		23	W	30·15	30·06	51	38	—	
		24	NW	30·19	30·15	43	28	—	5
		25	SE	30·20	29·98	39	24	—	
		26	SW	29·98	29·55	32	21	—	—
		27	NE	29·27	29·16	34	26	—	14
		28	NW	29·67	29·27	35	27	—	5
	Full M.	29	N	29·96	29·67	32	22	—	
		30	SW	30·08	29·96	30	25	—	
1812.		31	SW	30·08	29·88	35	31	· —	
1st Mo.		Jan. 1	SW	29·88	29·70	43	34	—	
		2	S	29·70	29·56	48	31	—	
		3	S	29·55	29·46	44	29	—	3
		4	N	29·46	29·37	38	33	—	41
		5	NW	29·76	29·47	37	29	60	26
				30·20	28·90	54	21	1·21	2·73

NOTES.—Twelfth Mo. 7, 8. Much wind; showery. 9. The wind at 9 a. m. E, yet sounds came freely from the westward, together with the clouds. *Nimbi:* rainbow; showers through the day; a brilliant twilight. 10. A gale of wind a. m., then fair. 12. A dripping mist. 14. Clear day; an extensive redness on the twilight.

1812. First Mo. 2. About 10 a. m. there having been no rain for some days, a few light clouds, just formed, and coming from the westward, suddenly exhibited a segment of a rainbow, terminating above and below at the edge of the mass of cloud. As the latter advanced by the north, and became denser, the arch increased, and became at length nearly complete; the eastern extremity descending towards the earth, with the usual appearance of rain under the clouds. The western end now began to fade, and was soon reduced to a pale white, which gradually pervading the whole, the bow disappeared, having lasted about ten minutes. It afforded an example of rain, formed and propagated in the atmosphere with such rapidity, as scarcely to give time for the previous appearance of buoyant particles, in the form of cloud.

NOTES T. F's *Journal, Clapton.—December* 7. Misty and cloudy, with showers towards night, and strong wind. 8. Hazy and cloudy, rain at night. 9. Much rain with wind. 10. Windy with some rain, fair by night. 11. Wind gentle from NE, a balloon launched at 10 a. m. indicated an upper current from the east: *Cirrocumulus,* &c. cloudy night. 12. Yellow fog, afterwards cloudy and windy, with small rain. 13. Wind and small rain. 14. Clear morning, rain at night. 15. Cloudy and hazy; small rain. 16. Fair; wind very high; light showers. 17. Morning cool; evening warmer with small rain. 18. Cloudy and hazy all day, with very little wind; starlight at times by night. 19. *Cirri* and haze: cloudy. 20. Small rain: wind by night. 21. Cloudy, windy, hazy: some rain. 22. Fair day: various clouds: at night a *corona lunaris* coloured. 23. Various clouds: the atmosphere finely coloured at sunset. 24. Cloudy and hazy. 25. Clear, and a few light *Cirri.* 26. White frost and cloudy sky: some *Cirri.* 27. Cloudy, frosty: snow and rain. 28. Some snow toward evening. 29. Cloudy. 30. Clear, a. m. cloudy, p. m. 31. Cold, frosty, and cloudy morning: then warmer.

1812. *January* 1. Complete thaw. 2. Lofty *Cirri,* then showers. 3. Clear and showers: windy night. 4. Foggy. 5. Showery.

RESULTS.

Prevailing Winds Westerly.

Barometer: Greatest height . . . . 30·20 in.

Least. . . . . 28·90 in.

Mean of the period . . . . 29·708 in.

Thermometer: Greatest height . . . .   54°

Least. . . . .   21°

Mean of the period . . . . 38·06°

Evaporation . . . . 1·21 in.

Rain (including several products of snow) 2·73 in.

On the Inequality in the Products of the Rain Gauge at different Heights, and on the proper Situation for this Instrument.

It is a fact long established, that two rain gauges, placed at different heights, afford unequal products; the lower commonly yielding more than the higher. The following table gives the results of observations on this subject made during twenty successive days, on which rain fell at Plaistow; the elevation or depression of the *mean temperature* for the 24 hours, and direction of the prevailing wind being added.

Table of the Products of Rain in the Gauges No. 1 and 2, with the Changes of Wind and Temperature.

	1811.	Wind.	M.Tem. higher.	M.Tem. lower.	Rain in No. 1.	No. 2.	REMARKS.
10th Mo.	Oct. 24	Var.		5°	5	8	Misty rain about midday ; little wind veering from SW to E.
	25	S		4	—	—	
	26	Var.	2°		45	50	Showers chiefly by night.
	27	SE	2		10	11	Rain by night.
	28	Var.		1	44	44	Clear a. m. with dew ; *Nimbi* ; vane SE, p.m. a heavy shower to S ; wind veered by S to NW ; then much cloud and rain.
	29	SW			18	18	Showers.
	30	Var.	2		8	14	Three currents in the air—see Journal under this date.
	31	W	3		13	14	Rain by night.
11th Mo.	Nov. 1	SW	6		5	11	Much cloud with a fresh breeze.
	2	SW		2	6	14	Cloudy ; much wind ; stormy night.
	3	SSW		4	6	8	Rain by night.
	4	W		2	—	—	
	5	SW		2	9	25	Stormy a. m. ; wet p. m.
	6	SW			31	50	Showery day ; *Cirrostratus* evening ; wet night.
	7	NE		3	6	7	Rain by night.
	8	S	1		16	19	Cloudy ; drizzling.
	9	Var.	3		29	34	
	10	SW		3	19	21	
	11	NW		3	1	3	Windy night ; *Nimbus* at sunset.
	12	NW	2		11	22	Windy night.
					2·82 in.	3·73 in.	

The upper gauge, No. 1, is fixed on the NW angle of a glass turret, or observatory, on the house top, having a small vane and a conducting rod a few feet to the S and SE, but no other commanding object near it. The whole of the amounts of rain given in the tables in the *Athenæum* during 1807, 1808, and part of 1809, were obtained

2

with this gauge. The gauge No. 2, the products of which I now prefer to register, is placed on a grass plot, about 70 feet from the west front of the house. Their difference in elevation is about 43 feet.

It appears, from the total result of these observations, that about one fourth of the rain which fell in twenty days was formed within 50 feet of the earth's surface.

In attending to the manner in which the rains fell, the cause of the frequent difference in the products of the gauges was, at times, obvious. When they were alike, the abundance and active appearance of the clouds in the higher atmosphere, together with the transparency of the lower, indicated that the whole supply might very well be derived from above. On the contrary, in several cases of excess in No. 2, the lower air was very turbid, showing that the decomposition of vapour was going on quite down to the surface of the earth ; or, in other words, that the raining clouds, though not distinguishable as aggregates, to us who were enveloped in them, actually swept or rested upon that surface.

On the first day, when the products were 5..8, the mean temperature was lowered 5°, probably by the effect of the gentle easterly current, which decomposed the vapour near the surface. On the 28th of the tenth month, when the results were large and equal, a southerly current appeared to prevail in the region of the clouds, with, probably, a NW wind above it; by which the vapour coming from the south was decomposed. This was accomplished at a distance from the earth, and the mean temperature was lowered 1°. These two cases may elucidate the phenomenon without a long train of reasoning.

If we admit, that a portion of the atmosphere, contiguous to the earth's surface, may be so cooled by a superior portion moving in a different direction, or with a different velocity in the same, as to become filled with a fine mist, which is ultimately resolved into clouds and rain, we shall perceive, that a set of rain gauges, placed at various heights within this portion, ought to collect less and less rain, as we ascend; since each stratum of air deposits its own redundant water, and transmits that of the higher ones.

But if the source of the rain be in a middle current, the lower part of which is above all the gauges, they ought all to afford like quantities; unless, indeed, the lower air be so dry, at the same time, as somewhat to lessen the bulk of each drop by evaporation; in which case (as is said to have happened in some instances) the products will be found *larger* as we ascend.

But there is another source of discordant results, which seems not to have been enough attended to. It exists in the deflection of the rain by accidental currents. On the 25th of the ninth month, finding in the gauge No. 2, 0·46 of an inch, while No. 1 had only 0·12 of an inch, I suspected that the wind, which came in squalls from the W, had a share in producing the difference. I took, therefore, two other gauges, No. 3 and No. 4; and on the 27th, placed No. 3 in the gutter, near and on a level with the W parapet of the house; and No. 4 about 20 feet in a line to leeward, at the same height, but in the valley between the roofs. It was then beginning to rain in moderately large drops; the wind fresh at SW. After two hours and a half, I found in No. 3, 0·08, and in No. 4, 0·11 of an inch; No. 1, on the ground, having also 0·11 of an inch. I removed No. 4 about 40 feet to leeward, near the E parapet, and got in an hour and a quarter from No. 1, 0·08, No. 2, 0·15, No. 3, 0·12, No. 4, 0·14 of an inch. The rain continued six hours, with a steady wind, and was at times heavy: near twice as much fell on the ground gauge as on that at the turret; and the results of the other gauges proved, that some part of the difference must be attributed to the wind between these. For it appears, that the stream of air, obstructed by the W front of the house (which has a contiguous building fronting S,) and rising in a curve, carried with it a part of the rain over the windward gauge, to let it fall on the leeward; hence the latter had more than its due proportion, the former less.

Thus *rain may be drifted as well as snow:* and it will be found difficult to affix a gauge to any part of a building, so that its products shall not be affected by partial currents, diminishing or overcharging them; and allowance must doubtless be made in the results of the foregoing table for this source of error.

On the whole, as the proper subject of calculation and comparison is the rain on the surface of the ground, *this* is the proper ordinary situation for the gauge; and it should be as remote as possible from all objects that may give rise to eddies in the stream flowing over it. As a further defence, both from these and from sudden frosts, the bottle, into which the rain enters from the funnel, should be placed in a box, sunk in the ground; above which there should be a cavity sufficiently large to contain the funnel, with its mouth on a level with the ground, and a free space of a few inches round it: the whole being laid with turf, both to keep it neat, and to break the spray in heavy showers. A grass plot in a garden affords the greatest facility for this arrangement: and the instrument thus placed is scarcely discoverable from a moderate distance.

ERUPTION OF ÆTNA.

An eruption of Mount Ætna took place on the 27th of October. On the same day a great quantity of very fine pulverized cinders fell in Messina, a distance of 50 or 60 miles, and some were carried to a much greater distance.

STROKE OF LIGHTNING.

An instance of the remarkable effects of lightning occurred October 28, at 6 a. m. at Berkeley, near Frome. A single flash only was seen, followed almost instantly by a tremendous clap of thunder, and attended with a heavy storm of rain and hail: it struck two oaks at a short distance from the front of Berkeley-house, one of which was completely shivered to pieces, and even in some measure rooted up: every limb, twisted and torn in various ways, was blown off; fragments and splinters, from a very small to a very large size, were scattered about to the distance of 60 or 70 yards.

METEOR.

November 7.—At a quarter before 9 p. m. a splendid meteor illumined the visible horizon of the metropolis. It appeared above the loose clouds, through which Jupiter was scarcely perceptible, and passed over an extent of about 60° in the third of a minute. Its greatest altitude, when first distinguished, was about 18 or 20°, and it disappeared in the NNE part of the horizon. Its diameter was about the sixth of a degree, its mass compact, its light dense, and the colour on the ground and among the clouds was blue, such as is produced by the combustion of Roman candles. Its whole appearance was that of a magnificent rocket.

Whitehaven.— On November 15, we were visited by one of the most awful tempests that has been witnessed here since the memorable month of January, 1796. Fortunately the tides were much lower last week than they were at the period alluded to; but the wind was far more violent than at that time, and from NNW, with little variation.—(PAPERS).

TABLE LXV.

	1812.		Wind.	Pressure, Max.	Min.	Temp. Max.	Min	Evap.	Rain, &c.
1st Mo.	L. Q.	Jan. 6	NW	29·77	29·68	38°	31°	4	
		7	N	30·05	29·68	38	33	—	19
		8	NW	30·19	30·05	37	26	—	
		9	NW	30·18	30·10	37	31	—	
		10	Var.	30·18	30·13	36	32	—	12
		11	N	30·13	29·89	38	32	—	25
		12	N	29·89	29·79	42	32	—	
		13	NW	29·98	29·79	40	35	—	1
	New M.	14	NW	30·08	29·98	41	33	14	
		15	W	30·18	30·08	45	28	—	
		16	W	30·20	30·17	39	32	—	
		17	N	30·25	30·20	38	31	—	
		18	W	30·25	30·15	43	34	—	5
		19	SW	30·15	29·97	47	36	—	3
		20	NW	29·97	29·88	43	29	—	
	1st Q.	21	NW	29·88	29·86	40	28	33	
		22	NW	29·96	29·86	41	31	—	
		23	NE	30·08	29·96	34	31	—	
		24	NW	30·10	30·06	39	27	—	
		25	W	30·06	30·00	41	39	—	3
		26	S	30·07	30·05	45	31	—	
		27	SW	30·05	29·87	47	31	28	
	Full M.	28	Var.	29·87	29·46	46	36	—	
		29	SE	29·34	29·28	45	40	—	12
		30	S	29·79	29·34	50	33	—	24
		31	SE	29·79	29·79	48	41	29	10
2d Mo.		Feb. 1	SE	29·69	29·67	47	42	—	4
		2	SE	29·64	29·34	50	40	—	8
		3	SE	29·69	29·45	47	42	—	2
		4	SW	29·58	29·45	49	42	32	1
				30·25	29·28	50	26	1·40	1·29

NOTES.—First Mo. 6. Very fine morning: wet evening: the night stormy with much snow. 7. Snowy morning, stormy day. 9. Snow fell through the night, to about three inches depth. 10. Little wind, changing to SW: a thaw. London was this day involved, for several hours, in palpable darkness. The shops, offices, &c. were necessarily lighted up; but, the streets not being lighted as at night, it required no small care in the passenger to find his way, and avoid accidents. The sky, where any light pervaded it, showed the aspect of bronze. Such is, occasionally, the effect of the accumulation of smoke between two opposite gentle currents, or by means of a misty calm. I am informed that the fuliginous cloud was visible, in this instance, from a distance of forty miles. Were it not for the extreme mobility of our atmosphere, this volcano of a hundred thousand mouths would, in winter, be scarcely habitable! 16. A dripping mist. 18. Misty morning. 19. Very cloudy: large lunar halo: stormy night. 22. Snowy evening. 23, 24. Lunar halo. 28. Windy night. 29. Windy morning: wet evening.

Second Mo. 2. Gloomy, with small rain at intervals. About half past 7 p. m. the wind rose and blew furiously from E and SE for about an hour and a half, the barometer falling a quarter of an inch: after abating, the wind rose again, and the night was stormy.

Notes T. F's *Journal, Clapton.—January* 21. Some *Cirri:* fair day. 22. Cloudy. 23. Fair. 24. Cloudy: fair: a fine coloured halo observed at Walthamstow. 25. Foggy: calm. 26. Foggy: then clear sky and misty horizon. 27. Hazy morning, windy night: corona, followed by halo. 29. *Cirri,* much coloured before sunrise. 30. Sun and mist: showers. 31. Cloudy, calm, and hazy.

February 1. Fair: much cloud. 2. Fair: *Cirrocumuli* and *Cumuli.*

RESULTS.

Winds from the N and W to the time of Full Moon, then from the Eastward.

Barometer: Greatest height . . . . 30·25 in.
Least . . . . 29·28 in.
Mean of the period . . . . 29·899 in.
Thermometer: Greatest height . . . . 50°
Least . . . . 26°
Mean of the period . . . . 38°
Evaporation . . . . 1·40 in.
Rain (including the products of snow) . . 1·29 in.

TABLE LXVI.

	1812.			Wind.	Pressure. Max.	Pressure. Min.	Temp. Max.	Temp. Min.	Evap.	Rain &c.
2d Mo.	L. Q.	Feb.	5	SE	29·58	29·54	47°	41°	—	59
			6	NW	29·86	29·54	47	35	—	1
			7	W	29·86	29·70	47	37	—	11
			8	NW	29·96	29·86	41	36	28	10
			9	N	29·98	29·95	43	38	—	
			10	E	29·97	29·87	45	26	—	
			11	E	29·87	29·60	48	33	—	2
	New M.		12	S	29·54	29·45	50	39	—	4
			13	W	29·77	29·59	44	38	—	10
			14	Var.	29·65	29·48	48	39	—	24
			15	NW	29·69	29·65	47	41	48	5
			16	NW	29·75	29·66	49	45	—	14
			17	W	29·80	29·46	50	40	—	2
			18	NW	30·06	29·80	46	38	—	
	1st Q.		19	S	30·06	29·97	53	34	—	
			20	S	29·97	29·84	54	42	—	
			21	S	29·84	29·58	54	43	—	28
			22	Var.	29·59	29·55	50	41	55	32
			23	NW	29·75	29·49	50	31	—	1·08
			24	NW	29·95	29·76	44	34	—	
			25	S	29·40	29·30	44	32	—	12
			26	Var.	29·76	29·40	42	30	—	
	Full M.		27	Var.	29·76	29·70	50	31	—	6
			28	S	29·70	29·65	47	31	—	1
			29	E	29·65	29·55	48	37	67	3
3d Mo.		March	1	E	29·85	29·65	48	33	—	
			2	NW	29·97	29·90	46	25	—	1
			3	E	29·90	29·86	44	38	—	22
			4	SW	29·87	29 80	52	35	—	3
			5	Var.	30·04	29·75	47	36	30	13
					30·06	29·30	54	25	2·28	3·71

NOTES.—Second Mo. 8. Rainy evening. 10. clear p. m.: a fine blush on the evening twilight. 11. Hoar frost. 12. Stormy night, 14. Rainy morning: very stormy day. 16. Wind boisterous all night, with rain. 17. Stormy night. 20 A very fine day: lunar halo at night. 21. Cloudy: a heavy shower of hail about half past 9 p.m.: night stormy. 22. About 9 a. m. came on a great storm of wind and rain, mixed with hail, which continued about an hour: on its ceasing, the clouds dispersed and the wind changed to W. About noon it became again stormy, continuing so at intervals till half past four, when it began to hail with great violence; this was followed by rain, and during the storm there were frequent flashes of lightning and some distant thunder. 23. Cloudy: a large lunar halo: wind high in the night with rain. 24. Very stormy morning: heavy rain about 3. a. m. with the wind very strong from NW. In an hour after, snow and sleet, with a freezing air: clear evening: the moon bright. 25. Very stormy. 27, 28. Hoar frost. 29. Misty morning.

Third Mo. Fine, with occasional clouds. 3. Hoar frost: night rainy. 4 W tmorning. 5. Wet night.

RESULTS.

Winds Variable.

Barometer: Greatest height .... 30·06 in.
Least...................... 29·30 in.
Mean of the period .... 29·738 in.
Thermometer: Greatest height............... 54°
Least...................... 25°
Mean of the period .... 41·73°
Evaporation .... 2·28 in.
Rain .... 3·71 in.

THUNDER STORMS.

February 22, p. m.—The neighbourhood of Windsor was visited with a most dreadful storm, accompanied with vivid flashes of lightning and very loud claps of thunder.

Plymouth, February 24, 2 p.m.—There has been so dreadful a storm of thunder, lightning, and hail, for these two hours past, as has not been seen or heard in this port for a series of years.

Thunder storms were very generally experienced throughout *France* in February, and, besides occasioning the loss of many lives, levelled many public buildings, houses, &c. On the night of the 23d, three vessels in L'Orient were sunk by thunder bolts.—The church of St. Pellerin, department of the Loire, was, on the 22d of February, struck by lightning, and burnt to the ground before the flames could be extinguished.

EARTHQUAKES IN NORTH AMERICA.

The American Papers abound with accounts of the recent earthquakes that have been experienced in different parts of the United States, (as it appears, in the month of December, and again in February.) Several islands in the Mississippi had totally disappeared, and the banks on both sides of that immense river had fallen in, to a prodigious extent.—(PAPERS).

6

TABLE LXVII.

1812.			Wind.	Pressure. Max.	Min.	Temp. Max.	Min.	Evap	Rain, &c.
3d Mo.	L. Q.	March 6	SW	29·88	29·84	56°	41°	—	
		7	NW	29·87	29·66	57	35	12	18
		8	NW	30·19	29·87	50	31	—	3
		9	NE	30·26	30·19	46	33	—	
		10	NE	30·26	30·20	44	30	—	
		11	NE	30·20	30·20	46	33	—	
		12	NW	30·20	29·96	44	34	—	12
	New M.	13	NE	29·99	29·96	45	34	—	6
		14	NE	29·99	29·87	44	26	—	4
		15	NE	29·87	29·76	42	31	48	1
		16	NE	29·77	29·75	35	31	—	
		17	NE	29·75	29·66	36	29	—	
		18	NE	29·66	29·40	39	26	—	
	1st Q.	19	E	29·40	29·30	39	29	—	14
		20	SW	29·24	29·10	50	40	—	8
		21	SE	29·54	29·24	54	39	30	18
		22	NE	29·74	29·54	53	39	—	2
		23	SE	29·74	29·27	42	40	—	67
		24	NW	29·64	29·27	40	32	—	16
		25	NE	30·27	29·64	42	24	—	1
		26	SE	30·35	30·17	46	30	—	
	Full M.	27	SE	30·20	29·46	51	41	36	16
		28	S	29·42	29·25	53	49	—	—
		29	SW	29·48	29·36	58	48	—	46
		30	SW	29·78	29·48	59	40	30	12
		31	E	29·53	29·48	47	40	—	10
4th Mo.		April 1	Var.	29·64	29·59	58	41	18	—
		2		29·70	29·58	—	—	—	
		3		29·68	29·58	55	43	17	26
				30·35	29·10	59	24	1·91	2·80

NOTES.—Third Mo. 9. A shower of hail p. m. 11. Hoar frost. 15. Frosty morning. 16. Wind very strong from NE all day. 17. Cold wind. 20. Snow in the morning, followed by rain. 22. Very wet night; high wind. 25. Snow: the barometer rising rapidly. 26. Very fine a. m.: barometer still rising. 27. Cloudy a. m.: a considerable depression of the barometer, with appearances indicating thunder. Late at night a shower of hail, with lightning. 28. Stormy, with showers. 29. Windy a. m.: at 2h. 30m. p. m., the temperature without being 54°, I found the vapour point in a room as high as 51°. In an hour after this it began to rain steadily, and there fell near half an inch in depth. 30. Much wind, at intervals changing to E. 31. Stormy from E and SE: cloudy: about 9 p. m. an extensive appearance of light in the clouds to the W, with rapid coruscations passing through them, in the manner of an aurora borealis. This phenomenon was apparently not more elevated than the clouds which then overspread the sky, and was certainly not produced by the reflection of a light situate below them: it continued 20 or 30 minutes.

RESULTS.

Prevailing Winds Easterly.

Barometer:	Greatest height	30·35 in.
	Least	29·10 in.
	Mean of the period	29·739 in.
Thermometer:	Greatest height	59°
	Least	24°
	Mean of the period	41·5°
	Evaporation	1·91 in.
	Rain	2·80 in.

PARHELIA.

On the 9th March, was seen at Carlisle, the beautiful phenomenon of two *parhelia*, or mock suns, in the heavens. They were first observed about ten o'clock, and appeared of variable brightness until near twelve, when they vanished.

STORMS, SNOW, &c.

In consequence of the great inclemency of the weather, the mail due on the 21st March, from the north of Scotland, had not arrived at Edinburgh, nor the Glasgow at Carlisle.

Between Appleby and Brough, the snow had so much drifted as to make the road impassable for a coach.

Between Sheffield and Manchester, and Bradford and Halifax, the snow had drifted from two to three yards deep.—The mails have been also greatly impeded in South Wales; the snow and the floods having made the roads impassable.

Plymouth, March 21.—It blew a most tremendous gale from SW the whole of last night and this morning.

Hull.—On Friday, March 20, we experienced a most tremendous gale throughout the day and night; accompanied with heavy showers of snow and sleet. The accounts from the coast of the effects of the gale are highly disastrous. *During the storm, the wind to the southward of the Humber, was from the SW, whilst to the northward it was strong from the NE and by E.*

Notes, &c. *(continued).*

Cold in Italy, &c.

The cold has been more severe in Italy this winter than for many preceding ones. At Venice, Turin, Naples, &c. the thermometer placed in the sun, was in January 3° below freezing point. In the southern parts of Germany the winter has been extremely severe, while in the north it has been very moderate.—(Pub. Ledger).

Earthquake in South America.

A supplement to the St. Thomas's Gazette, of the 9th of April, contains the following particulars of this dreadful catastrophe.

The 26th of March has been a day of woe and horror to the province of Venezuela. At 4 p. m. the city of Caraccas stood in all its splendour—a few minutes later 4,500 houses, 19 churches and convents, together with all other public buildings, monuments, &c. were crushed by a sudden shock of an earthquake. That day happening to be Holy or Maundy Thursday, and at that precise hour every place of worship being crowded, to commemorate the commencement of our Saviour's Passion, by public processions, which were to proceed through the streets a few minutes afterwards, augmented the number of sufferers to an incredible amount; as every church was levelled with the ground before any person could be aware of danger. The number taken out of one church amounting (two days after the disaster) to upwards of 300 corpses, besides those (it may be presumed) that could not yet have been dug out of such heaps of ruins, gives an idea of the extent of the calamity. The number of dead are differently stated, from 4 to 6, as far as 8,000. Horrible as this catastrophe appears, it would be matter of some consolation to know that the vicinity of that city offered some support or shelter to the surviving mourners, but the next town and sea port thereto, *viz.* La Guira, has in proportion suffered still more; and that appears to have been the case along its immediate coast; huge masses of the mountains have been detached from the summits and hurled down into the vallies. Deep clefts and separations of the immense beds of rocks still threaten future disasters to the hapless survivors, who are now occupied to bury and burn the dead, and to relieve the numerous maimed, perishing for want of medical aid, shelter, and other comforts.

It appears, by authentic accounts, that several other cities and towns had suffered by the earthquake which destroyed Caraccas and La Guira:—Cumana, New Barcelona, and Valencia, are nearly destroyed; Barquisimeto, Santa Rosa, and Caudare, totally destroyed; Arilaqua, sunk; and the inland town of St. Philip, with a population of 1,200 souls, entirely swallowed up.—(Papers.)

The writer of a letter inserted in the public papers, (dated from Jocame, two leagues from Caraccas, March 31,) gives the following account of the state of the atmosphere at the time. " My first idea (on feeling the shock, which is stated to have lasted about 15 seconds) was that the Silla, a mountain near Caraccas, had broken out into a volcano: but its peak, like all the surrounding hills, was unusually clear: nor had there been any sultry weather or violent winds:—on the contrary, the weather has long been and still continues to be, warm without being sultry; though without rain for a considerable time, except now and then a few drops—and with very heavy night dews."

———————

I may remark on the preceding statement, that probably it would have been happy for the city of Caraccas, had a volcano opened in its neighbourhood at this time. The celebrated *Humboldt*, whose opportunities of observing these phenomena have been most extensive, is clearly of opinion, that one and the same cause, deeply seated in the globe, produces both *earthquakes* and *volcanic eruptions:* he seems moreover to regard volcanoes as the *spiracula* through which the elastic vapours, occasionally disengaged in the bowels of the earth, find a vent into the atmosphere, without displacing the superincumbent strata. He considers the eruption in the island of St. Vincent, (for which see the Notes under next Table) as affording one example of this connection; as it began only 34 days after the present earthquake, and was long preceded by shocks, which were felt at the same time on the South American continent.

"We learned, says Humboldt, at Pasto, (120 miles N of Quito,) that a column of thick black smoke, which for several months in 1797, had been issuing from the volcano near that city, disappeared *at the very hour* when the cities of Riobamba, Hambato, and Tacunga were overthrown by an enormous shock, 60 leagues to the south of the volcano."—Voyage au Nouveau Continent: *Relation Historique, livre 2, chap. iv.:* which see throughout on this subject.

Supposing the prime agent in these phenomena to be *water*, penetrating to unusual depths through the inclined strata of mountainous countries, and thus meeting with masses of the highly oxidable bases of the earths, (for instance) which have not before undergone its action, there will then appear to be a close connexion between the recurrence of these effects and the variations of the atmosphere. For the most probable cause of such penetration, (though not the only assignable one) is excessive and continued *rain;* and this, not necessarily on the district which is the immediate seat of the earthquake, or eruption; but rather on some neighbouring mountainous tract, which by its position may be capable of collecting, and transmitting through subterraneous communications the requisite excess of water. Now the want of rain "for a considerable time" at Caraccas not only consists with, but even renders probable, an excess of it for the same period, in the more inland and elevated country.

z

TABLE LXVIII.

1812.			Wind.	Pressure. Max.	Min.	Temp. Max.	Min.	Evap.	Rain, &c.
4th Mo.	L. Q.	April 4	E	30·10	29·60	53°	36°	—	
		5	S	30·18	30·10	56	35	—	4
		6	S	30·10	29·93	53	42	25	
		7	E	30·02	29·88	50	37	—	10
		8	NE	30·15	30·02	49	25	—	
		9	NE	30·15	30·01	43	33	—	
		10	NE	30·01	29·98	47	36	—	
	New M.	11	W	29·98	29·89	44	37	43	
		12	NE	30·05	29·89	52	32	—	
		13	NE	30·05	30·05	51	33	—	
		14	NE	30·05	29·91	51	32	—	
		15	NE	29·91	29·73	53	30	—	
		16	N	29·73	29·64	49	27	—	—
		17	N	29·83	29·64	48	28	—	—
	1st Q.	18	N	30·09	29·83	51	29	87	
		19	Var.	30·09	30·02	50	33	—	
		20	NW	30·15	30·00	58	30	—	
		21	NE	30·15	30·07	58	37	28	
		22	N	30·01	29·97	54	30	—	
		23				52	32	—	
		24	W	29·94	29·86	52	36	—	
		25	SW	29·86	29·56	54	39	—	—
	Full M.	26	Var.	29·65	29·55	49	34	69	47
		27	NE	29·64	29 59	52	44	—	6
		28	NE	29·76	29·64	51	44	—	14
		29	NE	29·80	29·76	52	43	27	39
		30	Var	30·02	29·80	55	45	—	—
5th Mo.		May 1	E	30·02	29·92	59	43	—	
		2	NE	29·92	29·73	54	42	—	
		3	Var.	29·75	29·70	56	32	55	4
				30·18	29·55	59	25	3·34	1·24

NOTES.—Fourth Mo. 4. Cloudy a. m.: clear evening. 5. Much dew: barometer unsteady: heavy clouds through the day: a shower about sunset. 6. Much dew: grey sky, and the air nearly calm. 7. Lightly cloudy: little wind. 8. Cloudy a. m.: a shower p. m. 9. Brisk wind: cloudy. 10. Hoar frost. 11. Cloudy. 16. Slight showers. The *Cumulostratus* cloud has prevailed every day for a week past. 17. A little hail. 20. A few large drops. 23, 24. Occasional slight showers of hail, &c. 25. A rainy night. 26. Gentle showers of rain, yet not warm. 27. Misty morning: much dew: swallows appear. 28, 29. Cloudy: windy.

NOTES, &c. *(continued)*.

Fifth Mo. 1, 2. Cloudy: the cuckow heard. 3. About 1 p.m. a few drops of rain, attended with the smell of electricity in the air: the wind, which in the morning had been brisk at NE, died away, the canopy of the sky rose: the evening was calm, and dew fell.

RESULTS.

Prevailing Wind NE.

Barometer: Greatest height .... 30·18 in.
Least .... 29·55 in.
Mean of the period .... 29·902 in.
Thermometer: Greatest height .... 59°
Least .... 25°
Mean of the period .... 43·57°
Evaporation .... 3·34 in.
Rain .... 1·24 in.

It appears that a shower of *Meteoric Stones*, accompanied with the usual flashes and detonations, fell on the 10th of the Fourth month, about 8 p.m., at six leagues distance from *Thoulouse*.

ERUPTION OF THE SOUFRIERE MOUNTAIN, ST. VINCENT'S.

A letter of the 6th of May furnishes the following interesting particulars of this awful phenomenon:—" The Soufriere Mountain," says the writer, " had for some time past indicated much disquietude; and from the extraordinary frequency and violence of earthquakes, which are calculated to have exceeded 200 within the last year, portended some great movement or eruption. The apprehension, however, was not so immediate as to restrain curiosity, or to prevent repeated visits to the crater, which of late had been more numerous than at any former period, even up to Sunday April 26; when some gentlemen ascended it, and remained there for some time. Nothing unusual was then remarked, or any external difference observed, except rather a stronger emission of smoke, from the interstices of the conical hill at the bottom of the crater. About 2,000 feet from the level of the sea, (calculating from conjecture) on the south side of the mountain, and rather more than two-thirds of its height, opens a circular chasm, somewhat exceeding half a mile in diameter, and between 4 and 500 feet in depth: exactly in the centre of this capacious bowl, rose a conical hill about 260 or 300 feet in height, and 200 in diameter, richly covered and variegated with shrubs, brushwood, and vines, above half way up, and, for the remainder, powdered over with virgin sulphur to the top. From the fissures in the cone and interstices of the rocks, a thin white smoke was constantly emitted, occasionally tinged with a slight bluish flame. The precipitous sides of this magnificent amphitheatre were fringed with various evergreens and aromatic shrubs, flowers, and many Alpine plants. On the north and south sides of the base of the cone were two pieces of water, one perfectly pure and tasteless, the other strongly impregnated with sulphur and alum.

" A century had now elapsed since the last convulsion of the mountain, or since any other elements had disturbed the serenity of this wilderness than those which are common to the tropical tempest. It apparently slumbered in primeval solitude and tranquillity, and from the luxuriant vegetation and growth of the forest which covered its sides from the base nearly to the summit, seemed to discountenance the fact, and falsify the records of the ancient volcano. Such was the majestic, peaceful Soufriere on April the 27th; but we trod on ' *ignem suppositum cineri doloso*,' and our imaginary safety was soon to be confounded by the sudden danger of devastation. Just as the plantation bells rang 12 at noon, on Monday the 27th, an abrupt and dreadful crash from the mountain, with a severe concussion of the earth, and tremulous noise in the air, alarmed all around it. The resurrection of this fiery furnace was proclaimed in a moment by a vast column of thick black ropey smoke, like that of an immense glass house, bursting forth at once, and mounting to the sky; showering down sand, with gritty calcined particles of earth and favilla mixed, on all below. This, driven before the wind towards Wallibon and Morne Ronde, darkened the air like a cataract of rain, and covered the ridges, woods, and cane pieces with light grey coloured ashes, resembling snow when slightly covered by dust. As the eruption increased, this continual shower expanded, destroying every appearance of vegetation. At night a very considerable degree of ignition was observed on the lips of the crater ; but it is not asserted, that there was as yet any visible ascension of flame. The same awful scene presented itself on Tuesday; the favilla and calcined pebbles still increasing, and the compact pitchy column from the crater rising perpendicularly to an immense height, with a noise at intervals like the muttering of distant thunder. On Wednesday the 29th, all these menacing symptoms of horror and combustion still gathered more thick and terrific, for miles around the dismal and half obscured mountain. The prodigious column shot up with quick motion, dilating as it rose like a balloon. The sun appeared in total eclipse, and shed a meridian twilight over us, that aggravated the wintry gloom of the scene, now completely powdered over with falling particles. It was evident that the crisis was yet to come—that the burning fluid was struggling for vent, and labouring to throw off the superincumbent strata and obstructions which suppressed the ignivomous torrent. At night it was manifest that it had greatly disengaged itself from its burthen, by the appearance of fire flashing now and then, flaking above the mouth of the crater.

" On Thursday, the memorable 30th of April, the reflexion of the rising sun on this majestic body of curling vapour was sublime beyond imagination—any comparison of the Glaciers of the Andes, or Cordilleras with it, can but feebly convey an idea of the fleecy whiteness and brilliancy of this awful column of intermingled and wreathed smoke and clouds : it afterwards assumed a more sulphureous cast, like what we call thunder clouds, and in the course of the day a ferruginous and sanguine appearance, with much livelier action in the ascent, a more extensive dilatation, as if almost freed from every obstruction—in the afternoon the noise was incessant, and resembled the approach of thunder still nearer and nearer, with a vibration that affected the feelings and hearing : as yet there was no convulsive motion, or sensible earthquake.—Terror and consternation now seized all beholders.—The Charaibs, settled at Morne Ronde, at the foot of the Soufriere, abandoned their houses, with their live stock and every thing they possessed, and fled precipitately towards town. The Negroes became confused, forsook their work, looked up to the mountain, and, as it shook, trembled, with the dread of what they could neither understand or describe—the birds fell to the ground, overpowered with showers of favilla, unable to keep themselves on the wing; the cattle were starving for want of food, as not a blade of grass or a leaf was now to be found—the sea was much discoloured, but in nowise uncommonly agitated; and it is remarkable, that throughout the whole of this violent disturbance of the earth, it continued quite passive, and did not at any time sympathize with the agitation of the land. About 4 p. m. the noise became more alarming, and just before sunset the clouds reflected a bright copper colour, suffused with fire. Scarcely had the day closed, when the flame burst at length pyramidically from the crater, through the mass of smoke; the rolling of the thunder became more awful and deafening; electric flashes quickly succeeded, attended with loud claps; and now, indeed, the hurlyburly began. Those only who have witnessed such a sight, can form any idea of the magnificence and variety of the lightning and electric flashes; some forked, zigzag, playing across the perpendicular column from the crater—others shooting upwards from the mouth like rockets of the most dazzling lustre—others like shells with their trailing fuses flying in different parabolas, with the most vivid scintillations from the dark sanguine column, which now seemed inflexible, and immovable by the wind. Shortly after 7 p. m. the mighty caldron was seen to simmer, and the ebullition of lava to break out on the NW side. This, immediately after boiling over the orifice, and flowing a short way, was opposed by the acclivity of a higher point of land, over which it was impelled by the immense tide of liquid fire that drove it on, forming the figure V in grand illumination. Sometimes when the ebullition slackened, or was insufficient to urge it over the obstructing hill, it recoiled back, like a refluent billow from the rock, and then again rushed forward, impelled by fresh supplies, and scaling every obstacle, carrying rocks and woods together, in its course, down the slope of the mountain, until it precipitated itself down some vast ravine, concealed from our sight by the intervening ridges of Morne Ronde. Vast globular bodies of fire were seen projected from the fiery furnace, and, bursting, fell back into it, or over it, on the surrounding bushes, which were instantly set in flames. About four hours from the lava boiling over the crater, it reached the sea, as we could observe from the reflection of the fire, and the electric flashes attending it. About half past one, another stream of lava was seen descending to the eastward, towards Rabacca. The thundering noise of the mountain, and the vibration of sound that had been so formidable hitherto, now mingled in the sullen monotonous roar of the rolling lava, became so terrible, that dismay was almost turned into despair. At this time the first earthquake was felt: this was followed by showers of cinders, that fell with the hissing noise of hail during two hours.

" At three o'clock, a rolling on the roofs of the houses indicated a fall of stones, which soon thickened, and at length descended in a rain of mingled fire, that threatened at once the fate of Pompeii, or Herculaneum. The crackling and coruscations from the crater at this period exceeded all that had yet passed. The eyes were struck with momentary blindness, and the ears stunned with the glomeration of sounds."—Another letter says, " The stones that fell were as light as pumice, though in some places as large as a man's head. This dreadful rain of stones and fire lasted upwards of an hour, and was again succeeded by cinders from three till six o'clock in the morning. Earthquake followed earthquake almost momentarily, or rather the whole of this part of the island was in a state of continual vacillation. The break of day, if such it could be called, was truly terrific—a chaotic gloom enveloped the mountain, and an impenetrable haze, with black sluggish clouds hung over the sea. The whole island was covered with favilla, cinders, scoriæ, and broken masses of volcanic matter. It was not until the afternoon that the muttering noise of the mountain sunk gradually into a solemn silence."

The mountain continued to be agitated up to the 7th, when its more violent paroxysms gradually subsided; and it has since (to May 18) shewn scarcely any signs of commotion. By this dreadful calamity, the large rivers of Rabacca and Wallibon were dried (or choked) up, and in their places was a wide expanse of barren land. The melted minerals had formed a promontory which jutted out some distance from the main land, close to the post at Morne Ronde. The quantity of matter discharged from the crater is supposed to exceed twenty times the original bulk of this immense mountain. The soil, in many places, would be rendered incapable of vegetation. The extent of the planters' losses had not been ascertained; but it was thought to be very great.

Barbadoes.—The following extracts relate to the fall of the volcanic matter on that island. " I was lying in bed about six o'clock in the morning, (May 1,) when I observed my chamber more dark than usual. (Another account states, that the atmosphere at *four* was light and clear.) Some time after I arose and opened one of my windows, when I observed *to the north a dark thick cloud,* similar to the usual indication of a great deal of rain (which would have been very acceptable,) but at the same time I perceived a most remarkable *bright cloud to the southward,* so much so, as to reflect light on the houses. We had had what we conceived to be several rolling claps of thunder during the night, and the last was a quarter past seven, when an instantaneous total darkness ensued, and from that time till one o'clock I never saw so dreadful a phenomenon. During the time of the darkness we were assailed by immense falls of calcareous matter (as I think) to such a degree that it was dangerous to go out of the house. At first, what fell was a large black substance, very coarse; but it gradually became as fine as Scotch snuff, and in a few hours the streets and the tops of the houses were many inches thick in this matter. About half-past one o'clock, a small glimmering of light began to appear, and by half past two o'clock we could make out people in the streets. About half-past six last night, we saw like rays of fire in the southern quarter again. The whole island is in one complete sheet of this matter; the canes are all weighed down with it, and the poor cattle and horses must die for want, if we are not immediately relieved. We cannot see 20 yards before us for the immense volumes continually falling from off the tops of the houses; for so soon as it is dry, it is exactly like flour."

A letter of the 5th, says:—" We learn, by arrivals, that the explosion has not taken place at Dominica, St Lucia, or Martinique. The Neptune arrived this morning from *Bristol*. Capt. Powers says, *he met with the volcanic matter 500 miles to windward of this island;* in that case, it is likely to be from the Western Islands."—(PAPERS).

St. Vincent's (an island about 17 miles long, by 10 broad,) lies 20 leagues west, and somewhat south, of Barbadoes. The distance to which the *favilla* of this eruption was carried, and that in vast quantities, is truly astonishing. It reached Barbadoes about six in the morning on the fifth day of the eruption, and approached as it seems, *from the northward;* from which circumstance, as well as from its having been met with far in the Atlantic by a ship from Europe, the Barbadians might very well imagine it came from the *Azores.*—The facts are explicable on the supposition, that this matter was raised by the force of the explosion, as well as by the ascending heated column of air, quite into the superior current, which returns from the Equator to the Tropic, above the Trade wind: for *the latter*, blowing from Barbadoes to St. Vincent's, must have carried what was within its range towards *Grenada*, and we have no account of its even reaching that island, though nearer than Barbadoes to the volcano. By the superior current (which often moves with a velocity unknown in the lower atmosphere) it was probably carried over and beyond Barbadoes, appearing there at sunrise as a " bright cloud to the southward:" and falling soon after into the Trade wind, continued to be brought back to the island in the ordinary direction of that wind, until the supply from the volcano ceased.

TABLE LXIX.

	1812.		Wind.	Pressure.		Temp.		Evap.	Rain, &c.
				Max.	Min.	Max.	Min.		
5th Mo.	L. Q.	May 4	NE	29·86	29·78	63°	38°	—	
		5	E	30·01	29·86	64	40	—	
		6	E	30·01	29·98	60	42	—	
		7	E	29·94	29·86	58	45	70	
		8	SE	29·86	29·73	76	51	—	
		9	SW	29·78	29·68	72	53	90	—
		10	W	29·82	29·56	64	56	—	20
	New M.	11	NW	29·56	29·54	65	49	—	2
		12	SW	29·53	29·51	65	44	—	13
		13	S	29·56	29·50	60	40	44	15
		14	S	29·75	29·56	58	40	—	—
		15	NE	29·95	29·75	57	43	—	10
		16	N	30·00	29·95	62	45	36	2
		17	NE	29·95	29·87	53	45	—	7
	1st Q.	18	E	29·87	29·80	63	48	—	
		19	E	29·80	29·63	66	53	—	44
		20	SW	——		65	53	—	8
		21	Var.	29·94	29·63	61	45	—	60
		22	NW	30·18	29·98	52	35	—	
		23	E	30·27	29·98	61	40	32	
		24	SE	30·27	30·11	57	52	—	4
		25	SW	30·11	29·98	62	53	—	
	Full M.	26	SW	29·98	29·55	72	55	43	
		27	S	29·59	29·55	71	51	—	
		28	SE	29·69	29·59	69	54	—	14
		29	SW	29·84	29·69	72	53	—	23
		30	S	29·76	29·74	67	52	65	
		31	S	29·75	29·72	65	54	—	10
6th Mo.		June 1	SW	29·95	29·72	60	46	28	4
				30·27	29·50	76	35	4·08	2·36

NOTES.—Fifth Mo. 4, 5, 6. Much dew. 7. Windy. 8. Windy: *Cirrocumulus* and *Cumulostratus:* wind S above: thunder clouds: the evening twilight was luminous and coloured: the clouds dispersing, and scattered in loose flocks over the rich ground of the western sky, presented a striking appearance. 9. Shower very early: wind S, *Cirrus, Cirrocumulus:* evening, much wind. 10. Overcast a. m.: a gale from the W, with much cloud: showers: p. m. clear and pleasant. 11. A shower early: *Cumulostratus* prevails. 12. Showers. 13. A thunder shower, with hail about 3 p. m. 14. Showers. 15—17. Cloudy, windy. 18. Small rain a. m.: wind N, gentle: p. m. sunshine. 19. Wind

E a. m., pretty strong: clouds of different kinds, with haze above: p. m. thunder clouds: in the evening came on a violent thunder storm, which lasted several hours; it was chiefly to the S and W. The appearances were very similar to those of the destructive hail storm, which occurred here in the same month, and on the same day of the month, and nearly at the same time of the day in 1809: sheets of blue and white lightning came in quick succession, with an almost continual rolling of thunder. We had however no hail, (being only on the flank of the storm) but sudden and heavy showers of warm rain; which was of the same amount in the upper as in the lower gauge. At 11 p. m. wind NE, it still lightned far in the N. 20. Wind a. m. W, cloudy and misty. 23. About noon, during a shower, it thundered to the southward. 29. A little thunder to the SW about 4 p. m., with a few drops: wet night. 31. An electric shower about 9 a. m., *Nimbi:* windy night.

NOTES T. F's *Journal, Clapton.—May* 22. Cloudy morning, fair afternoon, with *Scud, Cumulus,* and *Cumulostratus.* 23. Various clouds in different heights. 24. Clouded sky with a little rain, a wavy, and in some places mottled appearance of the cloudy mass. 25. Small rain; fair evening. 26. Fine warm morning, *Cirrus,* &c., a sort of flying haze of a brownish colour here and there appeared in the afternoon. 27. Fine warm day and various clouds. 28. Clouds in two strata, some rain, fine sunset. 29. All the modifications appeared, and were followed by showers. 30. Some *Cirri* early, afternoon all the clouds were compact but rocky *Cumuli.* 31. Clouds in two strata, cloudy and rather windy by night.

RESULTS.

Winds Variable.

Barometer: Greatest height .... 30·27 in.

Least .... 29·50 in.

Mean of the period .... 29·810 in.

Thermometer: Greatest height .... 76°

Least .... 35°

Mean of the period .... 55·46°

Evaporation .... 4·08 in.

Rain .... 2·36 in.

TABLE LXX.

1812.		Wind.	Pressure.		Temp.		Evap.	Rain, &c.
			Max.	Min.	Max.	Min.		
6th Mo. L. Q. June	2		29·98	29·95	68°	53°	—	
	3	NE	30·02	29·98	65	46	—	
	4		30·04	30·00	72	50	38	
	5	E	30·08	30·04	70	49	—	
	6	NE	30·18	30·08	66	45	—	
	7	NE	30·35	30·12	70	44	52	
	8	NE	30·40	30·35	62	46	—	
New M.	9	N	30·40	30·15	66	51	—	
	10	Var.	30·27	30·17	65	43	36	
	11	NW	30·07	30·03	75	53	—	
	12	NW	30·03	29·93	74	48	33	
	13	SW	29·93	29·88	68	50	—	
	14	SW	29·88	29·81	69	49	—	
	15	SW	29·82	29·79	68	49	—	7
1st Q.	16	SW	29·79	29·58	65	48	—	23
	17	SW	29·78	29·58	52	46	—	39
	18	SW	29·58	29·49	59	53	—	40
	19	SW	29·49	29·34	63	49	1·15	9
	20	SW	29·53	29·33	60	46	—	21
	21	SW	29·66	29·32	60	46	—	16
	22	SW	29·83	29·66	60	43	37	1
	23	W	29·94	29·81	62	46	—	5
Full M.	24	SW	29·94	29·91	59	45	—	6
	25	Var.	29·91	29·60	63	50	—	25
	26	Var.	29·86	29·45	58	42	38	61
	27	Var.	29·86	29·78	63	46	—	22
	28	N	30·10	29·78	58	39	—	1
	29	SW	30·03	29·96	64	48	—	
	30	SW	29·86	29·70	62	52	60	5
			30·40	29·32	75	39	4·09	2·81

NOTES.—Sixth Mo. 3. A little rain at intervals. 4. A few large drops: *Cumulostratus* p. m. A shower to the SW. Wind E. 5. Much dew: clear with *Cirrus*. 6. Overcast, windy: then very fine, with red *Cirri* at sunset. 7. Cloudy morning: clear day afterward: brilliant orange twilight. 8. Cloudy: brisk wind. 9. Fair, with *Cumulus* and *Cirrus* above: at sunset the wind rose, with some appearance of *Nimbus*. 10. *Cumulostratus*, with a cold breeze all day. 11. Wind fresh at W a. m.: the maximum of temperature occurred at nine: the barometer fluctuating. *Cumulus* clouds, with very large plumose *Cirri* above, which showed red at sunset. The New Moon appeared (in a white crescent, becoming afterward of a gold colour) in the midst of a pretty luminous twilight. 12. Cloudy a. m.: barometer

5

still unsettled: evening twilight luminous and orange coloured: a *Stratus* began to appear at 9 p. m. 13. Misty a. m.: much dew. 15. Cool day: rather windy. 16. Rain last night: fair and cool. 17. Heavy short showers. 18. Fair, cloudy: rain by night. 19. The rainbow *twice* this morning. 21. Several hours rain a. m. Barometer fluctuating. 22. *Nimbi* a. m., fair p. m. 23. *Nimbi* through the day: thunder twice to the SW: the wind veered as far as to NW, but settled W. 24. Much cloud a. m.: calm air: showers. 25. *Cumulus*, with very elevated *Cirrus* in parallel bands E and W. A solar halo for above two hours after noon, the higher atmosphere filled with cloud: at sunset the wind, which had been SE and SW, came to NW. 26. Cold stormy morning, wind N. Thunder twice about 2 p. m.: rain almost from sunrise to sunset. 27. Sunshine a. m.: wind NW: a solar halo p. m.: wind SW: evening wet and stormy. 28. Wind N a. m.: a faint blush on the evening twilight. 30. Windy evening: rain at intervals.

RESULTS.

Winds Variable, the SW of longest continuance.

Barometer: Greatest height.............. 30·40 in.
           Least...................... 29·32 in.
           Mean of the period .... 29·881 in.
Thermometer: Greatest height.............. 75°
             Least..... .... 39°
             Mean of the period .... 55·87°
       Evaporation....................... 4·09 in.
       Rain.............................. 2·81 in.

Thunder Storm.

June 21.—South End, in Essex, and the neighbourhood for several miles round, were visited by a severe storm of thunder and lightning, accompanied by a shower of hail stones of uncommon size.

Eruption of Vesuvius.

Naples, June 15.—Vesuvius, which had been quiet for several years, has suddenly broken out. At nine in the morning of the 12th, loud reports proceeded from the bosom of the mountain, followed by an eruption of cinders and smoke. After this the mountain remained quiet for an hour. At 11 two fresh reports were heard, when the crater vomited fire and smoke, which covered the horizon. On the 13th and 14th, it was calm, but, at the instant we are writing, it is again in action, and its crater is covered with an immense column of smoke.

Agitation of the Sea.

Marseilles, June 24.—A singular phenomenon was witnessed here yesterday. On a sudden, a rush of water came from the sea into the port, forming a current so rapid that it drew every thing with it. The sea then retired all at once, leaving the harbour dry, and all the vessels aground. Almost at the same instant the sea returned by leaps and bounds, with extraordinary impetuosity, filling again the harbour, placing afloat the vessels, and inundating the quays. Afterwards every thing returned to its usual state.—(Papers).

TABLE LXXI.

1812.			Wind.	Pressure.		Temp.		Evap.	Rain, &c.
				Max.	Min.	Max.	Min.		
7th Mo.	L. Q.	July 1	SW	29·70	29·44	68°	52°	—	29
		2	Var.	29·56	29·40	65	47	—	53
		3	N	30·00	29·56	61	42	—	
		4	SW	30·05	30·01	61	42	45	
		5	SW	30·02	29·96	63	51	—	
		6	NW	30·27	30·02	67	50	—	
		7	N	30·29	30·27	72	51	43	
	New M	8	E	30·33	30·29	71	46		
		9	NE	30·33	30·29	73	50		
		10	N	30·39	30·29	72	51		
		11	N	30·29	30·16	69	54		
		12	NW	30·17	30·16	66	41		
		13	NW	30·19	30·16	64	52		
		14	Var.	30·19	30·17	64	46		4
		15	Var.	30·17	30·05	69	50		1
	1st Q	16	E	30·05	29·95	65	55	—	
		17	Var.	30·14	29·95	67	56	—	
		18	SE	30·10	30·00	75	56	37	
		19	SW	30·00	29·74	73	55	—	17
		20	W	29·85	29·70	75	50	—	34
		21	W	29·96	29·94	65	45	35	
		22	SW	30·09	29·96	63	42	—	15
		23	SW	30·09	29·94	65	52	—	
	Full M.	24	SW	29·94	29·78	62	58	—	26
		25	SW	29·79	29·78	71	57	55	3
		26	NW	29·85	29·79	68	49	—	
		27	Var.	29·66	29·60	61	48	—	1·00
		28	SW	29·66	29·65	64	50	—	
		29	W	29·80	29·66	63	49	55	22
				30·39	29·40	75	41	2·70	3·04

NOTES.—Seventh Mo. 1. Much wind: very cloudy: rain at intervals through the day and night. 2. Fair a. m.: thunder showers with hail, p. m. 3. Cloudy: a few drops of rain. 4. The wind veered gradually from N by E to SW. 5. Wind moderate. 22. Thunder and hail.

NOTES J. G.—*Stratford, July* 1. Rainy morning. 2. A shower of large hail between 5 and 6 p. m., with some thunder : evening showery. 5. Cloudy and fine : some rain in the evening. 12. Cloudy morning. 19. Some rain in the evening. 20. Some thunder in the afternoon : frequent lightning in the evening. 22. A heavy thunder storm about noon, accompanied with large hail. 27. Very rainy day. 29. A thunder storm about 3 p. m., with heavy rain mixed with hail.

RESULTS.

Winds Variable.

Barometer: Greatest height................30·39 in.
Least........................29·40 in.
Mean of the period............29·975 in.
Thermometer: Greatest height................   75°
Least........................   41°
Mean of the period............58·34°
Evaporation (in 21 days, the rest being lost
by accident)......................   2·70 in.
Rain ....   3·04 in.

In travelling at different intervals during this month between London and Folkstone, I have observed that the showers in great measure avoided the high chalky tracts, and followed the course of the rivers and moist valleys. The reverse distribution sometimes takes place.

Being on the heights above Folkstone, Kent, about 10 p. m. on the 14th of seventh month, I observed a brilliant meteor dart down in the southern sky, tending a little E in its fall; during which it was of a bright blue, but at the moment of extinction it became red and seemed to explode.

WATER SPOUT.

July 22.—The inhabitants of Cromer were gratified by the appearance of a water spout, which continued in sight for some minutes, when the mass or column of water collected between a cloud and the surface of the sea broke in the offing, without doing any injury to the shipping.

STORMS IN DORSETSHIRE.

On Wednesday the 29th July, a man was struck by lightning at Aston Pits, near Newport, in this county, and very dreadfully burnt. The same day there were very heavy storms of thunder and lightning in that neighbourhood.— During the present week much injury has been done to the growing crops in various parts of this county, and the vicinity of Pool, by violent torrents of rain.—(PAPERS).

TABLE LXXII.

1812.			Wind.	Pressure.		Temp.		Evap.	Rain, &c.
				Max.	Min.	Max.	Min.		
7th Mo.	L. Q.	July 30	NW	29·96	29·80	61°	46°		
		31	SW	29·96	29·80	64	51	20	8
8th Mo.		Aug. 1	Var.	29·86	29·80	61	53	—	2
		2	NE	29·86	29·80	64	54	—	10
		3	Var.	29·85	28·80	63	52	—	2
		4	NE	29·90	29·85	65	50	—	32
	New M.	5	SW	29·95	29·90	57	50	29	23
		6	Var.	30·00	29·95	63	47	—	41
		7	NW	30·00	29·95	61	49	—	
		8	NW	29·96	29·94	57	51	17	3
		9	NW	29·97	29·96	57	45	—	
		10	NW	29·97	29·96	58	53	—	2
		11	Var.	30·07	29·96	63	49	—	5
	1st Q.	12	NE	30·14	30·07	57	44	18	
		13	N	30·15	30·14	64	43	—	
		14	NE	30·15	30·12	67	49	—	
		15	E	30·12	30·07	65	50	—	
		16	E	30·06	30·05	68	54	34	
		17	SE	30·05	29·98	73	55	—	
		18	SE	29·98	29·76	78	58	—	
		19	SW	29·96	29·76	72	55	56	1
	Full M.	20	W	30·00	29·98	71	55	—	
		21	SW	29·97	29·94	69	57	—	
		22	W	30·05	29·97	68	53	—	
		23	SW	30·04	29·86	70	60	6	
		24	SW	30·10	29·86	66	47	—	5
		25	SW	30·10	30·04	69	52	—	
		26	NW	30·04	29·96	67	53	35	
		27	NW	29·99	29·96	69	51	—	
		28	NW	29·99	29·98	59	48	17	—
				30·15	29·76	78	43	2·92	1·34

NOTES.—Eighth Mo. 4. Wet afternoon. 5. Wet morning. 6. " The day was gloomy: about 4 p. m. a very heavy shower commenced, which continued for about 20 minutes, then abated for a short time, but increased again, and continued all the evening, with thunder and lightning: the barometer was nearly stationary." Such were the phenomena at the laboratory, where there fell 1·39 inch of rain. At Plaistow, two miles distant, there appears to have fallen only 0·41 inch of rain, and I find only this note, " Thunder in the afternoon." 13. Foggy morning: a *Stratus* at night. 14. The same.

17. The same: lunar halo. 18. Some lightning during the night. 21. Thunder between 1 and 2 p. m. 24. Bright moonlight. 28. The wind this night very high.

RESULTS.

Winds Variable.

Barometer: Greatest height . . . . 30·15 in.

Least. . . . . 29·76 in.

Mean of the period . . . . 29·968 in.

Thermometer: Greatest height. . . . . 78°

Least. . . . . 43°

Mean of the period . . . . 57·83°

Evaporation . . . . 2·92 in.

Rain . . . . 1·34 in.

METEORIC STONES IN INDIA.

The India Papers contain an account of the descent of two huge masses of stone in the neighbourhood of Lahore, accompanied by a series of explosions resembling the discharge of cannon;—a phenomenon, which had excited the utmost consternation throughout the country.—A meteoric stone also fell on the 6th of August, (1812?) near the British lines at Punderpoor. It weighed four pounds, was very heavy for its size, being impregnated with iron, and coated with a thin black crust, as if gunpowder had exploded round it. The ground where it fell was an open space, near the village of Kokurrgam, and by the rapidity of its descent, it was buried a foot deep in the earth. It was picked up by a native of rank, and was, with the greatest difficulty obtained by Captain M., as the possessor, *conceiving that it was of heavenly fabrication, had determined to repeat his prayers to it thrice a day.*——(P. LEDGER, *Oct.* 1813.)

Compare this with the worship paid by the Ephesians to the *Diopetous,* (that which fell from Jupiter) as mentioned Acts xix. v. 35: where note, that the word *image* is supplied in the translation, leaving it probable from the original, that it was in fact *a Meteoric Stone.*—L. H.

TABLE LXXIII.

	1812.		Wind.	Pressure. Max.	Min.	Temp. Max.	Min.	Evap.	Rain, &c.
8th Mo.	L. Q.	Aug. 29	NW	29·99	29·98	54°	52°	—	16
		30	NE	30·04	29·99	61	53	—	14
		31	N	30·15	30·04	60	48	26	
9th Mo.		Sept. 1	NE	30·18	30·15	59	48	—	
		2	SE	30·15	30·10	62	54	—	
		3	SW	30·10	29·95	62	53	—	
		4	SE	29·95	29·89	63	45	—	
	New M.	5	NE	29·96	29·89	65	46	28	
		6	NE	30·07	29·96	67	40	—	
		7	SE	30·09	30·07	67	41	—	
		8	E	30·07	29·97	69	50	—	
		9	SE	29·99	29·97	68	47	41	
		10	S	30·17	29·95	67	47	—	
	1st Q.	11	NW	30·28	30·17	68	42	—	
		12	W	30·26	30·22	69	44	—	
		13	Var.	30·18	30·14	71	42	—	
		14	Var.	30·18	30·07	69	40	53	
		15	NW	30·07	29·94	68	41	—	
		16	SW	29·94	29·82	72	47	—	
		17	Var.	29·96	29·82	62	41	—	15
	Full M.	18	NW	30·20	29·96	60	37	—	
		19	SW	30·14	30·12	60	46	48	
		20	SW	30·12	30·03	69	45	—	
		21	W	30·03	29·97	73	47	—	
		22	SW	29·95	29·91	68	50	—	84
		23	SW	29·95	29·93	57	44	57	
		24	NW	30·06	29·95	56	34	—	
		25	Var.	30·04	29·99	60	40	—	
		26	W	30·07	30·04	63	53	23	
				30·28	29·82	73	34	2·76	79

NOTES.—Eighth Mo. 30. Very showery. Between 4 and 5 p. m. a *tornado* (as it seems by the description given) crossed the village of Plaistow, in a direction from NE to SW, which left behind it considerable traces of its violence: a large quantity of wheat in sheaves was carried over a hedge into a neighbouring field: a fence was levelled, and about seventy *oak hurdles* torn out of the ground, some of which were seen tumbling over in the air, and fell at two hundred yards distance.

Ninth Mo. 12. Misty morning: much dew. 13, 14. The same in the evening, a dense *Stratus* reflecting on its surface with much brilliancy the orange colour of the western sky. 15. Hoar frost in the pastures: a *Stratus* at night as before, the wind coming about to the eastward soon after it was formed. 16. *Cirrus*, with *Cirrostratus* and *Cumulus*. 17. Rain most of the afternoon: a rich crimson tinge on the lower surface of the clouds at sunset. 18. At sunset the sky was extensively coloured with orange, surmounted by a distinct blush of red: the colour was reflected in the E horizon. 19. Much hoar frost. 22. Clear morning at first, but soon overcast, with rain. 25. Hoar frost. 26. *Cirrostratus*.

RESULTS.

Easterly Winds prevailed the fore part, and Westerly the latter part, of this period.

Barometer: Greatest height .... 30·28 in.
Least .... 29·82 in.
Mean of the period .... 30·040 in.
Thermometer: Greatest height .... 73°
Least .... 34°
Mean of the period .... 54·93°
Evaporation .... 2·76 in.
Rain .... 0·79 in.

FALL OF A PRECIPICE.

An avalanche occurred on the 4th of September, in the neighbourhood of Villeneuve, Switzerland. A part of the eastern chain of the Fourches, which had been sapped by a stream that ran at the base, suddenly fell with a terrific noise. About 30 cottages were buried beneath the ruins, and 12 of their inmates killed. The noise of the avalanche was heard at the distance of six miles.

EARTHQUAKE.

Shocks of an earthquake were felt throughout Italy on the 11th and 12th of September. At Florence, several houses, public edifices, and two churches were thrown down. The inhabitants, under the influence of extreme terror, fled into the country, and the greater part of the population abandoned their homes, and passed two days and nights in the fields.—(PAPERS).

TABLE LXXIV.

		1812.		Wind.	Pressure.		Temp.		Evap.	Rain &c.
					Max.	Min.	Max.	Min.		
9th Mo.	L. Q.	Sept.	27	SW	30·04	29·77	68°	59°	—	4
			28	SE	29·90	29·67	63	53	—	—
			29	NE	29·90	29·90	57	55	—	—
			30	E	29·90	29·75	59	55	20	—
10th Mo.		Oct.	1	W	29·92	29·75	62	48	—	28
			2	W	29·98	29·92	64	48	—	1
			3	SE	29·98	29·90	61	41	2	9
			4	SW	29·90	29·78	67	44	—	
	New M.		5	NW	29·78	29·31	69	55	—	23
			6	SE	29·32	29·12	64	41	—	22
			7	Var.	29·32	29·12	57	45	—	5
			8	SE	29·40	29·32	62	40	—	6
			9	NW	29·40	29·35	53	46	13	7
			10	S	29·35	29·28	55	43	—	32
			11	W	29·28	29·23	53	38	—	6
			12	Var.	29·23	29·16	51	37	—	8
	1st Q.		13	SE	29·16	28·83	57	45	—	60
			14	SW	29·14	28·83	51	42	3	10
			15	W	29·39	29·14	53	43	—	—
			16	W	29·56	29·39	56	34	—	—
			17	SW	29·56	29·05	55	39	—	29
			18	W	29·17	28·74	61	50	—	44
			19	SW	28·81	28·53	59	49	9	37
	Full M.		20	NW	29·54	28·81	67	42	—	—
			21	NW	29·74	29·50	53	37	—	
			22	SW	29·45	29·35	59	46	—	2
			23	W	29·92	29·45	54	41	10	
			24	SW	29·92	29·74	56	41	—	—
			25	SE	29·50	29·40	56	41	—	31
			26	SW	29·74	29·72	53	38	6	—
					30·04	28·53	69	34	63	3·64

NOTES.—Ninth Mo. 26. Windy. Some rain in the night. 27. Very foul sky a. m. 28—30. Rain at intervals in very small quantity.

Tenth Mo. 1. A thunder storm about 1 p. m., which was chiefly in the W, with heavy showers. 4, 5. Much dew. A storm of wind about midnight on the 5th. 6. Windy. 7. Misty morning: the trees dripping. 8. Rainbow, several times repeated between 8 and 9 a. m. Showers followed. 10. Rainbow, p. m. 11, 12. Rain in the night, misty morning. 13. *Cirrostratus* and *Nimbus* a. m., sunshine, and showers: a wet night. 16. Sunshine, with *Cumulostratus*. 17. Misty morning. 18. Squally during the night, with heavy showers.

19. Thunder and lightning about 2 p. m. Very heavy squalls with rain. 20. Sunshine a. m., much wind. 21. Clear and calm this evening. 22. Overcast, windy a. m. In the evening a wet squall, with some lightning. 24. No swallows have been seen since the 19th or 20th. 25. A few swallows appeared again to-day.

RESULTS.

Prevailing Winds Westerly.

Barometer: Greatest height .... 30·04 in.
Least....................... 28·74 in.
Mean of the period .... 29·468 in.
Thermometer: Greatest height............... 69°
Least....................... 34°
Mean of the period .... 51·46°
Evaporation .... 0·63 in.
Rain .... 3·64 in.

The evaporation was much greater during the above period than the amount here stated, as appears by observations at the laboratory. It was probably not less than two inches. The situation of the gauge had been changed.

HURRICANE AND EARTHQUAKE AT JAMAICA.

Kingston, Jamaica, October 17.—During the greater part of Monday, there was a considerable fall of rain, with a heavy swell from the southward, and strong indications of an approaching storm. Between the hours of six and seven of that evening the wind began to blow with great violence, and about midnight had increased to a perfect hurricane, the sea being at the same time dreadfully agitated, passing over the different wharfs, and sweeping every thing in its overwhelming course. The weather continued in this boisterous state until eight o'clock next morning, when the fury of the wind considerably abated, and copious torrents of rain fell during the remainder of that day.

October 24.—We are concerned to state, that the accounts received this week from various parts of the country, furnish melancholy details of the damage sustained in consequence of the late tempestuous weather.—At Salt-hill, (parish of Port Royal,) a piece of about 16 acres of land, with a small house on it, sunk down, *and was afterwards swept to the distance of* 3 *or* 400 *yards from its original situation,* without occasioning any injury to the house.

November 14.—On Wednesday morning, at 20 minutes past two o'olock, a smart shock of an earthquake was felt in this city and neighbourhood; and at ten minutes before six three most alarming and tremendous concussions immediately succeeded each other, accompanied by a most dreadful rumbling noise and crash, and continuing for upwards of 30 seconds. The shock was felt throughout the whole island, and many houses and plantations suffered severely.—The watchman at the dock yard, *a few minutes-previous to the shock,* *observed a large meteor,* which passed in a direction from the SE to the NW. The shock is stated to have lasted one minute and some seconds.—(PAPERS).

TABLE LXXV.

	1812.		Wind.	Pressure, Max.	Min.	Temp. Max.	Min.	Evap.	Rain, &c.
10th Mo.	L. Q.	Oct. 27	SW	29·40	29·24	52°	40°	—	26
		28	SW	29·75	29·40	51	32	—	
		29	W	29·80	29·74	49	33	15	
		30	SE	29·78	29·66	51	39	—	14
		31	W	29·94	29·78	54	41	—	
11th Mo.		Nov. 1	SW	29·94	29·87	55	50	—	—
		2	N	30·05	29·87	54	44	7	65
		3	SW			49	38	—	
	New M.	4	W	30·05	29·83	48	39	8	
		5	NW	29·83	29·80	49	30	—	
		6	SW	29·83	29·80	45	27	—	
		7	W	29·83	29·79	45	24	—	
		8	E	29·79	29·77	40	27	5	
		9	E			46	39	—	
		10	NE	30·15	30·03	45	33	—	
		11	SE			46	39	—	32
	1st Q.	12	E	30·03	29·58	45	39	—	38
		13	NE	29·58	29·20	51	44	—	55
		14	W	29·59	29·20	52	40	—	—
		15	SW	29·59	29·30	52	39	—	3
		16	NE	29·30	29·00	46	42	—	—
		17	NE	29·00	28·96	46	40	—	—
	Full M.	18	NW	29·66	28·96	42	32	7	13
		19	N	29·86	29·66	42	28	—	
		20	N	29·97	29·83	41	33	—	—
		21	NE	30·32	29·97	39	26	7	
		22	N	30·38	30·31	43	25	—	
		23	SW	30·31	30·08	44	26	—	
		24	SW	30·08	29·89	48	39	9	
				30·38	28·96	55	24	0·58	2·46

NOTES.—Tenth Mo. 27. Misty and overcast, a. m.: wet at noon: p. m. the barometer descended at the rate of a tenth of an inch per hour, the wind increasing in proportion, with much rain, the clouds sweeping the earth. The evening was very tempestuous; before midnight the barometer had risen again, and the weather was moderate. Many large trees were blown down. 28. Hoar frost: rather misty. At sunset, the sky exhibited a fine collection of coloured clouds, in the modifications *Nimbus* and *Cirrus*, with broad parallel bands of red in the haze above them. 28. Fair and calm. 30. *Cirrostratus* and *Cumulus:* the sky again beautifully coloured.

5

Eleventh Mo. 1. Cloudy. 2. Wet a.m. 5. Fine day. 6—11. Chiefly misty or cloudy, with hoar frost, and some very thick local fogs. 11. Overcast a. m. The *Cirrostratus* prevails, and sounds travel with the wind to an unusual distance : we hear the rattling of the carriages on the pavement in London, through a direct mean distance of five miles. This phenomenon is perhaps to be attributed to a thick continuous sheet of haze in the air above us, which acts as a sounding board. 12. Rain through the day. 13. Misty: rain: sounds are again distinctly heard from the city. 15. Fair: a *Stratus* at night. 16. Overcast; with an easterly gale. 18. Wet stormy day, night clear and calm. 20. Misty, much rime on the trees, which came off about noon in showers of ice. At 11 a.m. a perfect but colourless *bow* in the *mist:* near 4 p. m. there was a shower, in which the rainbow showed its proper colours. 22. Clear: the ground just sprinkled with hail balls. 23. Misty a. m. with rime; clearer p. m.: thaw in the night. 24. Clear morning.

RESULTS.

Winds for the greater part Westerly; though the Rain chiefly fell during an Easterly Wind.

Barometer: Greatest height................30·38 in.

Least......................28·96 in.

Mean of the period...........29·678 in.

Thermometer: Greatest height............... 55°

Least...................... 24°

Mean of the period...........41·31°

Evaporation...................... 0·58 in.

Rain.............................. 2·46 in.

A River changing its Course.

From the year 1750, the North Esk, in Kincardineshire, emptied itself into the sea upon the lands of Kirkside and Woodstone. About a twelvemonth ago, however, in consequence of an overflow in the river, and a very high tide, it excavated a new channel on the lands of Comieston, upon which, on an average, one half of its contents continued to run during last fishing season. A new revolution has now taken place in the course of the river, which, on Monday last, completely opened out its old channel upon the lands of Kirkside and Woodstone, by which a valuable property is restored to its former proprietors. This river had, several times previous to the first-mentioned date, undergone a similar change from the same causes ; and, upon one of these occasions, it gave rise to a long and expensive litigation between the new and old proprietors, which was, at last, decided in favour of the gentleman on whose grounds the river had begun to flow; all artificial means employed to obstruct or change the course of a river being declared illegal.—(Pub. Ledger, *Nov.* 2).

TABLE LXXVI.

	1812.		Wind.	Pressure. Max.	Pressure. Min.	Temp. Max.	Temp. Min.	Evap.	Rain, &c.
11th Mo.	L. Q.	Nov. 25	SW	29·89	29·80	48°	35°		
		26	E	30·07	29·77	47	43		26
		27	N	30·21	30·10	49	42		
		28	NE	30·10	29·92	47	38		
		29	SE	29·89	29·85	49	41		
		30	S	29·95	29·88	50	47		15
12th Mo.		Dec. 1	S	29·96	29·72	52	44		5
		2	NW	30·08	29·96	49	42		—
	New M.	3	E	30·11	30·09	49	45		—
		4	E	30·08	30·05	48	38		11
		5	E	30·22	30·08	44	33		
		6	NE	30·51	30·29	42	26		
		7	NE	30·51	30·41	35	23		
		8	NE	30·41	29·94	34	18		
		9	W	29·96	29·94	35	24		
		10	NW	29·89	29·78	34	29		—
	1st Q.	11	E	30·00	29·97	36	27		
		12	NE	29·97	29·79	32	24		
		13	NE	29·79	29·71	34	24		
		14	NE	29·71	29·66	35	28		
		15	E	29·66	29·20	34	28		
		16	E	30·20	28·98	34	28		—
		17	E	30·22	28·98	35	32		27
	Full M.	18	E	29·51	29·22	38	33		18
		19	E	29·57	29·47	38	35		
		20	E	29·76	29·57	36	31		
		21	NW	29·82	29·76	38	32		—
		22	Var.	30·02	29·82	42	33		—
		23	N	30·30	30·02	36	31		—
		24	N	30·46	30·30	35	32		3
				30·51	28·98	52	18		95

NOTES.—Eleventh Mo. 28. The sky, about sunset, was overspread with *Cirrus* and *Cirrostratus* clouds, beautifully tinged with flame colour, red and violet. 30. The sky again much coloured in the morning.

Twelfth Mo. 5. The weather, which has been hitherto mostly cloudy, with redness at sunrise and sunset, begins now to be more serene. 6. Hoar frost. 7. A little appearance of hail balls on the ground. 8, 9. Clear, hoar frost. 11. Snow this morning, and again after sunset. 13. An orange-coloured band on the horizon this evening; this phenomenon arises from reflection by the descending dew.

15. A gale from NE, unaccompanied by snow, came in early this morning. 16. a. m. The wind has subsided to a breeze, and there now falls (at the temperature of 27·5°) snow, very regularly crystallized in stars. 17. a. m. It snowed more freely in the night, and there is now a cold thaw, with light misty showers. 18. A little sleet, followed by snow. Ice has been formed in the night, by virtue of the low temperature which the ground still possesses. A wet evening. 21. A little rain, a. m. 22. A dripping mist. 24. Cloudy; a little rain; some hail balls in the night.

RESULTS.

Prevailing Winds Easterly.

Barometer : Greatest height . . . . 30·51 in.

Least . . . . 28·98 in.

Mean of the period . . . . 29·882 in.

Thermometer : Greatest height . . . . 52°

Least . . . . 18°

Mean of the period . . . . 36·68°

Rain and snow . . . . 0·95 in.

The Evaporation during this period was not ascertained; but it probably did not exceed half an inch.

COPENHAGEN, *December* 19.—The intense cold still continues, *in consequence of the east wind.* Reaumur's thermometer is 13 below Zero. From our custom house to the coast of Sweden, the Sound presents only one continued surface of ice. Should the frost continue a short time longer, it will soon be passable on foot. Some ships from our provinces, which are detained by the ice, have already sent home their crews.

FIGURE OF A DROWNED MAN IN THE ICE.

The curiosity of several persons in the neighbourhood of Halnaker, near Chichester, was last week excited by the figure of a man, apparently in a round frock, on the ice of a pond in the park; but no individual being missed in the parish, it was conceived to be an accidental impression on the ice, and was, by many, treated as a fancy. It being, however, last Tuesday, ascertained that a man named Richards, of the neighbouring parish of Charlton, was missing, the pond was searched, and he was found lying at its bottom, in 10 feet water, exactly under the figure. The ice was uniformly seven inches thick; but that which composed the figure was darker and more transparent.—(PUB. LEDGER, *Dec.* 21).

Meteorological Observations

MADE AT

TOTTENHAM, near LONDON,

IN THE YEARS

1813, 1814, 1815, 1816.

(First published Monthly in Thomson's Annals of Philosophy).

TABLE LXXVII.

	1812.			Wind.	Pressure.		Temp.		Evap.	Rain, &c.
					Max.	Min.	Max.	Min.		
12th Mo.	L. Q.	Dec.	25	N	30·46	30·40	35°	31°	—	—
			26	N	30·50	30·40	37	30	—	—
			27	N	30·52	30·48	36	29	—	1
			28	W	30·52	30·32	43	32	—	
			29	W	30·32	30·15	46	42	—	
			30	W	30·15	29·92	50	42	—	
1813.			31	W	29·81	29·75	44	40	—	—
1st Mo.	New M.	Jan.	1	W	30·09	29·81	45	38	—	—
			2	SW	30·26	30·09	44	36	6	—
			3	W	30·30	30·26	41	34	—	—
			4	SE	30·30	30·09	42	34	—	—
			5	SW	30·09	29·86	44	37	—	5
			6	SW	29·77	29·70	50	40	—	9
			7	NW	29·70	29·30	46	40	—	
	1st Q.		8	NW	29·62	29·30	48	28	—	
			9	NW	29·87	29·75	41	31	9	9
			10	NW	29·82	29·70	34	28	—	11
			11	SE	29·80	29·70	40	26	—	—
			12	SE	29·70	29·61	34	29	—	
			13	SE	29·58	29·53	38	34	5	16
			14	NE	29·74	29·53	38	33	—	
			15	NW	30·00	29·74	38	28	—	
	Full M.		16	E	30·20	30·00	44	29	—	
			17	SE	30·20	30·04	35	28	—	
			18	SE	30·14	30·04	31	30	—	
			19	E	30·26	30·14	33	31	—	—
			20	NE	30·27	30·26	34	30	—	
			21	NE	30·35	30·27	34	29	—	
			22	NW	30·50	30·35	36	23	15	
			23		30·38	30·22	35	30	—	8
					30·52	29·30	50	23	0·35	0·59

NOTES.—Twelfth Mo. 25. A very slight fall of snow. 27. A little snow last night. 30, 31. Windy night: small rain at intervals.

1813. First Mo. 1. Small rain at intervals. 3. Misty morning. 5. Windy. 6. Windy: small rain. 7. Very misty a. m., dark and cloudy p. m. About eight some lightning, which was soon followed by a shower. 9. Hoar frost: at 9 a. m. thick air, with *Cirrostratus* and *Cirrocumulus:* sounds came freely from the city, with the wind at SSW. Sleet and rain followed within an hour. 13. Overcast a. m., thin sleet and rain. 14. Cloudy. 19. A little snow a. m. 22. Clear

p. m. A fine red blush on the horizon at sunset. 23. Hoar frost: soon after noon a fine granular snow: after sunset a more plentiful spicular snow, adhereing to the trees and shrubs.

RESULTS.

Winds Variable.

Barometer: Greatest height .... 30·52 in.

Least .... 29·30 in.

Mean of the period .... 30·022 in.

Thermometer: Greatest height .... 50°

Least .... 23°

Mean of the period .... 36·25°

Evaporation .... 0·35 in.

Rain and snow .... 0·59 in.

STORM AT GIBRALTAR.

On the 29th December, a most violent storm came on at Gibraltar from the SE, in which many vessels and lives were lost.—The late gales at Gibraltar were more severe than any remembered to have been experienced. Twenty-eight vessels were lost in Catalonia Bay, and 20 in Gibraltar Bay. Many of the largest trees upon the rock were torn up by the roots, others entirely stripped of their limbs, *and some measuring upwards of* 18 *inches in diameter, broken short off in the trunk.*—(PUB. LEDGER).

TABLE LXXVIII.

1813.			Wind.	Pressure. Max.	Pressure. Min.	Temp. Max.	Temp. Min.	Evap.	Rain, &c.
1st Mo.	L. Q.	Jan. 24	NE	30·46	30·37	37°	24°	—	
		25	NE	30·47	30·45	36	29	—	
		26	NE	30·48	30·40	41	35	—	
		27	N	30·49	30·47	39	21	—	
		28	Var.	30·39	30·37	32	20	—	
		29	Var.	30·48	30·39	34	21	—	
		30	NW	30·48	30·44	42	30	—	
		31	NW	30·50	30·44	48	34	15	27
2d Mo.	New M.	Feb. 1	Var.	30·33	30·24	41	30	—	
		2	NW	30·37	30·32	41	36	—	
		3	NW	30·45	30·37	43	34	—	
		4	W	30·45	30·29	41	34	—	
		5	S	30·29	29·78	47	36	—	—
		6	SW	29·89	29·78	47	38	—	
		7	SW	29·98	29·79	48	37	—	—
	1st Q.	8	SW	29·66	29·63	52	44	—	—
		9	SW	29·88	29·66	51	35	—	36
		10	W	30·00	29·88	46	33	—	
		11	S	30·00	29·75	47	35	—	
		12	S	29·75	29·28	56	44	—	—
		13	SW	29·48	29·37	57	39	—	33
		14	SW	29·38	29·27	52	42	71	30
	Full M.	15	SW	29·34	29·27	52	41	—	18
		16	SW	29·44	29·34	48	41	—	—
		17	SW	29·37	29·30	52	43	—	27
		18	SW	29·88	29·37	52	41	—	—
		19	S	29·66	29·60	56	40	—	19
		20	SW	29·80	29·66	53	42	—	—
		21	SW	29·70	29·69	57	49	70	
				30·50	29·27	57	20	1·55	1·90

NOTES.—First Mo. 24. Light clouds and sunshine. 28. Rime on the trees: very misty a. m., clear p. m. 29. Hoar frost: the sky overcast. 30. Misty to the S a. m. A grey day. 31. Misty a. m. Heavy *Cirrostratus* clouds: rain at night.

Second Mo. 1. The *Cumulostratus*, which has not for a long time been exhibited, appeared to-day in large masses. 7. Showers and wind: at sunset, several large clouds of the modification *Nimbus*. 8. Stormy. 9. A violent thunder gust from the W about 2 p. m., by which considerable damage was done to the roofs and chimnies of houses, &c. This was followed by a series of heavy gales continuing (with a few short intervals of calm and pleasant weather) to the end of the period. The lunar halo appeared before several of these, of a large diameter; and, on the 18th, about 11 a.m. there was a brilliant rainbow. The River Lea has considerably inundated the adjacent lands.

RESULTS.

Winds in the fore part Northerly, with a very dry dense air, and low temperature: in the latter part Southerly, with a rare and moist atmosphere, and high temperature.

Barometer: Greatest height . . . . 30·50 in.

Least . . . . 29·27 in.

Mean of the period . . . . 29·957 in.

Thermometer: Greatest height . . . . 57°

Least . . . . 20°

Mean of the period . . . . 40·58°

Evaporation (at Stratford) . . . . 1·55 in.

Rain . . . . 1·90 in.

DEAL, *February* 16.—It last night blew excessively hard from the SSW, accompanied with a heavy sea.

February 18.—The wind the whole of last night and this day has blown tremendously from the west.

PORTSMOUTH, *February* 18.—The whole of last night and to-day it has blown very hard at SW, attended with very heavy squalls.—(PUB. LEDGER).

TABLE LXXIX.

1813.		Wind.	Pressure. Max.	Pressure. Min.	Temp. Max.	Temp. Min.	Evap.	Rain, &c.
2d Mo.	L. Q. Feb. 22	SW	29·80	29·69	57°	41°	—	
	23	NW	30·03	29·80	45	35	—	
	24	W	30·14	30·03	49	32	—	
	25	SW	30·13	29·83	50	35	—	71
	26	SW	29·90	29·70	52	35	50	8
	27	NW	30·33	29·90	46	32	—	
	28	NW	30·36	30·30	50	34	23	
3d Mo.	New M. March 1	SW	30·30	30·20	51	39	—	
	2	SW	30·20	30·02	47	35	—	—
	3	Var.	30·33	30·02	52	32	—	—
	4	SW	30·33	30·20	51	36	—	6
	5	W	30·34	30·20			—	
	6	W	30·40	30·34	53	35	—	
	7	NW	30·40	30·30	49	39	—	
	8	NW	30·30	30·20	54	43	—	
	1st Q 9	NW	30·20	29·89	52	36	56	—
	10	E	29·96	29·89	42	26	—	12
	11	NE	30·21	29·96	39	24	—	—
	12	NE	30·27	30·21	37	24	—	
	13	NW	30·27	30·20	40	29	—	
	14	SW	30·20	30·10	47	40	19	
	15	SW	30·18	30·10	53	43	—	
	16	Var.	30·18	30·09	51	32	—	
	Full M. 17	NE	30·09	29·96	56	32	—	
	18	NW	29·96	29·96	58	36	—	
	19	E	29·96	29·78	58	40	—	
	20	SW	29·96	29·78	56	35	—	14
	21	SW	29·96	29·84	53	42	31	
	22	SW	30·25	29·96	55	33	—	
	23	NW	30·30	30·28	50	34	—	
	24	W	30·28	29·98	47	39	18	35
			30·40	29·69	58	24	1·97	1·46

NOTES.—Second Mo. 24. Hoar frost. About 6 p. m. a very dark cloud came over, lowering with an arched base, as before thunder, and presently discharged a shower of large hail and rain, which was accompanied with a cold wind. 25. Fair a. m., wet and windy p. m., and night. 26. The same. 27. *Cirrus, Cumulus,* and *Cirrostratus* clouds appeared together: much wind: about 7 p. m. wind NW, a bright meteor passed from the zenith towards the N, declining a little westward. 28. Clear morning: wind moderate.

Third Mo. 1. Hoar frost, fair. 2, 3. Light showers. A *Nimbus* appeared S of the setting sun on the 3d, which went away southward. 9. Light showers. 10. Sleet a.m. At sunset a *Cumulostratus*, with a snowy appearance: some hail balls in the night. 11. A *Nimbus* was perceptible by 7 a.m., forming in the NE. There were some heavy (though transient) squalls of snow during the day. Abundance of snow fell on this and the following night to the southward, extending as far as the coast of France.

RESULTS.

Prevailing Winds Westerly, with a marked interruption by a current from the NE, occasioning snow about the middle of the period.

Barometer: Greatest height. . . . . 30·40 in.
Least. . . . . 29·69 in.
Mean of the period . . . . 30·109 in.
Thermometer: Greatest height. . . . .   58°
Least. . . . .   24°
Mean of the period . . . . 42·50°
Evaporation . . . . 1·97 in.
Rain . . . . 1·46 in.

EARTHQUAKE ON THE COAST OF DEVON.

March 21.—About six o'clock, the inhabitants of Exmouth were alarmed by the shock of an earthquake, which lasted for two or three seconds. The houses were shook, the people hurried from their beds, and the utmost alarm prevailed for some time throughout the town. The shock was felt in like manner at Sidmouth, Budleigh, Salterton, Starcross, and for many miles along the coast, but we have not heard of its having been attended by any ill consequences.—(PAPERS).

TABLE LXXX.

1813.			Wind.	Pressure.		Temp.		Evap.	Rain, &c.
				Max.	Min.	Max.	Min.		
3d Mo.	L. Q.	March 25	NW	30·37	30·28	47°	35°	—	
		26	NW	30·50	30·37	51	27	—	
		27	SW	30·47	30·43	55	32	—	
		28	NW	30·43	30·30	67	49	—	
		29	SW	30·30	30·10	66	53	—	
		30	Var.	30·10	29·89	58	47	—	1
		31	SE	29·89	29·18	57	42	40	
4th Mo.	New M.	April 1	SW	29·25	29·18	50	35	—	27
		2	W	29·45	29·25	52	35	—	9
		3	SW	29·74	29·45	48	27	—	3
		4	SW	29·85	29·74	54	29	—	
		5	SW	29·85	29·81	51	40	—	13
		6	SW	29·90	29·74	58	45	—	
	1st Q.	7	W	29·93	29·90	65	43	43	—
		8	E	29·97	29·87	69	37	—	
		9	SE	30·04	29·97	66	41	—	
		10	E	30·10	30·04	65	41	26	
		11	E	30·14	30·10	64	35	—	
		12	E	30·23	30·14	69	42	—	
		13	NE	30·34	30·23	66	35	—	
		14	E	30·20	30·10	66	42	36	
	Full M.	15	NW	30·20	29·96	68	42	—	
		16	NW	29·96	29·77	66	44	—	
		17	NW	30·10	29·77	68	41	—	—
		18	NW	30·13	30·10	56	42	53	
		19	NW	30·10	30·10	64	44	—	
		20	NW	30·10	30·05	64	40	—	
		21	N	30·09	30·05	57	32	—	
		22	NE	30·10	30·08	50	32	—	17
		23	NE	30·14	30·10	45	34	40	
				30·50	29·18	69	27	2·38	0·70

Notes.—Third Mo. 27. Hoar frost: large spreading *Cirri*. 28. Temperature 60° in the evening. 29. Overcast sky. 30. A veil of *Cirrostratus* a. m. The *Cumulus* afterwards shewed itself, and a slight shower ensued.

Fourth Mo. 1. Stormy, with rain. 2. Hoar frost: a sprinkling of opaque hail about sunrise. Several showers of this and some rain during the day. 3. Hoar frost: *Cumulus* a. m. Showers of snow, and of opaque hail p. m. 5. *Cirrostratus*, a. m. wet and windy. 10. *Cirrus* and *Cumulus* clouds: the wind increases in strength: the mornings have been misty of late, and there have been plentiful dews, in

1

consequence of the great difference between the temperature of day
and night. 15. Wind boisterous in the evening. 16. Cloudy a. m.
17. Slight showers. 20. From the 7th of this month we have had
summer like days and cold nights: the roads have become very dusty,
and the earth considerably dry. 21. Some clouds of a threatening
appearance from the NE in the evening, attended with depression of
temperature. 22. p. m. Hasty showers, mixed with hail; after which
steady small rain till evening. 23. Cloudy: several scanty hail showers
from large *Nimbus* clouds passing over. During the approach of one
of these, a slender, tapering, and somewhat twisted column, appeared
in front, detached from the main body, and reaching down to the
earth in the manner of a water spout. In a few minutes, by spreading
on all sides, it became incorporated with the rest of the shower.
This is not a very uncommon appearance, but I have seldom seen it
so perfectly exhibited.

RESULTS.

Prevailing Winds Westerly, interrupted (after the middle of the period)
by an Easterly current.

Barometer: Greatest height................30·50 in.
    Least......................29·18 in.
    Mean of the period...........30·005 in.
Thermometer: Greatest height............... 69°
    Least....................... 27°
    Mean of the period...........49·11°
  Evaporation........................ 2·38 in.
  Rain.............................. 0·70 in.

LARGE METEOR.

On Friday evening the 23d of April, between nine and ten o'clock, a fiery
meteor was seen at Brigg, (Lincolnshire,) in a northerly direction. When first
discovered, it was at a considerable height, and gliding downwards in an ob-
lique manner. It appeared about the size of a full moon, of the colour of
burning brimstone, and constantly emitting small balls of very brilliant red
light, which took an opposite direction, and soon disappeared; and by which
the meteor appeared completely exhausted before it reached the ground.
(PUB. LEDGER).

TABLE LXXXI.

	1813.		Wind.	Pressure.		Temp.		Evap.	Rain, &c.
				Max.	Min.	Max.	Min.		
4th Mo.	L. Q.	April 24	NE	30·14	29·95	—°	35°	—	
		25	NE	29·95	29·71	53	43	—	51
		26	SE	29·71	29·51	—	39	—	
		27	NW	29·51	29·36	48	45	—	52
		28	N	29·61	29·36	50	41	15	4
		29	NE	29·65	29·61	49	39	—	21
	New M.	30	E	29·65	29·63	—	—	—	—
5th Mo.		May 1	NE	29·72	29·63	59	45	—	25
		2	Var.	29·79	29·72	61	46	—	
		3	E	29·79	29·73	66	50	17	17
		4	Var.	29·86	29·73	65	51	—	24
		5	NW	29·91	29·87	64	49	—	—
		6	NE	29·87	29·69	68	50	—	16
	1st Q.	7	SE	29·70	29·69	69	45	—	8
		8	SE	29·70	29·57	68	49	27	—
		9	N	29·73	29·57	72	52	—	—
		10	Var.	29·90	29·73	69	47	—	8
		11	Var.	29·73	29·60	67	53	—	5
		12	SW	29·64	29·60	74	49	—	3
		13	SW	29·64	29·46	72	51	35	—
		14	Var.	29·47	29·39	68	52	—	9
	Full M.	15	SW	29·70	29·57	65	47	—	—
		16	SW	29·72	29·41	65	49	—	51
		17	NW	29·82	29·68	60	47	—	—
		18	Var.	29·80	29·78	60	49	45	25
		19	SW	29·84	29·56	69	50	—	—
		20	W	29·59	29·57	61	42	—	—
		21	W	29·64	29·59	59	39	—	—
		22	W	22·69	29·59	58	30	28	53
				30·14	29·36	74	30	1·67	3·72

Notes.—Fourth Mo. 24. Heavy *Cumulostratus* clouds through the day. 25. Rain nearly the whole day. 26. The *maximum* of temperature at 9 a. m., cloudy: clear at evening, with *Cirri*. 27. A wet day. 28. Wet morning: cloudy. 29, 30. Cloudy: much wind.

Fifth Mo. 1. The *maximum* of temperature at 9 a. m., wet. 2. Cloudy a. m. In the afternoon the sky cleared pretty suddenly, save that some dense *Cumulus* clouds remained in the NE, to the summit of one of which a *Cirrostratus* was observed for a considerable time adhering, which was at length incorporated with the larger cloud. The moon appeared with a pale golden crescent, the remainder of the

disk being pretty conspicuous. 3. Dense *Cumulus* clouds to the S,
with *Cirrus* and *Cirrocumulus* intermixed (as before thunder). A
shower of large drops about sunset. 4. Overcast sky a. m. About
6 p. m. (after some previous dripping) a thunder storm, the weight of
which fell to the E of us. A most brilliant rainbow, together with a
complementary one, was exhibited for about forty minutes. The
space included *within* the proper bow was very perceptibly *lighter*, and
that *without* it, extending to the complementary arch, as much *darker*
than the rest of the cloud. A nightingale sang with spirit in the
midst of the shower. 6. a. m. Much dew: p. m. a large *Nimbus* in
the N. *Cirrostratus* in the E, and *Cirrus* above, stretching from E
to W. The large cloud moved away by W into the S. A thunder
storm ensued in that direction, though nearly out of hearing, and
lasted till midnight; after which we had a sudden heavy shower.
7. a. m. Cloudy: p. m. (after a shower) clearer, but with indications of
more rain. 8. An appearance of much electrical action in the clouds
far to the S and SW. 9. A few drops of rain a. m., various modifications of cloud appeared this day. 10. *Nimbi:* dripping afternoon:
rainbow: fine evening. 13. Cloudy, windy. 14. Much wind. 15. The
same: calm night. 16—18. Much wind: showers.

RESULTS.

Winds Variable.

Barometer: Greatest height................30·14 in.
Least....... ....29·36 in.
Mean of the period............29·678 in.
Thermometer: Greatest height............... 74°
Least..................... 30°
Mean of the period............54·79°
Evaporation (the gauge being now placed
on the ground)................... 1·67 in.
Rain.............................. 3·72 in.

A gauge placed against the north wall of the house, at about seven
feet elevation, gave, for the evaporation in the same time, only 1·16 in.

Thunder Storms.

On the afternoon of Friday the 14th May, there was a dreadful storm of
thunder and lightning, about thirty miles north of Aberdeen.

May 17.—The town of Macclesfield was visited by one of the most awful
storms of hail, accompanied by loud thunder and lightning, remembered by any
of its oldest inhabitants—the lower part of the town was inundated.—(Pub.
Ledger).

TABLE LXXXII.

	1813.		Wind.	Pressure. Max.	Min.	Temp. Max.	Min.	Evap.	Rain &c.
5th Mo.	L. Q.	May 23	W	29·59	29·55	59°	49°	—	
		24	NW	29·70	29·55	60	49	—	
		25	NW	29·76	29·65	61	49	—	
		26	W	30·05	29·76	57	41	—	
		27	NW	30·10	30·05	60	42	—	
	New M.	28	Var.	30·10	29·96	72	42	—	
		29	NE	30·06	29·96	76	53	—	
		30	S	29·97	29·90	73	56	65	34
		31	NW	30·00	29·97	78	59	—	2
6th Mo.		June 1	E	29·97	29·83	85	50	—	
		2	NE	29·95	29·83	84	54	—	
		3	NW	30·17	29·95	72	51	—	
		4	NW	30·17	29·88	65	51	75	
	1st Q.	5	NW	29·88	29·82	58	48	—	
		6	NE	29·73	29·67	56	47	—	
		7	NE	29·80	29·73	71	43	—	
		8	NE	29·73	29·51	75	48	—	
		9	NW	29·53	29·43	75	54	48	—
		10	NW	29·83	29·53	75	48	—	
		11	S	29·83	29·74	76	52	—	
		12	SE	29·98	29·74	77	45	—	
		13	W	30·02	29·97	75	51	48	5
	Full M.	14	SW	30·07	29·78	67	52	—	—
		15	SW	29·98	29·78	67	46	—	—
		16	NW	30·04	29·98	63	44	—	—
		17	Var.	30·09	30·04	61	48	—	—
		18	N	30·15	30·09	61	39	—	—
		19	N	30·16	30·15	57	37	—	—
		20	NE	30·20	30·16	60	42	42	65
				30·20	29·43	85	37	2·78	1·06

NOTES.—Fifth Mo. 30. A shower p. m.: thunder to the westward. 31. Fine day: some thunder clouds appeared: the evening twilight was brilliant and tinged with orange—the new moon was conspicuous, and there fell much dew.

Sixth Mo. 1. *Cumulus, Cumulostratus,* and *Cirrostratus* clouds. The sunset was cloudy, with an orange tint. 2. *Cirrostratus* clouds, with haze to the S at sunset. At the same time there were *Cirri* in the N more elevated, and finely tinged with red. 4. Windy: cloudy till evening. 5. Clear a. m., afterwards cloudy and windy. 8. Windy: at sunset *Cumuli,* with the *Cirrostratus* attached: much orange in the twilight. 9. A shower early: cloudy, dripping. 10. p. m. Large elevated *Cirri*. 11. a. m. *Cumulostratus* clouds : p. m. *Cirri* in abundance lowering and thickening. 12. *Cirri*, tinged red in the morning early: before eight it was overcast, and rain fell. 13. a. m. Cloudy : a shower at evening. 14—20. Occasional showers, some of which were heavy.

RESULTS.

Prevailing Winds Northerly.

Barometer : Greatest height . . . . 30·20 in.

Least . . . . 29·43 in.

Mean of the period . . . . 29·889 in.

Thermometer : Greatest height . . . . 85°

Least . . . . 37°

Mean of the period . . . . 57·93°

Evaporation . . . . 2·78 in.

Rain . . . . 1·06 in.

Light of the Moon compared with the Evening Twilight.

Seventh Mo. 12, 1813.—A clear evening, with a gentle breeze from the west. The twilight being very brilliant and the Moon near the Full, I compared the intensity of their respective and opposite shadows, projected by a slender rod on white paper. At 9 p. m. the shadow of the twilight was very perceptible in the moonlight : but at 9 h. 20 min. the Moon cast an equally strong shadow into the twilight, which shadow appeared (as in the well-known experiment of the shadow of a candle in the daylight) of a bluish grey colour.

Thus it appears, that in certain conditions of the atmosphere favourable to refraction, the Sun affords us as much light an hour after he has set, as the Full Moon does an hour after she has risen.

TABLE LXXXIII.

1813.			Wind.	Pressure, Max.	Min.	Temp. Max.	Min.	Evap.	Rain, &c.
6th Mo.	L. Q.	June 21	NE	30·16	30·14	66°	46°	—	
		22	NE	30·18	30·14	67	41	—	
		23	NE	30·10	30·08	70	55	—	
		24	NE	30·15	30·08	64	46	—	—
		25	E	30·18	30·15	74	50	55	4
		26	NE	30·18	30·10	74	48	—	
		27	E	30·10	29·97	77	45	—	
	New M.	28	Var.	29·97	29·77	75	53	—	37
		29	SW	29·77	29·64	74	52	—	80
		30	W	29·64	29·54	65	52	40	75
7th Mo.		July 1	NW	29·75	29·64	70	53	—	16
		2	N	29·86	29·75	67	50	—	20
		3	NW	30·11	29·86	73	42	25	
		4	NW	30·18	30·11	70	50	—	2
		5	W	30·18	30·04	74	47	—	
	1st Q.	6	SW	30·04	29·74	78	56	35	
		7	S	29·65	29·57	78	57	—	
		8	W	29·59	29·55	75	55	—	—
		9	W	29·83	29·59	79	54	36	
		10	NW	29·91	29·83	78	51	—	
		11	NW	29·93	29·90	77	51	—	
		12	W	29·90	29·76	76	51	—	
	Full M.	13	SW	29·76	29·63	76	58	38	
		14	SW	29·63	29·60	71	58	—	
		15	NW	29·70	29·60	67	50	—	60
		16	NW	29·82	29·70	73	50	—	10
		17	NW	28·87	29·82	74	47	—	—
		18	W	29·87	29·73	73	50	43	
		19	W	29·73	———	73	—	—	—
				30·18	29·54	79	41	2·72	3·04

NOTES.—Sixth Mo. 21. Brisk wind through the day. 22. Wind more gentle: *Cumulostratus* and *Cirrostratus*. 24. A shower about 1 p. m. 25. The wind inclines to SE: clear twilight, somewhat orange coloured. 27. *Cirrus*, changing to *Cirrocumulus* and *Cirrostratus:* twilight somewhat opaque, but coloured. 28. Wind NE a. m.: the *Stratus* cloud appears to have prevailed in the night: slight showers at intervals during the day: at 7 p. m. several *Nimbi*, and some thunder to the SW, which with occasional lightning passed by S to NE: at 9 p. m. the air was so loaded with vapour as to deposit water on a glass vessel cooled only to 58·5°: it now began to rain heavily, ceasing at ten, with thunder and lightning still in the N. 29. *Cirrus, Cirrostratus,* and *Cumulostratus:* about 1 p. m. a heavy storm of rain and hail, with several electrical discharges. 30. In the forenoon heavy rain, ushered in by a peculiar hollow sound in the wind, then southerly: wet at intervals, p. m.: a part of this day's rain was taken by estimation, the gauge having been left under cover.

Seventh Mo. 3. After an appearance of two distinct orders of cloud during the forenoon, inosculation took place suddenly about one, and the *Cumulostratus,* with a brisk wind, prevailed till sunset. 4. A slight shower p. m.: from the 5th to the 9th several kinds of cloud prevailed, and occasioned at times considerable indications of rain, of which however a few drops only fell, the clouds still passing away to the N: in that quarter, on the evening of the 9th we had several distant *Nimbi,* with the usual appearances of a strong electric charge: a single flash of lightning, and some rain just discernible in the horizon, were the only results. 13. After repeated exhibitions of the *Cumulostratus,* which continued to pass over to the N, we had this night a few drops of rain. 14. Dripping at intervals: the dust laid. 15. A wet day. The vulgar notion, that rain on this day (by the church calendar given to St. Swithin) is followed by the same *daily for forty days,* if tried *at any one station* in this part of the island, will be found fallacious. 16. Thunder p. m. during a shower. 17. A slight shower p. m., dew on the grass. 18. A fine day: the *Cumulostratus* prevailed, and the evening was very clear, with dew. 19. Showers.

RESULTS.

Prevailing Winds Easterly to the New Moon, afterwards Westerly.

Barometer: Greatest height................30·18 in.
           Least.........................29·54 in.
           Mean of the period............29·875 in.
Thermometer: Greatest height.............. 79°
             Least....................... 41°
             Mean of the period...........61·69°
Evaporation........................ 2·72 in.
Rain.............................. 3·04 in.

Shadows of Clouds projected in a Hazy Atmosphere.

Seventh Mo. 16, 1813.—After pretty much wet yesterday, we had this day *Cumulus* and *Cumulostratus,* passing to *Nimbus,* with a thunder shower about 4 p. m. The sun then passed behind a group of dense clouds in the NW, darting those broad diverging beams of light separated by shadows, first downwards on the horizon and then up into the air, which Virgil so correctly defines in his first *Georgic:*—

> Aut ubi sub lucem densa inter nubila sese
> Diversi erumpent radii.—

It is obvious that these beams, should they be at any time extensive enough to pass the zenith of the observer, ought to converge again towards a point on the eastern horizon opposite to the sun's place in the western: which was in fact partially the case this evening.

Seventh Mo. 22. Travelling this evening between Winchester and Alton, with the setting sun behind me, I observed a considerable accumulation of dewy haze above the horizon before me, slightly reddened by the sun's rays; and in the midst of it several broad bars of shadow, *converging to a point on the horizon*, (conformably to the observation I made on the 16th inst.) and projected apparently by separate lofty dense clouds then remaining to the westward.

From long experience I am accustomed to connect the appearance of this elevated haze amidst clouds in warm weather with a thunder storm falling on the tract beneath it: and the reader will perceive, from the Notes, that this connexion probably existed in the present, as in the former instance.

Phenomena of the Evening Twilight.

Seventh Mo. 29, 1813.—This was a peculiarly fine evening for observing the phenomena of the twilight. The sun set, surrounded by a moderate glow of flame colour, and without a speck on the western sky. An extensive brightness was left along the horizon, which insensibly vanished upwards, and towards the NE and SW. Opposite to this, and quite distinct in its boundaries, appeared in the SE the reflection from the *dewy haze* (dew now falling in abundance) which resting on the horizon by an extensive violet or purplish base, rose diminishing, till it terminated in a rose-coloured apex. As this went off, an orange tint in the NW spread and deepened, and was suddenly surmounted by a fine blush of red, making out the remainder of a grand pyramid of coloured light, reaching half way to the zenith, and glowing every moment with increasing brightness: so that at half an hour after sunset, the landscape to the eastward was actually in a much stronger light than immediately after the disappearance of that luminary. By the time that this was in perfection, the opposite pyramid was no longer visible: and, the twilight beginning to go off, the *red* first vanished, the orange rather spreading upward, at the same time that together with the white light, or ground of the picture, it grew every minute fainter: it was not however wholly extinct at half-past ten, or two hours from the commencement at sunset.

Singular Meteor.

The same night at ten, after several small meteors (or shooting stars) a more considerable white one passed the zenith from SE to NW, in a kind of spiral or twisted line, wasting away as it went, leaving a long train of whitish sparks ; and, unless my sense was deceived by a strong association of ideas, *hissing* like a rocket in its course. The impression it left on my mind was altogether that of a solid ignited projectile.

Summer Lightning.

It is a popular error, very commonly entertained, that on fine summer evenings there is sometimes a harmless kind of lightning, without the usual accompaniments of dense clouds and electrical explosions. The mistake has originated in the great distance at which lightning may be perceived in a dark night.

Seventh Mo. 31. I perceived much faint lightning in the SE, although it was bright starlight, and not a cloud visible at the time. On communication with my brother, who was then at Hastings, he informed me, that they had on the above mentioned evening a heavy thunder storm in view for some hours, ranging, as he conceived, *in a line between Dunkirk and Calais on the opposite coast.* It is probable therefore that the greater part of the discharges, the faint light of which was perceptible at Tottenham, were actually made at the distance of 120 miles. I saw however one stroke with the usual linear zig-zag appearance, which I judged to proceed from the earth to the clouds, and which may have been *a returning stroke* far on this side of the storm.

THUNDER STORMS.

During a violent thunder storm, at Rolleston, Staffordshire, *ten head of deer* belonging to Sir Oswald Moseley, were killed by the lightning.

The neighbourhood of *Diss*, Norfolk, was visited by a short but severe storm of thunder and lightning, about two o'clock on the 16th of July.—(PUB. LEDGER).

TABLE LXXXIV.

		1813.		Wind.	Pressure. Max.	Pressure. Min.	Temp. Max.	Temp. Min.	Evap.	Rain, &c.
7th Mo.	L. Q.	July	20	E	29·51	29·49	75°	56°	—	
			21	NW	29·55	29·51	68	57	—	
			22	SW	29·55	29·50	75	58	—	
			23	SW	29·50	29·42	75	53	—	1·00
			24	W	29·53	29·40	74	55	48	77
			25	SW	29·54	29·40	73	53	—	4
			26	W	29·78	29·54	69	53	—	32
	New M.		27	W	30·07	29·78	75	51	32	10
			28	W	30·17	30·07	77	50	—	
			29	E	30·07	29·95	78	54	—	
			30	S	29·91	29·80	82	60	56	
			31	NW	29·95	29·91	78	53	—	
8th Mo.		Aug.	1	SW	29·96	29·93	78	59	—	5
			2	W	29·93	29·81	79	56	—	
			3	SW	29·78	29·72	76	51	38	—
	1st Q.		4	NW	29·75	29·55	73	49	—	—
			5	S	29·53	29·42	71	49	—	26
			6	NW	29·85	29·53	73	54	—	1
			7	W	29·92	29·91	74	49	—	
			8	SW	29·89	29·87	75	53	50	6
			9	NW	30·10	29·89	70	49	—	—
			10	Var.	30·10	30·07	76	57	—	—
			11	SE	30·07	29·98	79	51	30	
	Full M.		12	SW	29·96	29·89	80	54	—	
			13	NW	30·08	29·95	72	48	—	
			14	NW	30·08	29·83	72	57	42	—
			15	NW	29·83	29·78	76	51	—	
			16	NW	29·82	29·80	70	50	—	
			17	W	29·82	29·78	76	52	—	
			18	NW	30·05	29·82	75	47	46	
					30·17	29·40	82	47	3·42	2·61

NOTES.—Seventh Mo. 23. Continued heavy rain for above two hours p.m. with distant thunder. (This was about Bromley, in Kent, where a house was struck with lightning.) 24. A heavy thunder shower about 3 p.m.: rainbow. 25. Windy, cloudy, rainbow: broad diverging shadows on a coloured twilight, with *Cirrostratus* and haze to the S. 26. Fair a.m.: at noon began a steady rain with distant thunder: in the evening, several distinct *Nimbi*, in particular a well formed one in the NE. 27. *Cumulostratus* a.m.: rain p.m., with distant thunder: evening distant *Nimbi*, and a rainbow: much colour, with broad shadows in the twilight. 28. Much dew: *Cumulus*, with

Cirrus: at sunset a calm air, with large plumose *Cirri*, highly coloured. 29. A clear day, the wind passed from S to E: twilight brilliant, with dew: the New Moon shewed a well defined disc at 8h. 30m. p. m. 30. *Cumulus,* with *Cirrus* passing to the inferior modifications: in the evening, on the S horizon, *Cumulus,* mixed with *Cirrostratus* and haze, the twilight of a pink colour: it lightened frequently before ten at night very far in the SE, with the wind S.

Eighth Mo. 1. Rain at 5 a. m. succeeded by a close canopy of *Cumulostratus:* at sunset, *Cirrus* with *Cirrocumulus:* twilight opake, somewhat orange coloured. 2. Much the same phenomena as yesterday. 3. Some drizzling showers, with wind a. m.: sunset very dark, the sky being full of broken *Cumuli:* night windy. 4. Windy a. m., with *Cumulus,* which, p. m. inosculated with *Cirrostratus* above it. 5. Rain early, the wind S: in the evening (after several showers) clouds in various modifications, the wind W, with lightning to the S. 6. Much wind at NW, with *Cumulus:* a shower p. m. 8. Close *Cumulostratus* most of the day: rain, evening. 9. Wind brisk at NNW, a. m.: at noon, the upper clouds were perceived not to move with this wind, and at evening it fell calm: there were now in the sky rose coloured *Cirri* in stripes, from SE to NW, with *Cirrostratus* and *Cumulostratus* in a lower region: twilight orange, surmounted with rose colour. 11. A *Stratus* after sunset, with *Cirrostratus* remaining above: small scintillant meteors now appeared, falling almost directly down, and seeming to originate very low in the atmosphere. 13. *Cirrus* and *Cirrocumulus* abounded: there was a slight shower about noon. 14. Overcast, a little rain after sunset. 16. The *maximum* of the temperature for 24 hours past occurred at 9 a. m. 17. Overcast: windy.

RESULTS.

Prevailing Winds Westerly.

Barometer: Greatest height ....30·17 in.

Least........................29·40 in.

Mean of the period............29·799 in.

Thermometer: Greatest height................ 82°

Least........................ 47°

Mean of the period............63·88°

Evaporation........................ 3·42 in.

Rain............................ 2·61 in.

THUNDER STORMS.

July 23.—Between one and two, the city of *Glasgow* was visited by a heavy shower of rain, accompanied by tremendous peals of thunder.

Between three and four o'clock of the afternoon of July 24, *Scarborough* was visited with most tremendous thunder and lightning.—(Pub. Ledger).

[Description of a Hurricane, from a Bermuda Paper.]
" Nassau, (Bahama,) August 1, 1813.

" The dawning of Monday the 26th ult. exhibited a serenity, calculated to lull to sleep the fears of the most wary ; and the breeze freshening on the sky, it was hailed by all as a happy relief from the extreme sultry heat of the atmosphere.—

" At ten o'clock the wind increased, and continued increasing, accompanied by short showers of rain. It gained considerably in the course of an hour; at 11 it blew a strong gale, and some of the shipping in the harbour appeared uneasy at their anchorage ; but it was not until 12 o'clock that it attained the height which constitutes the commencement of an hurricane, and which soon became evident by its destructive effects upon the waters, and upon the shore. Some of the vessels in the harbour were driven from their moorings, and houses began to totter upon their foundations. At about half-past two o'clock the hurricane attained its greatest height, and its *acme* continued, without interval, until five, when it suddenly ceased ; and *in the space of half an hour succeeded a calm,* so perfect, that it can be compared only to that of death after the most dreadful convulsions.

" The Government-house, the greater part of the other public buildings, a great number of other houses, the wharfs, the orchards, and gardens were found either wholly or partly destroyed ; and all the vessels in the harbour (two excepted) were driven on shore or sunk.

" The inhabitauts of the colony, well knowing the nature of hurricanes, took every precautionary measure within their reach during the calm or lull, *to prepare for its second part, expected from the SW,* and which set in with great fury at about six o'clock, and continued until midnight, when it considerably abated, and soon after totally ceased. The SW storm differed from the north-eastern one, by appearing *in heavy blasts of a few minutes' duration,* repeated after lulls of equal length, until it so ceased : whereas the first storm *raged without intermission.* This last, however, nearly completed the general ruin, and it is believed that if it had raged another hour, scarcely a house would have remained standing in this city ; which, before the storm, was considered, in proportion to its size and population, to be one of the most wealthy and most flourishing in the world."

Hurricanes appear to have been felt about this time, through the whole extent of atmosphere between the two great divisions of the American continent. I have selected this from a number of reports in the papers, for the purpose of attempting to explain some peculiar and characteristic facts contained in the narrative.

The present storm seems to have consisted chiefly in a prodigious aggravation (for the time) of the velocity and force of the currents, which ordinarily proceed to, and return from the great equatorial stream. After an obstruction for some considerable time of this interchange of air, causing oppressive heat at the Bahamas, rarefaction suddenly ensues to the SW of these islands; a portion of the tropical air is rapidly elevated, and the air which supplies it place from the NE, from some cause not as yet apparent, flows in a *confined and accelerated* stream. That this stream, however, originates in the same cause with the ordinary trade wind, and is destined to supply the void caused by rarefaction to leeward, is evident, for it commences with the rising sun, assumes its greatest force at the time when the heat of the day is established, and ceases in the evening. This part of the hurricane blows with steady unintermitting violence, in which respect it agrees with our own north-easterly gales. When it is spent, *a stratum of air*, which had been interposed between it and the returning tropical current, suddenly shews itself; and, having been balanced between these opposite impulses, obeys neither, but appears (as it is) at rest as to horizontal movement. But this calm is of short duration: the superior SW wind comes down, and the moment it touches the earth, the storm is re-established. A difference however is now observable in the movement: for this inundation from the south, being urged by an unsteady force from behind, comes on in successive billows, like the tide making on the shore; and in this respect agrees precisely with our own southerly storms. At length, this current having spent on the earth the momentum it had acquired from the circumstances of the preceding day, it also ceases; and things return to their ordinary state, until the next occasion of obstructed, and subsequently accelerated, movement in the currents.

TABLE LXXXV.

1813.			Wind.	Pressure. Max.	Pressure. Min.	Temp. Max.	Temp. Min.	Evap.	Rain, &c.
8th Mo.	L. Q.	Aug. 19	N	30·14	30·05	71°	47°	—	
		20	N	30·14	30·11	70	40	—	
		21	NW	30·11	29·83	68	50	26	—
		22	NW	29·91	29·71	65	44	—	17
		23		30·23	29·91	67	46	—	
		24	NE	30·25	30·23	69	42	24	
		25	NE	30·25	30·23	70	52	—	
	New M.	26	NE	30·23	30·15	69	48	—	
		27	NW	30·15	30·10	68	51	25	—
		28	N	30·10	30·06	66	53	—	15
		29	NE	30·22	30·10	69	52	—	4
		30	NE	30·26	30·26	67	53	—	
		31	E	30·26	30·05	70	53	32	
9th Mo.		Sept. 1	SE	30·05	29·85	65	56	—	—
	1st Q.	2	SW	29·90	29·85	67	48	—	—
		3	S	29·95	29·85	75	58	—	
		4	S	29·85	29·75	73	60	—	—
		5	SW	29·75	29·25	70	56	—	69
		6	SW	29·49	292·7	68	52	61	—
		7	W	29·67	29·49	64	44	—	—
		8	NW	29·85	29·67	56	43	—	—
		9	NW	30·19	29·85	50	41	—	
	Full M.	10	NW	30·24	30·19	61	49	—	
		11	SW	30·24	30·10	70	58	55	
		12	SW	30·06	30·00	72	51	—	
		13	NW	30·06	30·05	63	51	—	18
		14	NW	30·18	30·06	64	42	—	
		15	SW	30·18	30·17	70	51	—	
		16	NW	30·29	30·17	72	50	40	
				30·29	29·25	75	40	2·63	1·23

Notes.—Eighth Mo. 19. *Cirrus* and *Cirrostratus* clouds : rather windy. 20. *Cumulostratus :* windy : *Cirrus* clouds, very red at sunset. 21. Windy. 22. Cloudy morning, small rain : showers and wind : rainbow about 6 p. m., the sky richly coloured, and the clouds evaporating. 24. *Cumulus* during the day : *Cirrus* at sunset : twilight brilliant and coloured, with traces of *Cirrocumulus* and of *Stratus.* 25. Overcast with *Cumulostratus :* twilight opake and coloured. 26. Windy a. m. : *Cumulostratus* clouds, the remains of which at sunset, glowed with a succession of crimson and purple tints on a full orange ground. 27. Windy a. m., a little rain. 28. Overcast : much wind, and at night rain. 29. *Cumulostratus* clouds chiefly: a shower or two: the twilight luminous, but opake and surmounted by a blush of red considerably elevated. 30. Cloudy a. m.

Ninth Mo. 1. Slight shower at evening. 2. Very cloudy morning. 5. Heavy rain after 6 p. m., lunar halo. 6. Rainy morning: high wind. 7. Much wind still: showery. 9. Very fine moonlight night. 12. Abundance of *Cirrocumulus*, gradually lowering, and arranged in close lines from SE to NW. 14. A shower in the evening. 16. About 5 p. m. a solar halo of short continuance: the sky at sunset was (as usual of late) much coloured: there was considerable diffused redness above the twilight, and some portions of the clouds seen against this, varied from the usual indigo colour to a pale olive green: an indistinct appearance of *Nimbus* in the E horizon.

RESULTS.

Prevailing Winds Northerly, with an interruption of some days continuance from the Southward, producing for the time a considerable depression of the barometer, together with elevation of the mean temperature, and rain.

Barometer: Greatest height .... 30·29 in.
Least........................ 29·25 in.
Mean of the period........... 30·009 in.
Thermometer: Greatest height .... 75°
Least...................... 40°
Mean of the period .... 58·44°
Evaporation...................... 2·63 in.
Rain .... 1·23 in.

TABLE LXXXVI.

	1813.	Wind.	Pressure. Max.	Min.	Temp. Max.	Min.	Evap.	Rain, &c.
9th Mo.	L. Q. Sept. 17	SE	30·22	30·11	72°	56°	—	
	18	SE	30·11	30·01	70	46	—	
	19	E	30·01	29·92	72	43	—	
	20	NE	29·90	29·89	66	45	—	
	21	NE	29·91	29·87	70	50	29	—
	22	N	29·96	29·90	68	48	—	6
	23	NE	30·01	29·96	64	54	—	14
New M.	24	NE	30·11	30·01	66	47	—	
	25	NE	30·11	30·03	62	51	20	
	26	NE	30·03	30·01	67	47	—	
	27	E	30·01	29·94	66	44	—	
	28	NE	———	29·94	60	47	—	—
	29	NE	30·12	———	62	47	21	
	30	NE	30·12	29·84	62	40	—	
10th Mo.	Oct. 1	NE	29·84	29·68	61	41	—	
1st Q.	2	NE	29·70	29·67	59	49	—	—
	3	SE	29·80	29·70	62	48	21	—
	4	SE	———	—	62	52	—	—
	5	W	29·78	29·65	66	54	—	
	6	SW	29·80	29·77	66	52	—	
	7	SW	29·52	29·47	65	51	—	
	8	W	29·73	29·46	60	54	—	2·50
	9	Var.	29·43	29·42	61	49	—	
Full M.	10	SW	29·44	29·15	65	45	—	4
	11	W	29·46	28·93	60	48	—	—
	12	SW	29·75	29·55	60	41	—	—
	13	SW	29·55	29·34	58	52	—	45
	14	NW	29·62	29·52	58	34	—	29
	15	S	29·58	29·21	51	33	28	47
			30·22	28·93	72	33	1·19	3·95

NOTES.—Ninth Mo. 19, 20. Breezes by day: much dew by night. 21. Cloudy quite to sunset: a few drops of rain. 22. A breeze a. m.: some clouds followed by a sudden shower p. m.: rain in the night. 23. Windy: showers. 24. Much wind. 25. Windy: cloudy. 26. Overcast a. m., clear p. m.: luminous twilight, with *Cirrus* and *Cirrocumulus*. 27. Morning twilight somewhat coloured but less luminous: forenoon overcast: p. m. clear: at sunset fascicular *Cirri*, arranged from W to E, the wind E, nearly calm: after these appearances, lightning far to the S, E, and SW. 28. Cloudy, with a few drops.

30. A pink twilight, with dense coloured *Cirri.* For three days past, a steady NE breeze, with pretty much sunshine.

Tenth Mo. 1. Overcast a.m., wind N: after sunset *Cirrocumulus* passing to *Cirrostratus,* a corona about the Moon, and a small meteor passing westward. 2. Overcast most of the day: a few drops p.m. 3. *Cirrus* with *Cumulostratus:* twilight opaque and orange coloured: the roads have of late become excessively dry, and the dust raised from them floats in great quantity in the air. 4. Early this morning began a steady rain which continued till after sunset. 5. Fine day: lunar halo at night. 6. Cloudy. 7. A considerable storm of thunder and lightning early this morning, followed by much rain. 8. Fair a.m., wet p.m. 9. Fine. 10. Wet with a fair interval about noon. 11. Wet a.m., fair p.m. 12. Fine a.m., showery p.m. 14. Fair. 15. Rainy.

RESULTS.

Prevailing Winds Easterly, and drying to the Moon's first quarter: after which they became Westerly, and brought much rain.

Barometer:	Greatest height	30·22 in.
	Least.....................	28·93 in.
	Mean of the period...........	29·752 in.
Thermometer:	Greatest height..............	72°
	Least.....................	33°
	Mean of the period...........	55·28°
	Evaporation	1·19 in.
	Rain	3·95 in.

The rain of the 4th of tenth month having put a period to a fair season of some weeks continuance, I availed myself of a journey I made immediately after it, to endeavour to ascertain by inquiry how far it had extended. I found it had rained on that day from morning to night through the whole distance from London to York; as likewise further north, quite to the Tyne, and across the island from Cheshire to Northumberland. It having been likewise a very wet day on the south coast, the conclusion seems to be that the whole of England, at least, was on this occasion irrigated at once, by the introduction of a current from the Atlantic, which now *displaced* completely the easterly breeze previously deflected from the NE to the SE; both currents on this occasion probably depositing the excess of water with which they were charged.

WEATHER AT GIBRALTAR.

Gibraltar, October 5.—This has been the coolest summer ever remembered, the rains have fallen early; and the weather now is not warmer than in England in the autumn.—The mornings, for these some days have been cold enough to render a fire agreeable, with a *NW* wind, and the most delightful, pure, and serene sky that can possibly be imagined.

TABLE LXXXVII.

1813.			Wind.	Pressure.		Temp.		Evap.	Rain, &c.
				Max.	Min.	Max.	Min.		
10th Mo.	L. Q.	Oct. 16	W	29·20	28·74	51°	38°	—	19
		17	SW	29·01	28·64	52	42	—	8
		18	NW	29·55	29·44	58	35	—	
		19	Var.	29·66	29·55	50	27	—	
		20	E	29·66	29·50	52	42	12	14
		21	Var.	29·76	29·50	54	43	—	
		22	NE	29·89	29·45	53	40	—	10
		23	E	29·91	29·86	52	51	12	
	New M.	24	NE	29·86	29·82	57	45	—	
		25	NE	30·12	29·82	53	37	—	—
		26	NE	30·12	29·85	47	36	—	
		27	NE	29·79	29·77	41	34	17	6
		28	N	29·90	29·79	46	33	—	
		29	E	29·90	29·70	47	27	—	—
		30	Var.	29·70	29·00	52	30	—	50
		31	W	29·25	29·00	56	30	—	—
11th Mo.	1st Q.	Nov. 1	W	29·68	29·25	49	31	—	
		2	NW	29·83	29·41	49	35	8	9
		3	SW	30·32	29·83	51	29	—	
		4	W	30·34	30·32	45	27	—	
		5	S	30·32	29·95	47	29	—	
		6	E	29·95	29·70	45	35	14	6
		7	SW	29·70	29·37	53	41	—	17
	Full M.	8	SW	29·42	29·37	58	43	—	22
		9	W	29·62	29·42	51	42	—	—
		10	W	29·60	29·48	57	43	—	10
		11	SW	29·65	29·51	54	44	17	—
		12	W	29·49	29·45	54	32	—	35
		13	SW	29·58	29·49	46	29	—	
		14	W	29·58	29·17	43	33	3	8
				30·34	28·64	58	27	0·83	2·14

NOTES.—Tenth Mo. 16. Fine morning: wet p. m.: lightning in the evening. 17. Showery. 18. Fine a.m., shower p.m. 19. Hoar frost: fair day after misty morning. 20. Cloudy a.m., much wind at E. 22. About half-past 7 p.m. a bright blue meteor appeared in the N, and passing to westward with a steady and rather slow motion, became extinct: there were traces of *Cirrostratus* clouds, which increased afterward. 23. Maximum of temperature at 9 a.m.: cloudy, with a breeze. 24. Overcast. 25. Cloudy a.m., clear p.m.: during the twilight there was abundance of red haze, first in the E horizon over clouds in that quarter, then at a considerable elevation in the W: it ended more clear and orange coloured. 26. Cloudy at intervals. 27. Windy a. m. 28. Hoar frost, which being examined, was found to consist not of spiculæ attached to the herbage, but of the drops of dew frozen clear and solid. 29. Hoar frost: *Cumulostrati* followed by *Nimbi:* one of the latter approaching from the E at 4 p. m., exhibited a double rainbow, on a ground of purple. 30. Spicular hoar forst: very misty: clear at noon; *Cirrostratus* and a little rain p. m.: wet and stormy night. 31. Windy a. m.: *Cirrus* with *Cumulus:* a shower about four, with a fine bow.

Eleventh Mo. 1. Hoar frost. 2. Granular hoar frost: very clear sunrise: clouds at noon: rain p. m., very windy night. 3. Sunshine, a. m. wind NNW, a shower p. m. 6. Calm clear weather, with hoar frost these three days, the wind now rising. 7. Small rain at intervals: a *solar halo* of large diameter p. m. 8. Fair with wind a. m., but before 4 p. m. dark *Nimbi* and rain beginning: being then on the south side of London, I was surprised with a flash of lightning and a sharp peal of thunder: about half-past six, a glimpse of a meteor passing to the W: it was said to lighten after this time. 9. Windy. 11. Stormy night. 12. Windy: cloudy p. m., wet. 13. Hoar frost. 14. The same, with a crystallized rime on the shrubs.

RESULTS.

Winds Variable: the Easterly prevailed in the former, the Westerly in the latter part.

Barometer: Greatest height................30·34 in.
　　　　　　Least........................28·64 in.
　　　　　　Mean of the period...........29·625 in.
Thermometer: Greatest height...............　58°
　　　　　　　Least......................　27°
　　　　　　　Mean of the period..........43·41°
　　Evaporation........................　0·83 in.
　　Rain ....　2·14 in.

THUNDER STORMS, &c.

Between three and four o'clock on Tuesday morning (Nov. 9), Brighton was visited by a dreadful storm of wind and rain, accompanied by some severe strokes of thunder and vivid flashes of lightning.

Extract of a letter from an Officer serving in the Mediterranean:—" On the 2d of September, the Hibernia was struck by lightning, which *set fire to the main-top-mast and fore-rigging,* but the rain falling in torrents at the time, shortly extinguished it: five men were knocked down, but none killed. The Swiftsure, Union, Ocean, and Leopard, *all had their main-top-masts shattered to pieces.* Not a life was lost on board any of them, although many men were struck down and slightly hurt."—(PUB. LEDGER).

2 F 2

TABLE LXXXVIII.

1813.			Wind.	Pressure. Max.	Pressure. Min.	Temp. Max.	Temp. Min.	Evap.	Rain, &c.
11th Mo.	L. Q.	Nov. 15	NW	29·38	29·20	47°	34°	—	
		16	NE	29·10	29·02	45	35	—	13
		17	NW	29·46	29·02	43	30	—	—
		18	W	29·65	29·46	44	32	—	
		19	SW	29·72	29·65	51	37	10	15
		20	S	29·86	29·72	56	49	—	
		21	S	29·88	29·86	55	47		
	New M.	22	NW	29·93	29·88	55	37	—	
		23	E	29·98	29·93	45	38	7	
		24	NE	30·08	29·98	45	33	—	—
		25	N	30·08	30·05	43	37	—	
		26	NE	30·05	30·02	45	31	—	
		27	NE	30·02	29·88	43	28	—	
		28	E	29·95	29·85	37	34	—	
		29	E	29·90	29·85	40	25	—	
		30	SE	29·85	29·50	35	29	—	
12th Mo.	1st Q.	Dec. 1	E	29·50	29·42	37	33	—	
		2	SE	29·42	29·09	40	37	12	12
		3	SE	29·37	29·09	43	40	—	—
		4	NE	29·44	29·37	44	40	—	—
		5	N	29·76	29·44	44	36	—	—
		6	N	29·78	29·77	44	36		20
	Full M.	7	NE	29·76	29·72	44	41	2	3
		8	NE	29·86	29·72	44	40	—	—
		9	NE	30·05	29·86	44	39	—	—
		10	NE	30·18	30·05	43	37	4	14
		11	NE	30·18	30·05	42	37	—	
		12	SE	30·05	29·82	37	29	—	
		13	NW	29·93	29·82	37	26	2	
				30·18	29·02	56	25	0·37	0·77

Notes.—Eleventh Mo. 16. A stormy night, after some rain.
17. Much wind a. m.: snow from 2 to 3 p. m., which remained
through the night. 18. A fine day: at sunset much rose colour, with
Cirrostratus. 19. Overcast at 8 a. m., the barometer falling: misty
and drizzling. 22. Grey morning: fair day. 23. Misty, calm.
24. Overcast, drizzling. 25—27. Fine days: at sunset on the 27th,
the wind being SE, the smoke of the city, in passing away in a body,
swelled up into several distinct heaps, each of which inosculated at
its summit with a small cloud : this union was clearly the result of mu-
tual attraction, and it continued for great part of an hour : the clouds

which at this time overspread the sky in great numbers were the remains of larger ones, which had in part evaporated, and now resembled
Cirrostrati: those which were attached to the smoke became sensibly
denser than the rest. 30. This month exhibited an unusual proportion
of fine days.

Twelfth Mo. 2. Cloudy: a steady breeze: some light snow and
sleet: rain in the night. 3. Overcast: some rain. 4, 5. Cloudy,
misty: rain at intervals. 8. Small rain a. m.

RESULTS.

Prevailing Winds Easterly.

Barometer: Greatest height . . . . 30·18 in.
Least. . . . . 29·02 in.
Mean of the period . . . . 29·728 in.
Thermometer: Greatest height . . . . 56°
Least. . . . . 25°
Mean of the period . . . . 39·63°
Evaporation . . . . 0·37 in.
Rain . . . . 0·77 in.

The latter part of this period has been remarkable for a general
prevalence of the diseases commonly attributed to obstructed perspiration. The detail of these belongs properly to the medical reports:
it may, however, be suitable here to point out the circumstances
which appear to have contributed to this effect. I say *appear*, because
there may exist modifications of the air capable of exciting disease,
to the discovery of which none of the present means of examination
are competent.

The wind, during the time alluded to, came in a moderate stream
from the eastward; the barometer, which had been depressed, rising
gradually. The sky was almost constantly overcast with *Cirrostratus*;
beneath which the atmosphere was perceptibly full of diffused water,
of the density of *dew*, quite down to the earth. The sun's rays thus
intercepted, the temperature varied little from 40° by day or night;
and evaporation was nearly at a stand. Electricity in such circumstances could not accumulate: hence, though it drizzled at intervals,
the air never got cleared by showers.

Let us now apply these facts to the case. In health, the matter of
perspiration is thrown out on the skin in a fluid state, in quantity
and with a force proportioned to the state of the circulation.
Here it has to evaporate in the common way of fluids; but it
will do this very slowly, even at the common temperature of the skin,
in an air already loaded with moisture. In such an air, too, the whole

muscular system being relaxed, the heart and arteries act with less force. If it be, at the same time, only moderately cold, and void, in great measure, of light and electricity, it will want that exciting action on the nerves, which results from the *sudden* loss of heat, as well as from the application of the two latter *stimuli*. In a word, such an air, succeeding to a dry and clear air, will be a *sedative*: it will counteract, by constant, though insensible, effects, the healthy energies even of the strong. The usual *sponging* operation on the skin being withheld, at the same time that the *vis a tergo* is lessened, we need not wonder that the matter of perspiration should stagnate in the fine extremities of the cuticular arteries, or be thrown on some internal secreting surface; that the skin should take on a state of spasm, and that fever and local inflammation should ensue. Thus, without supposing in this case any occult quality in the air, we may trace a combination of qualities which may be reasonably thought to have made it productive of disease.

THUNDER STORMS, &c.

The Barfleur of 98 guns, Sir E. Berry, was struck by lightning, the latter end of October, in the Mediterranean. The fluid struck the foretop-gallant-mast, which it shivered to pieces; descended the foretop-mast and foremast, and proceeding through all the decks, tore up part of the lead at the light-room door, which is situated close to the magazine!—The destruction of the ship, however, was providentially averted.

November 17.—A Correspondent at Bridgewater says: " Wednesday last we were visited by a tremendous thunder storm: the lightning struck the spire of the church, and shivered it to pieces, the windows were broken, and the tiles were forced off the houses in every quarter. At Weston Zoyland, a short distance from this, the lightning much damaged the tower of the church; the windows were broken, and the church soon filled with thick smoke, which smelt so sulphureously, that the inhabitants throughout the town were greatly alarmed; the battlements were entirely thrown down, and some of the stones thrown to a distance of 60 yards.

PLYMOUTH, *November* 17.—It has blown a very heavy gale all last night and to-day from NW.

SCILLY, *December* 5.—On the 1st instant we experienced the most violent gale ever remembered, from the SE, but it only lasted a short time. Three vessels in the Roads parted from their cables—no other accident.

(PUB. LEDGER).

INSTANCE OF EXTRAORDINARY DROUGHT.

The island of Aruba, says a letter from Curaçao, of November 3, is at present in a most deplorable condition, on account of the scarcity of rain. *Within the last nine months only five showers have fallen;* and they have been so gentle, as to have had no visible effect on the parched soil. The trees on the island are withered and no longer bear; all traces of vegetation are gradually disappearing; and the efforts to raise corn, or esculent roots, have proved fruitless. A dreadful mortality had taken place among the cattle, as the few springs from whence they had been supplied were exhausted, and there was neither fodder nor water to keep them alive. The principal inhabitants had abandoned the island.

A letter from Curaçao, dated November 5, says:—" During the last ten months there has been no rain in this colony; a famine among the horned cattle has been the natural consequence. More than 1700 head have been lost on some plantations. The lower classes and the slaves have been exposed to the prospect of famine.—(PUB. LEDGER, *February*, 1814).

REPORTED LOSS OF A CARAVAN BY THE KAMSIN, OR HOT WIND OF THE DESERT.

The following is an extract of a letter from Smyrna:—" We have received intelligence of a dreadful calamity having overtaken the largest caravan of the season, on its route from Mecca to Aleppo. The caravan consisted of 2,000 souls—merchants and travellers from the Red Sea and Persian Gulph, pilgrims returning from performing their devotions at Mecca, and a numerous train of attendants, the whole escorted by 400 military. The march was in three columns.—On the 13th of August last they entered the great Arabian Desert, in which they journeyed seven days, and were already approaching its edge. A few hours more would have placed them beyond danger; but alas! they were not permitted to return in safety. On the morning of the 23d, just as they had struck their tents, and commenced their march, a wind rose from the NE, and blew with tremendous violence. They increased the rapidity of their march to escape the threatening danger; but the fatal Kamsin had set in. On a sudden, dense clouds were observed, whose extremity obscured the horizon, and swept the face of the desert. They approached the columns, and obscured the line of march. Both men and beast, struck by a sense of common danger, uttered loud cries. The next moment they fell, beneath its pestiferous influence, lifeless corpses. Of 2,000 souls composing the caravan, not more than 20 escaped this calamity.—They owed their safety to the swiftness of their dromedaries."—(PUB. LEDGER, *December*, 1813).

TABLE LXXXIX.

1813.			Wind.	Pressure.		Temp.		Evap.	Rain, &c.
				Max.	Min.	Max.	Min.		
12th Mo.	L. Q.	Dec. 14	NW	29·93	29·87	35°	25°		
		15	NW	29·87	29·49	38	26		
		16	SW	29·49	29·35	49	40		—
		17	S	29·35	29·14	51	50	2	19
		18	SW	29·16	29·14	54	44		—
		19	SW	29·28	29·16	50	35		
		20	W	29·57	29·28	45	31		
		21	NE	29·50	29·44	46	32	5	12
	New M.	22	SW	29·77	29·50	47	36		
		23	SW	29·89	29·77	47	38		—
		24	SW	30·02	29·89	53	46		—
		25	SW	30·28	30·02	51	41	1	9
		26	NW	30·49	30·28	41	28		
		27		30·49	30·38	31	25		
		28		30·35	30·31	30	24		
		29		30·36	30·35	30	19		
	1st Q.	30	NW	30·36	30·32	32	22		
1814.		31	N	30·32	30·17	35	22		
1st Mo.		Jan. 1		30·17	29·86	31	20		
		2		29·86	29·71	32	28		
		3		29·71	29·60	33	29		
		4	E	29·60	29·20	33	25		
		5	NE	29·12	29·08	33	32		
	Full M.	6	N	29·52	29·12	34	15		
		7	NW	29·66	29·60	28	11		
		8	NW	29·65	29·62	31	12		
		9	NW	29·79	29·65	29	8		
		10	NW	29·89	29·79	26	21		
		11	SE	29·89	29·54	25	15		
		12	N	29·88	29·49	27	15	6	1·28
				30·49	29·08	54	8	0·14	1·68

NOTES.—Twelfth Mo. 14, 15. Hoar frost: clear days. 16, 17. Cloudy: rain at intervals: bees quit the hive in unusual numbers for the season. 18. Windy: some rain. 19. Very misty a. m. 20. Hoar frost. 21. The same, followed by wind and rain. 23—25. Misty: drizzling rain at intervals. 26. A clear morning, with much dew: the barometer rising fast: *Cirrus* and *Cirrostratus:* orange coloured twilight. 31. Since the 26th we have had a succession of thick fogs, with a calm air, or at most a breeze from the NE: yesterday the air cleared a little, and to-day has been fine: a display of *Cirrus* clouds, with much red in the morning and evening sky: the peculiar smell of electricity has been perceptible of late, when the air cleared up at sunset.

1814. First Mo. 4. The mists, which have again prevailed for several days, and which have rendered travelling dangerous, are probably referable to the modification *Stratus*. The air has been, in effect, loaded with particles of freezing water, such as in a higher region would have produced snow. These attached themselves to all objects, crystallizing in the most regular and beautiful manner. A blade of grass was thus converted into a pretty thick *stalagmite*: some of the shrubs, covered with spreading tufts of crystals, looked as if they were in blossom; while others, more firmly incrusted, might have passed for gigantic specimens of white coral. The leaves of evergreens had a transparent varnish of ice, with an elegant white fringe. Lofty trees, viewed against the blue sky in the sunshine, appeared in striking magnificence: the whole face of nature, in short, was exquisitely dressed out in frost work. When the sun, at length, broke through, and loosened the rime, it fell unmelted, and lay in heaps under the trees; after which a deep snow, brought by an easterly wind, reduced the whole scenery to the more ordinary appearances of our winter. 5. Snow early, and during the day; the wind increasing in force from the NE. 6. A dark morning. Snow falling in some quantity to-day, with the temperature at the surface 33° or 34°, presented an amusing phenomenon, which was pointed out by my children. Instead of driving loose before the wind, it was collected occasionally into a ball, which rolled on, increasing till its weight stopped it: thousands of these were to be seen lying in the fields, some of them several inches in diameter. 9. A somewhat misty morning: the snowy landscape, visible to the distance of about a mile only, exhibited a bluish tint. A thermometer, *placed on the snow*, fell this night to 6°: and I am informed that at Croydon a temperature of *5°* was observed by a thermometer at the usual elevation from the ground: the time 11 p.m. 11. Very red sunrise: a steady breeze at SE till evening. Minimum temperature on the snow 11°. 12. *Cumulus* and *Cirrostratus* clouds. Minimum temperature on the snow 12°. The River Lea is now firmly frozen, and the Thames so much encumbered with ice that navigation is scarcely practicable.

RESULTS.

Winds Variable: a considerable interval calm.

Barometer: Greatest height................30·49 in.
Least.......................29·08 in.
Mean of the period...........29·757 in.

Thermometer: Greatest height............... 54°
Least.......................   8°
Mean of the period...........32·36°

Evaporation....................... 0·14 in.
Rain, (with melted snow and rime)...... 1·68 in.

There was an eruption of Mount Vesuvius, on the 25th and 26th of the twelfth month.

TABLE XC.

	1814.		Wind.	Pressure. Max.	Min.	Temp. Max.	Min.	Min. on the Snow.	Rain &c.
1st Mo.	L. Q.	Jan. 13	NE	30·13	30·03	30°	14°	8	
		14	NE	30·03	29·61	26	19	14	
		15	E	29·61	29·30	31	20	21	
		16	NE	29·63	29·25	32	22	19	—
		17	N	29·25	29·14	30	11	8	—
		18	E	29·14	29·07	36	30		—
		19	NE	29·42	29·07	34	28		—
		20	NE	29·76	29·42	33	14	6	—
	New M.	21	Var.	29·76	29·72	25	14	11	—
		22	N	29·82	29·71	32	8	6	—
		23	N	29·77	29·69	35	15	11	—
		24	Var.	29·88	29·77	33	24		1·25
		25	Var.	29·88	29·60	36	20		—
		26	SW	29·60	29·22	36	33		12
		27	W	29·16	29·11	39	33		20
		28	Var.	29·11	28·54	40	28		—
	1st Q.	29	Var.	28·94	28·22	41	32		86
		30	W	29·32	28·94	40	25	20	
		31	NW	29·80	29·32	38	26		—
2d Mo.		Feb. 1	NW	30·01	29·90	36	26		6
		2	N	30·04	30·02	41	24	22	
		3	N	30·01	29·98	33	19	2	
	Full M.	4	W	30·15	30·11	32	19	12	
		5	SW	30·11	29·68	38	29		8
		6	NW	29·61	29·51	44	33		4
		7	W	29·65	29·48	40	32		—
		8	SW	29·70	29·36	50	35		5
		9	SW	29·94	29·70	47	40		
		10	SW	29·98	29·94	49	42		
		11	S	29·95	29·90	50	35		
				30·15	28·22	50	8	2	2·66

NOTES.—First Mo. 13. Much wind last night: very fine day: *Cumulus* and *Cirrostratus.* 14. Somewhat cloudy a. m. 15. Overcast with *Cirrostratus:* light breeze. There being no evaporation to-day, the surface of the snow is a little warmer than the air. 16. Overcast: a slight thaw, from the warmth of the earth: at evening snow and frost again. 17. A clear day: *Cirrus* and *Cirrostratus* in the evening, with a low *Nimbus*, or two, forming. At sunset, the temperature being 18°, there was a bed of haze in the E, like that which accompanies the fall of *dew*, but much more dense, and singularly coloured. It was chiefly of an indigo blue, but passed below into opake white,

NOTES, &c. *(continued)*.

and above into a faint red, more transparent. 18. A snowy morning:
small snow and sleet through the day. 19. A snowy day. 20. Snow:
the wind strong at NE. 21. At 8 a. m. beginning to snow again. In
the drifts the snow is many feet deep. This day proved fine and
calm: the wind, at times, SW. 22. A little snow in the morning:
fine day: strong breeze, with various clouds. 23. Snow, morning
and evening. There was a fine exhibition of clouds: as *Cumuli*, well
formed, but with little colour; the superior modifications, including
the *Cirrocumulus;* and several distinct *Nimbi*. 24. The sky as yesterday: snow, a. m. the wind brisk: about 2 p. m. a squall, with plenty
of snow. 25. A fine elevated grey sky, a. m.: then, *Cirrus* and *Cirrostratus*, the wind veering by N to W. It has had lately much tendency to this quarter at night. 26. A hollow wind at SW: snow,
followed by small rain. 27. A misty thaw: a little rain at intervals:
wind and rain in the night. 28. Misty a. m., with rain: the wind W:
after this a little snow: then fair, and frosty after sunset. 29. Stormy:
the wind SE with snow early: then steady rain, followed by more
snow. The *minimum* of pressure occurred after 6 p. m. It was not
confined to a small space of time. As the barometer began to rise,
the wind came round by SW to NW, and blew with great violence till
near morning. 30. A fine day: strong breeze, with clouds: at sunset
a sprinkling of opake hail. 31. Fair day: but at 7 p. m. a sudden
heavy snow shower.

Second Mo. 2. Hoar frost. 3. A very fine sunny day, followed
by bright moonlight. A thermometer, at an intermediate height, between the standard one and that on the snow, gave 14° as the minimum.
The latter instrument at 8 a. m. had risen only to 4°, and was thickly
covered with *rime*, although at night there had been an appearance of
strong evaporation. 4. Sunshine, and fair night. About 9 p. m. a
colourless halo, of the largest diameter, with *Cirrus* clouds. At half
past ten a small coloured halo, with *Cirrostratus*. The temperature
on the snow being 12°, the wind W, a breeze, I now exposed in a
metallic dish, close to the thermometer, 2600 grains of snow (which
had become hard by freezing) in two or three masses. At half
past eight, or in ten hours, the temperature having risen to 28°, it
had lost 27 grains in weight. 5. Crimson sky at sunrise: hollow
wind: snow and sleet. 6. A gale from SW, with showers of rain:
at evening it cleared up, and blew from NW. 7—11. *Cirrostratus*
and *Cirrocumulus:* red sky at sunrise: gales of wind, and a few
showers.

RESULTS.

Winds Northerly in the fore part of the period, variable in the middle, and in the conclusion Southerly.

Barometer: Greatest height................ 30·15 in.

Least........ ....28·22 in.

Mean of the period ....29·591 in.

Thermometer: Greatest height................ 50°

Least...................... 8°

On the snow, least............. 2°

Mean of the period ....31·31°

Evaporation (from water containing a little salt)............. .... 0·32 in.

Rain and melted snow................ 2·66 in.

The Thames Frozen Over.

January 15.—The masses of ice and snow had accumulated in such quantities at London Bridge, on the upper side yesterday, that it was utterly imposspossible for barges or boats to pass up.

January 20.—During the whole of this week, that part of the Thames below Windsor Bridge, called Mill River, has been frozen over, and has been crouded with persons skaiting.

February 3.—The confidence of the public in the safety of the passage over the frozen surface of the Thames was yesterday increased. All the avenues from Cheapside to the different stairs on the banks of the river were distinguished by large chalked boards announcing " A safe footway over the River to Bankside ; " and, in consequence, thousands of individuals were induced to go and witness so novel a spectacle, and many hundreds had, what we cannot help terming the fool hardiness, to venture on the fragile plain, and walk, not alone over, but from London to Blackfriars' Bridge. Several booths, formed of blankets and sail-cloths, and ornamented with streamers and various signs, were also erected in the very centre of the river, where the visitors could be accommodated with various luxuries. In one of the booths, the entertaining spectacle of a *sheep roasting* was exhibited.

February 7.—Friday several printers brought presses and pulled off various impressions, which they sold for a trifle :—

e. g. " Printed to commemorate a remarkably severe frost, which commenced December 27, 1813, accompanied by an unusual thick fog, that continued eight days, and was succeeded by a tremendous fall of snow, which prevented all communication with the Northern and Western Roads for several days. The Thames presented a complete field of ice between London and Blackfriars' Bridges, on Monday the 31st of January 1814.—A Fair is this day (February 4, 1814), held, and the whole space between the two Bridges covered with spectators."

This " field of ice " was, indeed, a very rugged one, consisting of

1

masses of drift ice of all shapes and sizes, covered with snow, and cemented together by the freezing of the intermediate surface. The deceitfulness of the latter caused (as is too common on such occasions) the loss of some lives by drowning. The following passage, announcing the opening of the river soon after, is worthy of preservation on account of the spirit in which it is written.

February 11.—We are happy to see the lately perturbed bosom of Father Thames resume its former serenity. The busy oar is now plied with its wonted alacrity, and the sons of Commerce are pursuing their avocations with redoubled energy. Cheerfulness is seated on the brow of the industrious labourer : those who were reduced to receive alms as paupers, again taste the sweets of that comparative independence with which labour crowns the efforts of the industrious. What a fruitful source of congratulation does this prospect afford ! nor can the contemplative mind dwell on the subject without feeling gratitude to that beneficent Being, who, in a time of such calamity, opened the hearts of the benevolent to administer, from their abundance, to the necessities of their poorer brethren, and thus add cement to the bond by which all mankind are linked together.

The mischief done on the river, during the late frost, is greater than can be remembered by the oldest man living. Among the craft alone, it is calculated to amount to upwards of 10,000*l.* independent of the damage sustained by the cables, tackle, &c. of the shipping.

January 11.—The quantity of snow which has fallen in the upper parts of Hampshire, and on the Hind Head, is very great, lying in many places fifteen feet deep.—(PUB. LEDGER).

Similar accounts of the accumulation of snow, more especially on the western side of the island, as also of the freezing of the large rivers, &c. abounded in the public prints, within two weeks from this date.

In Dublin a fog enveloped the city at the same time that it prevailed here, and was so extremely dense, that a great number of serious accidents were the consequence.

Dublin, January 13.—From the uncommon depth of the snow, the streets appeared yesterday almost deserted : none were to be seen abroad but an occasional messenger, a walking physician, a letter carrier, or a bank runner, scrambling through the snow to their several vocations.

February 21.—A letter from *Heligoland* mentions, that the intense frost had there, as in England, been preceded by thick fogs, and heavy falls of snow. The latter was 10 and 12 feet deep. The frost, which had lasted six weeks, had on the 8th, every appearance of continuance.—(PAPERS).

TABLE XCI.

	1814.		Wind.	Pressure, Max.	Pressure, Min.	Temp. Max.	Temp. Min.	Evap.	Rain, &c.
2d Mo.	L. Q.	Feb. 12	S	29·94	29·92	48°	39°	—	
		13	SE	29·91	29·87	46	37	—	10
		14	NE	30·07	29·91	41	29	—	
		15	NE	30·13	30·07	38	29	—	
		16	NE	30·40	30·13	39	28	—	
		17	NE	30·42	30·39	33	19	—	
		18	NE	30·39	30·25	39	30	—	—
		19	NE	30·41	30·25	40	23	—	
	New M.	20	Var.	30·41	30·35	31	18	—	
		21	SE	30·30	30·25	34	19	—	
		22	E	30·30	30·22	32	21	—	
		23	SE	30·22	30·15	32	18	—	
		24	E	30·20	30·12	33	18	—	
		25	SE	30·12	30·02	34	21	—	
		26	NE	30·08	30·00	35	24	—	
	1st Q.	27	SE	30·00	29·83	39	26	—	
		28	SW	29·83	29·12	41	30	25	3
3d Mo.		March 1	Var.	29·07	29·05	45	31	—	—
		2	SW	29·05	28·97	45	31	—	—
		3	E	29·28	29·05	42	30	—	
		4	NE	29·59	29·28	35	31	—	
		5	NE	29·88	29·59	34	28	—	—
	Full M.	6	NE	29·88	29·77	34	28	—	
		7	E	29·85	29·77	32	21	—	
		8	NE	29·85	29·76	33	26	—	
		9	NE	29·76	29·66	34	27	—	
		10	NW	29·66	29·58	35	29	—	—
		11	NE	29·70	29·45	41	32	—	
		12	NE	29·99	29·70	39	21	—	
		13	N	30·18	29·99	38	30	28	82
				30·42	28·97	48	18	0·53	0·95

NOTES.—Second Mo. 12. Cloudy. At half past 10 a. m. after a peculiarly unpleasant atmosphere, with a breeze from SE, we had a continued *shower* of fragments of *burnt paper*, descending from an elevation which the eye could not appreciate : it was found to originate from the burning of the Custom House, (distant in a right line about five miles S), at which there happened an explosion of gunpowder. 13. Misty morning. 14. *Cumulus* clouds, beneath *Cirrocumulus* and *Cirrostratus*. 15. Fine morning : *Cumulus :* breeze : evening twilight clear orange, and the stars brilliant. 16. Cloudy morning : strong breeze : fine clear day. 17. Fine day : *Cumulus* with *Cirrus*. 18. Much hoar frost: fine morning: some rain, evening. 19, 20. Hoar frost. 21. The same: the moon visible, and very well defined at 6 p. m. 22, 23. Hoar frost. 24. The same: *Cirrus* clouds, in parallel stripes from N to S all day: minimum temperature at the laboratory, Stratford, 15°. 25—27. Hoar frost: *Cirrus*, in stripes from N to S: lunar halo. 28. *Cirrostratus* appears: the stones grow moist, and the wind has a hollow sound.

Third Mo. 1. Damp and cloudy: hollow wind: sleet, p. m. 2. Rain and sleet at intervals. 3—8. Snow at intervals: the country has become again white with snow. 9. Snow, more plentiful, in the night. 10—12. Snow at intervals: which, in northern exposures, lies to some depth.

RESULTS.

Prevailing Winds Easterly.

Barometer: Greatest height . . . . 30·42 in.

Least. . . . . 28·97 in.

Mean of the period . . . . 29·889 in.

Thermometer: Greatest height . . . . 48°

Least. . . . . 18°

Mean of the period . . . . 31·93°

Evaporation (the water in the gauge being most of the time solid) . . . . 0·53 in.

Rain and melted snow. . . . . 0·95 in.

STORMS.

Monday morning, February 21, about four o'clock, the town of Uttoxeter was much alarmed by a most violent storm of wind and hail, which was immediately succeeded by a blaze of lightning, particularly remarkable for its duration. The report which accompanied it resembled the firing of three cannons in quick succession.

PLYMOUTH, *March* 13.—Arrived the Insolent brig, of 14 guns, Captain Brazier, from Passages, having thrown her guns overboard, and sustained considerable damage in her masts and rigging, in a tremendous gale, which commenced in the Bay of Biscay on the 28th ult. and continued to the 4th inst.

DESTRUCTIVE INUNDATION IN THE EAST INDIES.

Extract of a letter from Surat :—" Accounts have been received from Gurry, which mention a dreadful calamity, but, alas! of too frequent occurrence in this country. On the 12th of February, the Nerbudda unexpectedly, and during the night overflowed its banks, and swept away upwards of 15 villages. The calamity was so sudden that the inhabitants, their houses, furniture, and cattle, shared one common fate. It is difficult to compute the number of human lives lost; but it is supposed to exceed three thousand."—(PAPERS).

TABLE XCII.

		1814.		Wind.	Pressure. Max.	Pressure. Min.	Temp. Max.	Temp. Min.	Evap.	Rain, &c.
3d Mo.	L. Q.	March	14	NE	30·34	30·22	36°	30°	—	
			15	NE	30·42	30·34	37	30	—	
			16	NE	30·42	30·32	40	29	—	
			17	NE	30·32	30·23	39	28	—	
			18	NE	30·23	30·06	37	29	—	
			19	NE	30·06	29·78	35	30	—	
			20	SE	29·78	29·65	49	35	—	
	New M.		21	SE	29·65	29·60	45	41	12	38
			22	SW	29·75	29·60	55	35	—	—
			23	SW	29·75	29·73	55	34	—	—
			24	SW	29·61	29·59	48	40	—	—
			25	W	29·61	29·60	54	35	—	—
			26	SW	29·80	29·60	55	41	—	
			27	W	29·84	29·80	59	36	—	—
	1st Q.		28	Var.	29·84	29·50	59	40	25	10
			29	SE	29·75	29·50	52	35	—	
			30	SW	29·87	29·75	60	31	—	36
			31	SE	29·87	29·60	54	45	—	36
4th Mo.		April	1	SW	29·60	29·34	60	42	—	15
			2	W	29·37	29·31	56	41	—	6
			3	SW	29·58	29·37	57	32	—	
	Full M.		4	NW	29·70	29·58	58	34	—	
			5	Var.	29·90	29·70	59	38	—	
			6	W	30·00	29·90	61	38	—	
			7	SE	30·20	30·00	64	40	—	
			8	NE	30·20	30·16	63	32	—	
			9	N	—	—	—	—	—	
			10		30·20	29·93	63	43	—	
			11	NE	29·93	29·83	62	36	83	
					30·42	29·31	64	28	1·20	1·41

Notes.—Third Mo. 14—19. Dull cloudy weather: the latter a misty day. 20. More clear and spring-like, after a misty morning. *Cirrus,* passing to *Cirrocumulus* and *Cirrostratus* 21. Rainy: the morning opened with *Cirrostratus* lowering. 22—25. Variable spring-like sky: large *Cumulus* clouds, inosculating with *Cirrocumulus* and *Cirrostratus:* the *Nimbus* appeared occasionally: the air nearly calm, with little evaporation, the mean temperature considered. 27. Over-cast a. m., with *Cirrostratus:* a few drops of rain: p. m. large *Cirri:* sunshine and clouds. 28. Fair a. m., with *Cirrus* and other light clouds. 29. Overcast: *Cirrostratus,* with *Cumulus* and large *Cumu-lostratus* clouds. 30. The evaporation is now considerable: wind and rain by night. 31. Fine morning: wet forenoon: fair afternoon: thunder clouds appear.

Fourth Mo. 1. Cloudy a. m., with a few drops: a fine day: the *Cumulus* inosculates with the superior clouds: wind and rain by night. 2. Windy morning: squalls with rain, p. m.: a *Nimbus* on the NW horizon, with much wind at sunset. 3. Windy a. m.: *Cumulus* with *Cirrostratus:* p. m. *Cirrus* only, with a brisk evaporation. 4. Morning overcast with the lighter modifications. 5—11. Fair weather: generally misty mornings, with much dew, and clear days: the roads are become already quite dusty by the brisk evaporation.

RESULTS.

Winds Variable.

Barometer: Greatest height .... 30·42 in.
Least........................ 29·31 in.
Mean of the period........... 29·842 in.
Thermometer: Greatest height .... 64°
Least....................... 28°
Mean of the period .... 44·14°
Evaporation...................... 1·20 in.
Rain .... 1·41 in.

The frost may be said to have gone off in the first week of the present period, and the mean temperature having steadily advanced since, the latter week has been seasonably warm.

LEGHORN, *April* 5.—On the morning of the 30th, we had three shocks of an earthquake here. No lives were lost, but several buildings were damaged.

(PUB. LEDGER).

TABLE XCIII.

	1814.		Wind.	Pressure.		Temp.		Evap.	Rain, &c.
				Max.	Min.	Max.	Min.		
4th Mo.	L. Q.	April 12	N	29·83	29·76	71°	39°	—	
		13	SE	29·76	29·62	74	42	—	—
		14	S	29·62	29·48	70	43	—	
		15	SE	29·44	29·42	69	46	—	—
		16	S	29·42	29·23	65	47	—	30
		17	S	29·54	29·23	62	40	—	
		18	SW	29·59	29·43	66	39	—	
		19	SE	29·65	29·59	64	43	—	
	New M.	20	SE	29·65	29·64	68	48	78	15
		21	SW	29·73	29·64	62	39	—	13
		22	NW	29·96	29·73	57	35	—	
		23	SW	29·96	29·74	59	42	—	25
		24	NW	29·80	29·74	55	41	—	3
		25	NW	29·86	29·80	51	40	—	
	1st Q.	26	NW	30·10	29·86	55	33	37	13
		27	NW	30·12	30·09	56	41	—	3
		28	SW	30·09	30·04	52	43	—	12
		29	SE	30·00	29·98	60	46	—	—
		30	SW	30·10	30·00	64	42	—	
5th Mo.		May 1	Var.	30·12	30·10	68	48	—	
		2	E	30·05	30·00	60	44	—	
		3	E	30·00	29·83	63	41	—	
	Full M.	4	SE	29·83	29·70	57	36	70	
		5	Var.	29·70	29·28	53	40	—	—
		6	Var.	29·70	29·28	63	45	—	66
		7	NE	29·95	29·70	67	45	—	—
		8	W	30·14	29·95	69	43	—	
		9	NE	30·27	30·14	57	35	—	
		10	NE	30·42	30·27	55	33	30	
				30·42	29·23	74	33	2·15	1·80

NOTES.—Fourth Mo. 12. *Cirrus* clouds, passing to the intermediate modifications : the dust is become so dry and light that the air is filled with it : this is in part to be ascribed to electrical action. 13. Various clouds appeared a. m., including the *Nimbus:* and a few drops of rain fell : thunder clouds p. m.: a coloured twilight. 14. Clouds of various modifications. 15. A few drops a. m.: a swallow appeared on the wing: p. m. steady rain. 16. Much wind: showers. 17. Steady rain in the morning: wind S, hollow and murmuring : after this large *Cumulus* clouds (beneath *Cirrus*), which in the evening rapidly evaporated or dispersed: an *Aurora Borealis* ensued. (See the Note).

18. Wet morning and evening: much *Cumulus* appeared to-day, intermixed with *Cirrostratus* in the region of its base, an appearance very unusual. 19. *Cirrostratus* a. m.: day fine: a *lurid* sunset, the disk showing enlarged, through spots and lines of *Cirrostratus*. 20. Clouds a. m., beneath an elevated haze, in which were discernible streaks from N to S: rain in the night. 21. The clouds inosculating a. m.: rain ensued in the evening. 22. Windy: *Cumulostratus:* a brisk evaporation. 23. Shower p. m.: rain in the night. 24. Windy: small rain. 25. Overcast: showery. 26. After a shower, various clouds through the day, and a rainbow: at sunset a beautiful *Cumulostratus* in the SE, reflecting the splendour of the twilight. 27. Cloudy: rain to the SW. 28, 29. Wet mornings. 30. A grey elevated sky.

Fifth Mo. 1. Windy. 2. Overcast a. m.: then fair, with *Cirrus*. 4. Cloudy morning. 5. The same: a windy and very wet day: the rain mixed at intervals with sleet and hail. 6. Showers, attended with the union of clouds in different strata. At sunset an appearance of extensive rain in the E, with groups of *Cumulus* and *Cirrostratus* before it. In the N and W these had, intermixed and adhering, a transparent brown-red haze, distinguishable from the substance of the cloud, and which gave a pink tinge to the twilight, elsewhere of the usual lemon colour. At the same time a *Stratus* arose in the meadows. 7. A misty morning: while the earth presents the aspect of spring, the sky of late is quite autumnal: showers again passed by in the E. 8. Cloudy morning: rain in the S and E: *Cumulus* clouds, uniting with *Cirrus* above. 9. Overcast sky.

RESULTS.

Winds Variable.

Barometer: Greatest height..............30·42 in.
Least......................29·23 in.
Mean of the period...........29·770 in.
Thermometer: Greatest height............... 74°
Least...................... 33°
Mean of the period...........51·39°
Evaporation....................... 2·15 in.
Rain.............................. 1·80 in.

AURORA BOREALIS.

Fourth Mo. 18. The *Aurora Borealis*, of late years a very unfrequent visitant in these parts, appeared last night, with no great degree of splendour, but with the usual characteristic marks of this phenomenon. About 11 p.m. when my attention was first called to it, there was a body of white light, in part intercepted by clouds, extending at a moderate elevation from the N to the NW, with a short broad streamer rising from each extremity. After this it became an arch, composed of similar vertical masses of fibrous light, which moved along in succession, preserving their polarity and curved arrangement. One large streamer in particular, went rapidly through nearly the whole length of the arch from W to E, in which direction the rest chiefly moved. Some of these masses were rather brilliant, and one exhibited colours. After some cessation, and a repetition of this appearance, carried more towards E and W, the light settled in the N, and grew fainter: in which situation, at midnight, I ceased to observe it.

THUNDER STORMS, &c.

During a thunder storm at *Mantua*, on the 20th of March, a flash of lightning penetrated the theatre. Four hundred people were in the house, two of whom were killed, and ten were struck senseless, but afterwards recovered. The electric fluid melted the brass wire, also several gold and composition ear rings and watch keys, without hurting the wearers, and split the diamonds of two ladies of rank.

Friday, during a most tremendous storm of thunder and lightning, an electric bolt fell on the summit of a closely compacted rick of hay, the property of Mr. Jordan, of Cheltenham, which it perpendicularly pierced, penetrating several feet in the earth. The aperture it made at the summit was about three feet in circumference, but it lessened in its progress downwards to about six inches, and was much diminished at the base of the mow. A sulphureous effluvium arose from the rick, which was diffused for several yards round.

(PUB. LEDGER, *April* 14).

Admitting the facts here related (and there is nothing on the face of the accounts to lead us to dispute them) they present two curious instances of the surprisingly varied operations of the electric fluid. To split a *diamond*, even with a knife and hammer, in the direction of the *laminæ* of which it consists, is an easy operation; and this cristallized and combustible substance being a non-conductor, we ought to expect such an effect occasionally, from a powerful stroke of electricity : but to perforate instantaneously a closely compacted *hay rick* is a more difficult task, and, considering the elastic nature of dried grass, would probably baffle the force of the heaviest ball from a cannon. The most probable solution of the fact seems to me to be this; that the substance of the hay was dispersed into the air by a *returning stroke* proceeding from the earth at the base of the hay rick to the clouds above. The true date of this occurrence is probably the first of the month. It will be seen that I noted thunder clouds on the preceding day.

SUPPOSED EARTHQUAKE.

We have heard, from most respectable authority, that a convulsion of the earth, exactly similar in effect and appearance to an earthquake, was sensibly perceived about ten minutes before eight o'clock, on Thursday night last, at Knill Court, Harpton, Norton, and Old Radnor, Radnorshire; at Knill Court, the oscillation of the house was plainly perceptible, and felt by all the family, and that too in several apartments, and was accompanied with a peculiar rumbling noise. At Harpton, a severe storm of thunder and lightning was experienced the same night, and at the same time.—(PUB. LEDGER, *January* 27).

I insert this report, under some uncertainty as to the date (though it appears to have been some time in the first month of this year) for the sake of the following remark. I do not apprehend that these local tremors of the ground in the time of thunder storms are to be classed with real earthquakes. I have stood at the distance of six or seven miles from the extremity of a most extensive and violent thunder storm, visible from Plaistow, and have sensibly felt the ground shake under my feet at the time of the nearer discharges, owing, as I conclude, to the circumstance of the electrical action taking place between the clouds and the thick *substratum* of indurated clay on which the country hereabouts reposes. Such strokes as penetrate but a little below the surface I suppose to excite a lateral tremor proportionally less extensive.

The reader will find two cases forward, in my own observations, of a nature similar, I conclude to the present, under the dates of Tenth month 7, and Twelfth month 14, 1816.

TABLE XCIV.

1814.			Wind.	Pressure. Max.	Min.	Temp. Max.	Min.	Evap.	Rain, &c.
5th Mo.	L. Q.	May 11	NE	30·42	30·28	57°	34°	—	
		12	NE	30·28	30·08	58	40	—	—
		13	N	30·08	29·97	52	42	10	21
		14	N	29·97	29·93	55	42	—	
		15	NE	29·94	29·89	59	34	—	
		16	SE	30·06	29·94	62	40	—	
		17	NE	30·09	30·06	68	43	—	
		18	E	30·06	30·04	68	40	40	
	New M.	19	SE	30·04	29·85	70	41	—	
		20	SE	29·85	29·70	62	42	—	
		21	SE	29·70	29·48	60	36	—	
		22	SE			52	39	—	
		23	NW			54	41	—	—
		24	NW			46	41	—	—
		25	NW	29·87	29·57	58	31	—	—
	1st Q.	26	NE	29·87	29·83	63	33	—	
		27	NE	29·83	29·58	66	39	—	
		28	NE	29·65	29·56	70	39	—	1·75
		29	NW	29·96	29·65	70	42	—	
		30	NW	30·01	29·96	63	48	—	
		31	SW	30·01	29·94	67	46	1·00	
6th Mo.		June 1	SE	29·94	29·79	58	47	—	—
		2	E	29·80	29·73	53	46	—	—
	Full M.	3	E	29·75	29·65	55	50	—	—
		4	NE	29·90	29·75	55	45	—	77
		5	NE	30·07	29·90	54	44	—	
		6	NE	30·07	30·00	55	43	—	
		7	NE	30·00	29·91	58	40	—	
		8	NE	29·97	29·89	67	41	—	
		9	NE	30·03	29·97	70	36	55	
				30·42	29·48	70	31	2·05	2·73

NOTES.—Fifth Mo. 13, 14. Cloudy: much wind. 15. Cloudy and windy: the *Cumulostratus* prevails. 16, 17. Misty mornings. 18. A breeze a. m. from the NE, with *Cirrus, Cirrocumulus,* and *Cirrostratus* clouds. 19. A clear morning. 24. A very wet day and night. 25. A *Stratus* on the low grounds at night. 26. Ice on the pools of standing water this morning to the thickness of $\frac{1}{10}$ of an inch.

Sixth Mo. 2. A slight shower last evening: and this morning small rain. 8. Much wind: cloudy.

RESULTS.

Prevailing Winds Easterly.

Barometer: Greatest height .... 30·42 in.

 Least .... 29·48 in.

 Mean of the period .... 29·910 in.

Thermometer: Greatest height .... 70°

 Least .... 31°

 Mean of the period .... 50·50°

 Evaporation .... 2·05 in.

 Rain .... 2·73 in.

The minimum of the barometer for this period is somewhat uncertain, from the loss of three days observations about the time of its occurrence: the depression was followed by frost, as is frequently the case when much rain has fallen. The first of the present month was a contrast to the same of last year, when the thermometer rose to 85°; yet it is observable that the same low temperature occurred then also four days afterwards. On the whole this period has been more changeable than we should have expected with an easterly current prevailing, and full 6° colder than the corresponding one of 1813, in which westerly winds predominated. Birds of song are remarked to have been less heard than usual; a circumstance perhaps ascribable to their number having been reduced by the severity of the winter.

COLD IN RUSSIA.

An extract from a letter, dated Moscow, May 13, (25th), says:—" The weather here is worse than the oldest person alive recollects at this season. It snows generally every day, more or less, and sometimes the whole day: *last night the ice was half an inch thick.* Every body is alarmed for the consequences in regard to the agriculture of the country. The roads are next to impassable."

By accounts from Archangel, of the 17th of June, we learn, that the last winter had made greater inroads into the summer of that northern latitude than ever had been known in the memory of man. The ice of the Dwina had not broke up till the 24th of May, and even in the middle of June the White Sea was full of drift ice.—(PUB. LEDGER).

ICE ISLANDS IN THE ATLANTIC.

The following is an extract of a letter from Halifax, dated the 31st of May, 1814:—" The convoy under the Spencer, bound to Quebec, on the 14th of May, in lat. 44° 18′ N, long. 50° 50′ W, fell in with upwards of 20 large islands of ice, some of which were 80 feet above the surface of the water, and about two acres in extent. In the afternoon of the same day, the convoy met a field of ice, computed at 20 *miles extent, and about* 30 *feet above the water's level,* some parts being considerably higher; most happily these were discovered in clear weather, and in the day time. For several days prior, and many subsequent to the 14th May, the fogs were so intense that one ship could not discern another, within the range of half cable, so that many of the convoy would have been wrecked, had the 14th proved as foggy. Happily, the wind was fair, and the atmosphere clear and frosty, and before night the convoy was considered out of danger.—(PUB. LEDGER).

TORNADOES.

[*From an American Paper.*]

I communicate the following, chiefly on account of one or two circumstances attending those phenomena of Nature, which I do not recollect to have noticed in any description I have seen.

Two Tornadoes passed in this vicinity on Saturday last, (June 4), attended with their usual destructive effects on the timber, and razed the few buildings in their course to the foundations, destroying fences, corn, &c. In crossing the Ohio River, the water was taken out, and fish of every description were thrown upon the land.

The courses of the two were nearly parallel and simultaneous, about 15 or 20 miles apart, proceeding from SW to NE. One passed through the prairies on Little Wabash, and was beyond our sight. The other passed in plain view, distant, on the first appearance, about three or four miles, and, from the levelness of the country, was visible for many leagues in its progress. Its shape was much like that of a cone, or a sugar loaf, with the small end downwards, or rather like a speaking trumpet, its upper part flaring considerably as it joined the cloud above. It was as black as pitch, and appeared to boil like that substance over a furnace. The cloud above was also very black. The extent at the bottom of the Tornadoes has been ascertained to have been between half a mile and a mile.

A singular circumstance, observed by myself, and others, who were within half a mile, was this—that the most vivid flashes of lightning were seen to pass between the heavens and earth in quick succession, just in front and rear, and sometimes through the body of the Tornado (or perhaps around it, in a line from it to our eyes), and still no peals of thunder were heard from them. I do not think I ever saw so broad and vivid flashes before in my life. Indeed, I never before had so deliberate and fair a view of a phenomenon of this nature.

It has occurred to me, and I submit it to philosophers to decide, whether the extreme velocity of the air within the whirl, did not prevent the vibrations (or undulations) by which sound is conveyed, from being communicated to the tranquil air without? Were not the vibrations carried round and round within the Tornado and there expanded? This opinion is confirmed to me, by the recollection of what I have heard persons say, who have been in Tornadoes, that there is a continual loud, thundering sound—which I think is produced by the electrical concussions within the whirl. They hear it continually—those out of it hear nothing, even from the fiercest flashes of lightning.

I will mention but one other circumstance—the hail stones which fell in these Tornadoes were as large as a man's two fists. They were tried to be put into a pint tin cup, and would not go in. Hail of this dimension may be formed, by being long borne up and driven round through a moist medium by the whirling wind, before being let down to the ground : whereas, by descending in a direct line, or nearly so, it can never become so large.

S. GRISWOLD.

Shawnoetown, Illinois Territory, June 9, 1814.

In another part of this work I may revert to this interesting description, and make some observations on the nature of these terrible whirlwinds. In the mean time it must be apparent that, whatever be their primary cause, the effects are most extensive and complicated. To the utmost mechanical force of air in motion, we see joined an extremely rapid condensation of its moisture, attended with a proportionate developement of electrical phenomena. The great breadth of atmosphere through which the whole extends is another important consideration : and it would be but a superficial view of the subject that should lead us to ascribe all this to the accidental meeting of two opposite currents in the lower atmosphere.

TABLE XCV.

1814.		Wind.	Pressure. Max.	Pressure. Min.	Temp. Max.	Temp. Min.	Evap.	Rain, &c.
6th Mo.	L. Q. June 10	E	30·03	29·80	69°	42°	—	
	11	E	29·80	29·70	69	48	—	—
	12	E	29·94	29·70	74	48	—	—
	13	E	30·01	29·94	67	55	—	—
	14	W	30·01	29·82	85	53	50	1·00
	15	S	29·87	29·75	78	52	—	—
	16	NW	30·00	29·92	71	43	—	—
	New M. 17	NW	30·04	29·92	68	52	—	—
	18	NW	30·02	———			—	—
	19	NW	———	29·74	66	42	26	38
	20	NW	29·74	29·58	65	48	—	
	21	NW	29·83	29·77	59	49	—	7
	22	NW	30·07	29·83	60	49	—	—
	23	NW	30·20	30·07	62	47	—	2
	1st Q. 24	NE	30·27	30·24	60	41	—	—
	25	NW	30·24	30·04	63	49	30	—
	26	NW	30·01	29·99	62	48	—	—
	27	NW	30·01	29·94	65	49	—	8
	28	NW	29·94	29·84	74	47	—	
	29	S	29·86	29·80	71	55	20	
	30	NW	29·86	29·81	74	47	—	
7th Mo.	July 1	SW	29·90	29·82	70	46	—	
	Full M. 2	NW	29·94	29·90	74	42	—	
	3	NW	29·95	29·93	79	50	33	
	4	NW	29·95	29·94	75	58	—	—
	5	NW	29·96	29·94	81	50	—	6
	6	SW	29·96	29·88	79	49	—	
	7	SW	29·82	29·78	73	63	—	—
	8	SW	29·78	29·67	69	62	—	—
	9	SW	29·67	29·67	72	63	50	13
			30·27	29·58	85	41	2·09	1·74

NOTES.—Sixth Mo. 10. A shower in the night. 11. Misty morning, with a smell of electricity in the air: *Cumulus* and *Cirrus* prevail. 12. Showers. 13. A *Stratus* in the meadows: thunder clouds afterwards appeared in the horizon; and, at 11 p. m. lightning. 14. A thunder storm very early, with much wind, and heavy rain mixed with hail. 15. Wet morning: evening cloudy and windy. 21—23. Windy, and overcast, with *Cumulostratus* and *Cirrostratus*. A fire in the grate has been again acceptable. 24. A shower early: this evening the cloudy canopy passed off, with a definite boundary to the S, where, at the same time, it appeared to be raining. 25. Again over-cast, with *Cumulostratus,* and windy : much honey dew on the lime trees. 26. A shower this evening. 27. The same. 28. Cloudy, calm : vane at N : *Cumulostratus,* very heavy in the middle of the day, but which cleared off in the evening. 29. Misty morning : clouds as yesterday. 30. Much sun to-day.

Seventh Mo. 1. *Cirrostratus* and *Cirrocumulus,* with some appear-ance of the *Stratus* at night. 2. a. m. Plumose *Cirri* of uncommon beauty, with *Cumuli* beneath them : the latter prevailed through the day. 3. Clear morning : the *Cumulostratus* again prevails. 4. Large ill-defined *Cirri* filled the sky this morning : various denser clouds succeeded. 5. Rain by half past six; then cloudy and fair : at even-ing the sky cleared, with the wind W, and an orange twilight. 6. Loose driving *Cumuli :* brisk wind : clear twilight, with little colour. 7. Much wind : *Cirrus* passing to *Cirrostratus,* with *Cumulus* beneath, the whole formed to windward : in the evening a few drops of rain. 8. Wet, a. m.: fair, p. m.: windy night. 9. Windy, cloudy : dripping at intervals : the quickthorn hedges are in many parts quite stripped of their leaves by caterpillars.

RESULTS.

Wind chiefly from the North-West.

Barometer : Greatest height . . . . 30·27 in.

Least. . . . . 29·58 in.

Mean of the period . . . . 29·90 in.

Thermometer : Greatest height . . . . 85°

Least. . . . . 41°

Mean of the period . . . . 60·01°

Evaporation . . . . 2·09 in.

Rain . . . . 1·74 in.

TABLE XCVI.

		1814.		Wind.	Pressure. Max.	Min.	Temp. Max.	Min.	Evap.	Rain, &c.
7th Mo.	L. Q.	July	10	S	29·83	29·67	67°	44°	—	16
			11	SW	30·03	29·83	76	51	—	17
			12	NW	30·05	30·02	74	54	—	—
			13	NW	30·02	29·77	70	48	—	
			14	NW	29·77	29·68	69	48	—	—
			15	W	29·63	29·62	74	52	36	—
			16	NE	29·87	29·63	66	50	—	7
	New M.		17	SE	29·87	29·85	73	48	—	14
			18	SW	29·84	29·83	75	48	—	15
			19	NW	29·83	29·62	74	57	—	
			20	SW	29·70	29·56	74	56	—	2
			21	NW	29·87	29·70	78	56	40	—
			22	NW	30·07	29·87	69	50	—	
			23	SW	30·15	30·09	82	52	34	
	1st Q.		24	SW	30·09	29·83	78	55	—	
			25	SE	———	———	84	61	—	
			26	NW	30·05	29·90	80	60	—	
			27	SE	30·05	29·98	83	70	—	
			28	SE	29·98	29·89	91	66	1·08	—
			29	SW	30·05	29·90	71	56	—	—
			30	NW	30·10	30·05	72	65	37	
			31	W	30·10	29·98	79	54	12	17
8th Mo.	Full M.	Aug.	1	NW	30·10	29·98	80	59	—	
			2	NW	30·10	30·05	76	64	—	
			3	SW	30·14	30·05	79	55	—	—
			4	SW	30·14	29·98	77	62	73	
			5	SW	29·98	29·86	76	58	—	—
			6	W	30·05	29·98	69	57	—	—
			7	SW	29·98	29·80	70	60	57	23
					30·15	29·56	91	44	3·97	1·11

NOTES.—Seventh Mo. 10. Rain came on gradually this morning, and continued the whole forenoon, after which appeared the *Cumulus* with *Cirrostratus:* the twilight was luminous, with faint horizontal streaks. 11. Wet afternoon: then *Cirrostratus.* 12. Clear morning: after which different strata of clouds inosculating, followed by a slight shower: a little of the *Cirrocumulus.* 13. *Cumulostratus* through the day, changing at evening to *Cirrostratus:* a strong breeze. 14. Clear morning: then *Cumulostratus,* with a breeze. 15. Various modifications of cloud: the day at length overcast, with one or two very slight showers, and more rain in the night. 16. *Cumulostratus* a. m.:

slight showers p. m.: the wind variable: a *Stratus* at night. 17. Misty morning, after which various clouds, with the wind E: inosculation followed, and a heavy shower in the evening: an electrical smell was perceived at different intervals to-day. 18. Windy: showers through the day. 19. *Cumulostratus*, after some sunshine: at evening the lighter modifications prevailed, including *Cirrocumulus*, in a turbid sky. 20. Wind S a. m., hollow and threatening rain: some showers followed, after which the clouds separated, shewing several modificacations, distinct, and well formed. 21. Fine morning, with *Cumulus*: groups of thunder clouds formed p. m., chiefly to the E: but at sunset the electrical character gave place to *Cirrostratus* and wind. 22. After a clear morning the *Cumulostratus*, which has so long predominated, with its usual attendant, a strong breeze of wind. 23. A nearly serene day. 24. Clear day: a breeze from SE. 27. Some lightning at night. 28. (At Tottenham, the thermometer rose to 92·5°: the observations for the latter half of this period were made at the Laboratory, Stratford). Frequent vivid lightning in the evening. 29. Some lightning this morning, with rain: a strong breeze from the SW all day.

Eighth Mo. 3. A few slight showers in the evening. 7. Day showery, with brisk wind.

RESULTS.

Prevailing Winds Westerly.

Barometer: Greatest height . . . . 30·15 in.
        Least . . . . 29·56 in.
        Mean of the period . . . . 29·918 in.
Thermometer: Greatest height . . . . 91°
        Least . . . . 44°
        Mean of the period . . . . 65·5°
Evaporation . . . . 3·97 in.
Rain . . . .   1·11 in.

HAIL STORM.

The storm (says a Stamford paper) which alarmed this town and neighbourhood on Thursday night the 28th ult. was felt with much greater severity in some parts of Leicestershire. Its course seems to have been from SW to NE. At Reasby, between Leicester and Melton Mowbray, it began before ten o'clock, with such a wind and hail, followed by such thunder and lightning, as horror struck every person in the parish. The hail stones were bigger than hens' eggs; many window frames have been beaten in at Reasby, Syston, Thrussington, and Hoby, and fields of corn totally destroyed. Of course the demolition of glass in windows facing the direction of the storm has been general in those places.—(PUB. LEDGER).

TABLE XCVII.

	1814.		Wind.	Pressure.		Temp.		Evap.	Rain, &c.
				Max.	Min.	Max.	Min.		
8th Mo.	L. Q.	Aug. 8	W	29·62	29·61	73°	50°	—	5
		9	W	29·82	29·62	69	52	—	—
		10	W	29·97	29·82	72	48	—	13
		11	W	30·05	29·97	73	56	75	1
		12	SW	30·05	29·97	77	56	—	
		13	SW	29·89	29·77	74	51	27	—
		14	NW	29·76	29·74	70	49	—	
	New M.	15	SW	29·83	29·80	70	52	—	
		16	S	29·75	29·75	67	46	—	40
		17	W	29·90	29·84	71	57	58	
		18	W	29·98	29·92	76	53	—	—
		19	NW	29·99	29·88	70	43	—	—
		20	NW	29·93	29·92	70	49	34	
		21	SW	29·87	29·77	70	53	—	2
	1st Q.	22	SW	29·79	29·75	76	55	—	4
		23	SW	29·64	29·54	79	52	—	
		24	N	29·57	29·40	63	55	39	1·21
		25	SW	29·60	29·40	71	51	—	28
		26	NW	29·78	29·70	67	43	—	—
		27	NW	29·87	29·85	66	43	—	
		28	N	30·00	29·91	65	37	55	
		29	NW	30·09	30·07	68	46	—	
	Full M.	30	W	30·17	30·10	73	57	—	
		31	NW	30·24	30·17	71	58	30	
9th Mo.		Sept. 1	E	30·24	30·20	68	58	—	
		2	NE	30·20	30·10	72	49	—	
		3	NE	30·18	30·10	65	45	32	
		4	NE	30·15	30·10	65	41	—	
		5	N			65	52	—	
		6	NW			65	54	42	17
				30·24	29·40	79	41	3·92	2·31

NOTES.—Eighth Mo. Rain last night: a stiff breeze, a. m.: rain, p. m.: a large *Nimbus* in the SW, and a transient bow in the evening. 9. The breeze continues, with *Cumulus* and *Cumulostratus:* after a drizzling shower or two, a calm evening, with *Cirrocumulus.* 10. The night was overcast: *Cumulus* a. m.: and *Cumulostratus*, with repeated showers p. m., passing off very heavy to the E at sunset. 11. Morning cloudy: evening close, with a *Stratus* forming at sunset. 12. Cobwebs on the grass with the dew: cloudy a. m.: clear at noon: evening calm: a fine plumose *Cirrus* in the N. 13. A few drops of rain this evening. 15. Drizzling rain: the wind veered to S in the night:

heavy showers from nine to twelve, when it cleared up, the wind turning to the westward. 17. Fine morning, with a light wind, WNW: *Cirrocumulus* prevails, with other clouds. 18. *Cirrus* and *Cirrocumulus :* light showers: at sunset the lower edges of the clouds exhibited a deep red. 19. *Cumulostratus*, with some light showers. 20. Morning calm: much dew: the first feeling of autumnal cold: at sunset a group of dense *Cirri* in the NNW, casting shadows into the atmosphere: a *Stratus* in the marshes. 21. Towards evening some light showers from low driving clouds. 22. *Cirrostratus* in the N horizon, a. m.: and at sunset a singular one in the W. 23. Some rain in the night: *Cumulostratus*, with showers: close and warm. 24. *Cumuli*, hung like curtains round the horizon: little air stirring: *Cirri* formed above, moving briskly from the NW, with the vane at N: soon after 11 a.m. it began raining, continuing to rain incessantly through the day, and at intervals through the night also. 25. A magnificent display of *Cumulostratus* this morning, followed by repeated showers: p.m. a thunder shower, the wind changing to W. 26. Some rain during the night: wind brisk and cool: *Cumulus* and *Cumulostratus :* the sky at sunset ruddy orange, reflecting a bronze hue from the eastern clouds. 27. The night was serene: a *Stratus* this morning, with a plentiful dew: *Cirrocumulus* appeared alternately with *Cumulostratus :* a few large drops of rain at 2 p.m.: evening calm: clear sunset. 28. Morning almost cloudless: a fresh breeze from N: much dew and cobwebs on the grass: *Cirrocumulus* in the evening, beautifully illuminated by the setting sun. 29. Misty morning: dew and cobwebs: fine day. 30. Grey morning, calm, cloudiness in the north.

RESULTS.

Prevailing Winds Westerly.

Barometer: Greatest height................30·24 in.

Least............... ....29·40 in.

Mean of the period...........29·88 in.

Thermometer: Greatest height................ 79°

Least...................... 41°

Mean of the period...........60·20°

Evaporation........................ 3·92 in.

Rain .... 2·31 in.

I am indebted for most of the observations contained in this period to the kindness of my brother, William Howard, by whom they were made, in my absence, at Tottenham. It appears there was a tremendous *hail storm* at Coventry, on the 24th of eighth month.

TABLE XCVIII.

	1814.		Wind.	Pressure. Max.	Pressure. Min.	Temp. Max.	Temp. Min.	Evap.	Rain, &c.
9th Mo.	L. Q.	Sept. 7	SW	29·94	29·67	62°	50°	—	37
		8	NW	30·05	29·94	65	50	—	
		9	N	30·06	30·05	64	49	—	—
		10	N	30·07	30·06	62	47	—	—
		11	N	30·15	30·06	65	38	—	
		12	NE	30·15	30·13	62	33	—	
	New M.	13	E	30·14	30·10	63	34	37	
		14	E	30·14	30·12	65	34	—	
		15	E	30·12	30·05	64	36	—	
		16	E	30·00	29·98	66	43	—	
		17	E	30·01	29·95	68	37	—	
		18	SE	30·05	30·01	71	39	—	
		19	SE	30·02	29·97	72	44	—	
		20	SW	29·97	29·76	75	52	—	—
	1st Q.	21	SW	29·76	29·70	68	44	56	16
		22	SW	29·79	29·75	65	40	—	1
		23	S	29·75	29·59	65	48	—	—
		24	S	29·57	29·52	70	60	—	—
		25	SW	29·77	29·74	70	44	—	35
		26	SE	29·74	29·65	66	55	—	—
		27	W	29·76	29·75	66	48	—	31
		28	N	29·97	29·76	64	44	25	—
	Full M.	29	N	29·97	29·94	61	54	—	—
		30	NE	30·00	29·94	64	46	—	
10th Mo.		Oct. 1	NE	30·06	30·00	54	40	—	
		2	NE	30·16	30·06	54	37	—	
		3	NE	30·20	30·16	57	37	—	
		4	NE	30·20	30·03	60	39	—	
		5	NE	30·03	29·79	59	31	43	
				30·20	29·52	75	31	1·61	1·20

NOTES.—Ninth Mo. 7. Showery: wind veered to NW: in the evening *Nimbi,* with large *Cumuli:* rain in the night. 8. Overcast, windy: calm evening. 9. Cloudy: a few drops. 10. Windy, with *Cumulus* and *Cirrostratus:* a slight shower at 8 p. m. 11. Windy: fair day: *Cumulus* with *Cirrostratus,* densely grouped at sunset. 12. Much dew: orange sky at sunset: from 11 to 15 inclusive, hoar frost in the mornings, and the *Stratus* by night. 16. Brisk wind: calm at night. 17. A very wet mist this morning: the day was fine, with large *Cirri.* 18. Fine: a group of *Cirri* obscured the setting sun. 19. Misty morning: clear day. 20. *Cirrus* in streaks from N to S, mixed with haze lowering and passing to *Cirrocumulus* and *Cirrostratus*, the motion being from E to W: a shower of rain followed these appearances. 21. Brisk wind: changeable sky. 22. Dew this morning: the sky overcast with *Cirrostratus* and haze: p. m. windy, with a shower. 23. Misty a. m. with *Cirrostratus :* showers, with wind followed. 24. Blustering at S a. m.: temperature at the maximum at nine. Cloudy, damp, and close. In the evening pretty much lightning to the S and W: at nine the storm came hastily over us, giving a few discharges of blue lightning, with thunder and heavy rain. 25. A small steady rain, succeeded by calm sunshine. 26. Dew a. m.: afterwards cloudy and windy, with showers. 27. Cloudy a. m.: a *Stratus* at night. 28. Overcast: a slight shower, succeeded by *Stratus*. 29. Windy: overcast: a few drops. 30. Windy.

Tenth Mo. 1—5 inclusive. Clear weather, with brisk drying winds.

RESULTS.

Winds Variable : the SW uniformly brought rain ; the opposite current, fair weather.

Barometer:	Greatest height................	30·20 in.
	Least.......................	29·52 in.
	Mean of the period............	29·945 in.
Thermometer:	Greatest height...............	75°
	Least.......................	31°
	Mean of the period............	53·79°
	Evaporation	1·61 in.
	Rain...............................	1·20 in.

TABLE XCIX.

	1814.		Wind.	Pressure.		Temp.		Evap.	Rain, &c.
				Max.	Min.	Max.	Min.		
10th Mo.	L. Q.	Oct. 6	SW	29·73	29·63	59°	39°	—	10
		7	NW	29·83	29·73	57	32	—	—
		8	N	30·01	29·83	55	31	—	
		9	NW	30·16	30·01	52	24	—	
		10	N	30·16	29·91	56	34	—	
		11	S	29·91	29·68	54	41	—	
		12	SW	29·68	29·57	59	44	—	—
	New M.	13	SE	29·57	29·40	64	50	32	1
		14	SW	29·49	29·31	67	48	—	13
		15	SW	29·48	29·40	59	42	—	10
		16	SW	29·68	29·48	56	38	—	—
		17	S	29·68	29·28	58	43	—	—
		18	SE	29·40	29·13	58	36	—	14
		19	SW	29·25	29·03	54	42	—	80
		20	W	29·68	29·25	55	30	—	—
	1st Q.	21	W	29·73	29·68	55	33	30	
		22	SW	29·55	29·50	59	47	—	4
		23	NW	29·74	29·54	57	34	—	
		24	S	29·74	29·09	53	39	—	95
		25	NW	29·50	29·09	52	32	—	6
		26	NE	29·70	29·50	50	38	—	—
		27	W	29·81	29·70	53	32	—	
	Full M.	28	NE	29·82	29·75	52	32	17	
		29	SE	29·84	29·66	52	37	—	13
		30	Var.	29·90	29·87	53	39	—	
		31	NE	29·93	29·87	55	47	—	
11th Mo.		Nov. 1	NE	29·96	29·93	50	43	—	4
		2	NE	29·96	29·93	44	38	—	
		3	NE	29·93	29·83	47	33	10	
				30·16	29·03	67	24	89	2·50

NOTES.—Tenth Mo. 6. Misty morning: the trees dripping: the wind veered from NW to SW, with *Cirrostratus*, then *Cumulostratus*, and finally rain. 7. Clear a. m.: much dew p. m.: a sudden shower, with hail. 8. Hoar frost: fine day, with *Cumulostratus*. 9. Hoar frost: *Cumulus:* fine day: the wind a strong breeze: clear orange sky in the twilight. 10. Hoar frost: clear: the temperature of last night was destructive to various tender garden plants, and brought down the leaves from the mulberry trees. 11. Wind S a. m.: the sky turbid, and streaked from N to S. 12. Overcast with *Cirrostratus:* it appears to have rained in the night. 13. a. m. Large *nimbiform*

3

Cirri, passing to *Cirrocumuli:* the wind E, very gentle, the vane remaining at SW: a smart breeze from this quarter ensued. 14. The lighter modifications of cloud, with much wind, a. m.: driving showers p. m. 15. Fine a. m.: then wind and some showers. 16. Clear: windy. 17. The sky veiled with *Cirrostratus.* 18. Much wind: small rain at intervals: then showers. 19. Windy: wet a. m.: clear at noon: p. m. rain again, by inosculation of the clouds: rainbow: wet, stormy evening. 20. Rather windy a. m., with *Cirrostratus* and *Cumulus:* large *Cirri* and *Cirrocumuli* above the wind: a few drops of rain. 21. Hoar frost: calm. 22. Overcast: wind, followed by rain in the night. 23. *Nimbi* to N and S p. m.: a rainbow in the former. 24. A fine day: *Cumulus,* with *Cumulostratus,* and extensive beds of the lighter clouds above, which appeared to descend, passing to the westward: the evening was overcast, and there followed a tempestuous night: the wind was mostly S or SE, changing in the morning to SW. The barometer had been lower than the minimum here noted. 25. Showers a. m.: cloudy p. m.: clear night. 26. Hoar frost: somewhat misty: a little rain followed. 27. A fine day, after a cloudy morning: large distinct *Cumulostrati,* in an air nearly calm. 28. Misty morning: minimum temperature about seven: the drops of dew frozen: a fine day. 29. *Cirrostratus,* overcast sky, a. m.: showery, with wind, p. m. 30. The lighter modifications appeared till noon, in elevated lines stretching NE and SW, the wind being NW.

Eleventh Mo. 1. Cloudy: a steady breeze from NE, and small rain. 2. The breeze continues. 3. The same: overcast morning: fine day.

RESULTS.

Wind for the most part Westerly: but during the latter part of the period Easterly.

Barometer: Greatest height .... 30·16 in.
     Least................... ..... 29·03 in.
     Mean of the period .... 29·662 in.
Thermometer: Greatest height .. .. 67°
      Least........... .... 24°
      Mean of the period .... 46·43°
  Evaporation .... 0·89 in.
  Rain.................... .... 2·50 in.

KINGSTON, JAMAICA, *October* 27.—On the 17th instant came on dark gloomy weather, with heavy rain. On the 18th it blew *from the N and NW,* one of the most tremendous gales of wind, attended with torrents of rain (which lasted three days), that has been experienced here these 25 years.—(PUB. LEDGER).

TABLE C.

	1814.		Wind.	Pressure.		Temp.		Evap.	Rain, &c.
				Max.	Min.	Max.	Min.		
11th Mo.	L. Q.	Nov. 4	NE	29·83	29·74	44°	33°	—	—
		5	NE	29·80	29·71	42	32	—	45
		6	N	29·84	29·77	47	30	—	
		7	SW	29·77	29·32	49	31	—	1
		8	SW	29·37	29·23	46	32	—	9
		9	NW	29·89	29·37	42	29	—	
		10	NW	30·28	29·89	40	26	—	
		11	SW	30·28	29·78	47	31	18	
	New M.	12	SW	29·85	29·78	52	38	—	8
		13	NW	29·86	29·80	49	39	—	—
		14	SW	29·85	29·75	53	43	—	—
		15	SW	29·78	29·28	54	45	—	—
		16	W	30·00	29·28	49	32	—	18
		17	SW	30·00	29·82	53	35	—	
		18	SW	29·82	29·37	52	42	—	17
		19	NW	29·48	29·37	47	28	22	
	1st Q.	20	NE	29·57	29·48	43	34	—	20
		21	N	29·69	29·57	38	20	—	
		22	NW	29·67	29·65	36	19	—	
		23	NW	29·81	29·67	35	26	—	
		24	Var.	29·81	29·48	46	35	—	16
		25	S	29·48	29·18	50	42	—	32
		26	W	29·59	29·18	50	32	5	7
	Full M.	27	Var.	29·68	29·30	44	35	—	68
		28	SW	29·64	29·44	45	36	—	27
		29	W	29·44	29·15	48	34	—	
		30	N	29·45	29·12	44	32	—	
12th Mo.	Dec. 1		NW	29·73	29·45	43	30	—	—
		2	W	29·86	29·73	38	30	—	3
		3	E	29·86	29·50	38	28	6	32
				30·28	29·12	54	19	0·51	3·03

NOTES.—Eleventh Mo. 4. Misty morning: cloudy: shower at night. 5. Overcast a. m.: wet afternoon and evening. 6. The drops of dew were frozen, but so clear as not to appear like hoar frost: a fine day, with large *Cirri*, and afterwards *Cirrostratus* and *Cumulus*: a streaked orange sky at sunset. 7. Hoar frost, followed by cloudy sky: some rain after dark. 8. Clear morning: hoar frost: about 4 p. m. the sky became very dark, and there was a storm of rain, attended by a single loud explosion of electricity. It is remarkable that on the same day of the month last year, at the same hour, and in the same direction (SW), we had a similar storm. 9. *Cirrus*, with the compound modifications: several shooting stars this evening: windy afterwards. 10. A steady breeze from the NW: *Cirrostratus* appeared over the Thames at sunrise: ice a quarter of an inch thick, and permanent: a brilliant twilight; first, lemon colour surmounted by purple; then, by the blending of the two, a rich deep orange. 11. Hoar frost. I observe this often continues an hour longer on some tufts of *saxifraga cœspitosa* than any where else in my garden: the plant grows on close spongy masses of fibre, which are bad conductors of heat. 12. It was stormy last night; and is said to have thundered and lightened about 3 a. m. Wet this morning; but the day fine: twilight orange, delicately varied with dusky horizontal striæ. 13. Cloudy and raining at 8 a. m., but fine afterwards, with a smart breeze. 14. Cloudy. 15. *Cirrostratus* a. m.: then *Cirrus nimbiformis*, depending and curling beneath: a turbid sky ensued, and wind in the night. 16. A wet stormy morning. 17. Much wind by night. 18. Small rain a. m.: wet and stormy at night. 20. Hoar frost: a little rain at intervals. 21. Overcast. 22, 23. Hoar frost: misty. 24. The same: near sunset a hollow wind from SW, and rain in the night. 25. Wet day. 26. Some rain a. m. 27. Clear morning, with evaporation: but before noon cloudiness was coming on from the W, and it rained steadily from sunset till late at night. 28. Wet again in the night. 30. Large elevated *Cirri*, with *Cirrostratus*.

Twelfth Mo. 1. Brisk wind at N: in the night a slight sprinkling of snow. 3. Wet morning.

RESULTS.

Prevailing Winds Westerly.

Barometer: Greatest height................ 30·28 in.
Least..................... ..29·12 in.
Mean of the period .... 29·635 in.
Thermometer: Greatest height............... 54°
Least....................... 19°
Mean of the period .... 39·05°
Evaporation.................. .... 0·51 in.
Rain............................. 3·03 in.

On the 6th of November, about six in the morning, two violent shocks of an earthquake were felt at *Lyons*, in the direction from west to east, preceded by a loud clap of thunder, unaccompanied by lightning. Both before and after the report the rain descended in torrents; the weather had been the night preceding very tempestuous.—(PUB. LEDGER).

There was a considerable fall of snow at Edinburgh on the 9th, a circumstance rather unusual so early in the season. At the beginning of last week a great deal of snow also fell on the NW of Devon, some of the hills were covered, and on Dartmoor, several inches deep.—(PUB. LEDGER, *November* 17).

TABLE CI.

1814.			Wind.	Pressure. Max.	Min.	Temp. Max.	Min.	Evap.	Rain, &c.
12th Mo.	L. Q.	Dec. 4		29·50	29·34	44°	36°	—	
		5		29·94	29·50	41	33	—	
		6		30·04	29·91	38	28	—	
		7		30·04	29·60	45	33	—	—
		8	S	29·60	29·25	52	45	—	44
		9	SW	29·70	29·25	52	25	—	6
		10		29·37	29·35	51	29	11	53
	New M.	11	W	29·63	29·37	54	48	—	33
		12	SW	29·65	29·50	56	53	—	17
		13		29·72	29·50	56	41	—	3
		14	SW	29·56	29·54	53	42	—	7
		15	SW	29·68	29·40	55	43	—	3
		16	SW	29·92	29·40	55	37	—	10
		17	SW	29·78	29·68	55	44	—	5
		18	SW	29·82	29·67	56	51	33	2
	1st Q.	19	SW	30·05	29·67	55	30	—	11
		20	N	30·08	29·98	38	28	—	
		21	NE	29·98	29·73	39	31	—	
		22	E	29·73	29·46	37	32	—	
		23	E	29·69	29·67	38	32	—	
		24	E	29·67	29·62	33	29	—	—
		25	NE	29·61	29·58	33	30	20	—
	Full M.	26	NE	29·62	29·45	35	32	—	16
		27	SE	29·45	28·94	41	35	—	1·14
		28	NE	29·70	28·94	40	33	—	—
		29	SE	29·78	29·70	43	34	—	2
		30	S	29·90	29·71	49	33	—	9
1815.		31	E	30·18	29·90	44	31	—	—
1st Mo.		Jan. 1	N	30·43	30·18	42	30	4	
				30·43	28·94	56	25	0·68	3·35

NOTES.—Twelfth Mo. 4—6. Slight hoar frosts: ice on the ponds: windy at intervals. 7. Occasional small rain. 8. The maximum temperature of the last 24 hours this morning, with a hollow SW wind: an unsteady rain followed, of many hours' continuance, with much wind at S. 9. Windy: the barometer appears to have stood through the night at the minimum noted. 10. Hoar frost: *Cirrostratus:* overcast a. m.: after one, the day wet and stormy: much wind in the night. 11. Cloudy: rain before nine, and at intervals through the day. Though it was very cloudy, the bees came out in great numbers, as observed about the same time last year. 12. Temperature this

morning 54°, with low driving clouds, and a gale at SW. 13. Early
this morning the wind was violent: it moderated about 3 p. m.: and
the clouds, after a squall, assumed the *Cumulostratus:* much water
out in the marshes. 14. Red sunrise: *Cirri* over the sky, with their
extremities drawn out westward: p. m. it blew hard again from SW,
followed by rain at night. 15. Much wind and cloud a. m. The
millers, who are said to have remarked that for two years past our atmosphere has been calmer than usual, will now probably complain of
an excess of wind. The *Cumulostratus* appeared after 3 p. m., and
the evening and night were calm. 16. Early this morning the wind
rose again, and blew with great violence, with rain at intervals, till
about 3 p. m., when the same change ensued as yesterday. 17. Overcast: windy: some rain p. m.: a gale through the night. 18. The
tempestuous weather continues. 19. A shower about 8 a. m., and
again in the evening, after which light clouds were observed to pass
the moon's disk with less and less velocity. 20. Hoar frost: a fine
and nearly calm day: swarms of gnats in the air, and *gossamer* floating: the dew froze on the grass by half past 4 p. m.: there was a
corona round the moon, and a *Stratus* at night. 21. Hoar frost.
22. Dry air: steady breeze: *Cirrus*, passing to *Cirrostratus:* a lunar
halo of large diameter. 23, 24. Cloudy: steady breeze. 25. Some
snow in the night, followed by a little more in the day. 26. Snow
more plentiful a. m.: a thaw. 27. Rain: sleet and snow: rain again.
28. The maximum temperature early this morning: much rain.
29. Overcast sky. 30. Misty. 31. Hoar frost: large *Cirrus* clouds:
a few drops of rain: p. m. *Nimbus.*

    1815. First Mo. 1. Hoar frost: a rose-coloured sky, with *Cir-
rostratus* at sunset.

RESULTS.

Winds in the fore part Westerly and violent: in the latter, Easterly
and moderate: in both attended with rain: but one clear day in
the period.

Barometer:	Greatest height	30·43 in.
	Least.	28·94 in.
	Mean of the period	29·665 in.
Thermometer:	Greatest height	56°
	Least	25°
	Mean of the period	40·13°
	Evaporation	0·68 in.
	Rain	3·35 in.

STORMS OF WIND.

Newcastle, *December* 23.—The late hurricane, which commenced in this
neighbourhood on Friday morning (December 16), has been generally felt

throughout the whole country. In the south, the storm commenced on the 11th instant, and the shipping on the SW coast has suffered most dreadfully from its effects. The gale from that time *has moved gradually northwards,* and has left sufficient evidence of its violence in all the places within its vortex.

WHITEHAVEN.—The night of December 15 was very tempestuous. Numbers of chimneys in this town were blown down; several houses suffered greatly in their roofs, and some were nearly unroofed.

At Lancaster the tide began to flow about two hours earlier than set down in the tide table; and the coach from Ulverston, crossing the Lancaster Sands, was obliged to make for the shore above Silverdale. The tide continued to flow about half an hour later, and was between six and seven feet higher than mentioned in the table.

The Hull Paper says, the gale, or rather hurricane, of December 10, was the longest and most severe that has been experienced on this coast for many years. The Humber for twelve hours was covered with white spray; the waves washed over all the vessels at anchor, and many of them were forced from their anchorage.

GREENOCK, *December* 16, half-past 3 p.m.—Since our last, the wind continued blowing strong from the S and SW. Last night it greatly increased, and until now has continued to blow with unremitting violence. In the morning there was a considerable deal of thunder and lightning, accompanied by excessive falls of rain.

LIMERICK, *December* 17.—Within the memory of the oldest inhabitant of this city there has not been so tremendous a hurricane as was experienced here yesterday morning; it commenced with the wind at WSW, about three o'clock in the morning, being then about low water, raising the tide (the spring tides falling off) to an height never before known in this port.

A Portsmouth Paper of December 18, says—" A very strong gale of wind, from the W and SW, has prevailed here nearly the whole of the present week; in consequence of which, we are sorry to state, some lives have been lost, and much anxiety occasioned. The gale attained its utmost fury on Wednesday (December 14): during the greater part of that day it was found impracticable to proceed to Spithead from the port.

WEYMOUTH, *December* 23.—This week we have experienced one of the most tremendous gales of wind at SE ever known at this port.

FALMOUTH, *December* 28.—Put back the Duke of Kent packet, which sailed from this port, with mails for Lisbon, on the 5th instant, having experienced very heavy gales of wind, in one of which, on the 21st, they were under the necessity of cutting away their mainmast (then in lat. 44½ long. 11½).

LIVERPOOL, *February* 8, 1815.—We this day learn, that the Star, Disbrow, from New Brunswick for this port, met with a severe gale on the 21st of December, in lat. 40 long. 20, in which her dead lights were stove in, bulwarks carried away, a passenger and four seamen washed overboard; the captain and one man died on the wreck; and the mate and five of the crew were taken from the foretop, by a Portuguese schooner, where they had been seven days.

The late tempestuous weather has been succeeded, in the neighbourhood of Bristol, by a heavy fall of snow, which covers the country in that part to the depth of three inches, and continued on Thursday (December 22), without any appearance of a change. Fortunately there was not any wind, and the snow had not drifted.

No place in the empire has suffered so much from the late storm as Adare, the beautiful seat of the Hon. W. Quin, in the county of Limerick. It has lost about 700 trees, of which above 500 were full grown timber, of very great size, and ornamental to the grounds, torn up by the roots.—(PAPERS).

LARGE HAIL AND HOT WIND IN NORTH AMERICA.

Extract of a Letter dated Cincinnati, November 9, 1814, from Dr. N. Crookshank to Dr. Peter Wilson, Columbia College.

On the 4th of June last, about meridian, a dark cloud appeared in or near the SW point of the horizon, having the usual appearances of electricity, as was known, by the hemispherical or convex appearances of various parts on the superior sides of the different shelves composing it ; while the the lower part appeared parallel to the earth. Some light clouds were seen to move with great rapidity from the NE, and appeared to meet the former ; when both seemed to rise perpendicularly several degrees, so as to attain an extraordinary height. I then predicted hail, which presently fell, of uncommon size. Several stones were picked up after the shower [which ended in rain], too large to put into a cup four inches in diameter. Others were picked up in the time of the fall, thirteen, fourteen, and even fifteen, inches in circumference : yet, strange to tell, no material damage was done, though the width of the shower was five miles, and twenty or thirty in length ; the tract of the largest hail, in centre of the former, about a mile. The larger pieces appeared to be aggregated of numerous others, which were likewise composed of smaller ones. Some, however, of more than ordinary size, appeared single, as if they had been snow balls immersed in water, and re-frozen. The largest of those I saw resembled the section of a large hen's egg. About five the same afternoon, a hot or very warm current of air passed over, of such temperature and composition, as to threaten those who breathed it with instant death. Their only remedy was by stooping down near the earth. It actually did prove fatal to the leaves, and parts of leaves of many vegetables, by which means its traces were discovered above sixty miles, along and near the Ohio river.

This current of hot air happened entirely out of the tract of the hail, to the southward and eastward of it.

(From a Paper on Meteorology, read before the Literary and Philosophical Society of New York, December 8, 1814. By JOHN GRISCOM).

TABLE CII.

	1815.		Wind.	Pressure. Max.	Min.	Temp. Max.	Min.	Evap.	Rain, &c.
1st Mo.	L. Q.	Jan. 2	NE	30·45	30·42	35°	32°	—	
		3	NE	30·42	30·16	33	30	—	
		4	N	30·16	30·06	33	30	—	
		5	NE	30·06	30·05	31	30	—	
		6	N	30·05	29·95	34	25	—	
		7	NW	29·95	29·48	35	24	—	
		8	W	29·99	29·48	34	25	—	
		9	Var.	30·02	29·76	42	26	—	
	New M.	10	NW	29·76	29·52	44	35	13	
		11	NW	29·88	29·52	40	32	—	
		12	NW	30·15	29·88	40	24	—	
		13	SW	30·15	29·75	40	28	—	28
		14	N	30·17	29·75	43	32	—	
		15	NE	30·30	30·17	36	27	—	
		16	Var.	30·25	30·20	37	30	—	
		17	NW	30·25	30·13	41	32	19	
	1st Q.	18	N	30·13	29·96	36	27	—	
		19	E	29·96	29·68	34	22	—	
		20	NE	29·80	29·77	28	26	—	
		21	NE	29·80	29·75	35	29	—	
		22	N	29·85	29·80	35	28	—	
		23	Var.	29·85	29·73	37	17	—	
		24	Var.	29·61	29·56	29	20	—	
	Full M.	25	N	29·56	29·41	30	22	—	
		26	SE	29·41	28·96	39	23	—	
		27	E	28·96	28·88	39	30	—	
		28	SW	29·03	28·95	42	36	—	57
		29	Var.	29·24	29·03	43	34	—	
		30	SE	29·29	29·22	41	35	—	
		31	S	29·36	29·34	45	38	17	22
				30·45	28·88	45	17	49	1·07

NOTES.—First Mo. 2. Slight hoar frost : misty : cloudy. 3. Breeze fresh, and drier air : slightly clouded. 4. The same. 5. The same : a little snow. 6. The same. 7. A bank of *Cirrostratus* in the SE a. m.: during the forenoon these clouds passed over in flocks, and becoming denser at night, there fell a little rain or sleet. 8. A fine day, with *Cirrus:* bright starlight night. 9. Clear morning, but a *Cirrostratus* over the marshes, and *Cirrus* above in lines from NE to SW: at midday a little snow : windy evening. 10. *Cirrostratus* and *Cumulostratus:* maximum of temperature for the day this morning. 11. Brisk wind : *Nimbi* to the S, succeeded by *Cumulus* and *Cirrus:*

a little snow after dark. 13. Barometer falling a. m.: hoar frost: red lowering *Cirrus* and *Cirrostratus* clouds, in lines from N to S: these indications were followed by rain after 2 p. m. 15. Cloudy: smart breeze: SE in the night. 16. The sky overcast with *Cirrostratus* a. m.: this modification continued through the day, with a dry air. 17. Cloudy: windy night. 18—23. Snow fell at intervals during these six days, often in regular and beautiful crystallizations. 24. Much rime on the trees and shrubs a. m.: misty from *Cirrostratus:* temperature 18° about nine: wind easterly in the night. 25. Somewhat misty: overcast: snow. 26. Cloudy a. m.: snow p. m. in crystals of all sizes, varying from the simple unison of six prisms in a minute star to broad feathery flakes of the most regular compound structure: it was nearly calm during this time ; so that the crystals escaped the derangement consequent on being driven about in their descent. 27. Snow and sleet p. m.: a thaw, interrupted by a little frost in the evening: maximum of temperature in the night. 28. Wind and rain in the night. 30. Misty a. m.: the trees dripping: rain.

RESULTS.

Winds Northerly till near the close of the period.

Barometer: Greatest height............... 30·45 in.
             Least...................... 28·88 in.
             Mean of the period........... 29·77 in.
Thermometer: Greatest height.............. 45°
             Least...................... 17°
             Mean of the period.......... 32·66°
Evaporation .... 0·49 in.
Snow and Rain .... 1·07 in.

THUNDER STORM IN THE NETHERLANDS.

The 11th of January, 1815, was a remarkable day for the whole country from the North Sea to the provinces of the Rhine, on account of a tremendous storm of thunder and lightning, during which the steeples of many churches, in places far distant from one another, were set on fire nearly at the same hour. The tempest took its course over Arnheim, Utrecht, Bois-le-Duc, and several other places. (Details of about a dozen of these buildings damaged or burn. follow). It is remarkable, that only the highest steeples were every where struck, and that this tempest confirmed in a striking manner the accuracy of previous observations, that storms of thunder and lightning, though of rare occurrence in winter, are generally most mischievous in that season.

(GERMAN PAPER).

TABLE CIII.

	1815.		Wind.	Pressure. Max.	Min.	Temp. Max.	Min.	Evap.	Rain, &c.
2d Mo.	L. Q.	Feb. 1		29·45	29·36	48°	39°	—	
		2	E	29·62	29·45	46	36	—	
		3	W	29·62	29·39	49	39	—	13
		4	SW	29·80	29·39	54	36	—	
		5	SE	29·92	29·74	50	38	—	
		6	SW	29·74	29·55	47	42	—	4
		7	Var.	29·70	29·55	52	36	—	—
		8		29·62	29·49	50	38	15	15
	New M.	9	SE	29·66	29·63	47	37	—	
		10		29·63	29·43	46	40	—	4
		11	Var.	29·40	29·28	51	39	—	11
		12	SW	29·51	29·37	50	48	—	21
		13	SW	29·60	29·40	52	32	—	
		14	E	29·65	29·40	50	37	—	—
		15	Var.	29·65	29·56	51	42	—	—
		16	SW	29·56	29·50	54	35	18	16
	1st Q.	17	NW	30·26	29·50	52	33	—	2
		18	SW	30·26	30·16	51	34	—	
		19	SW	30·16	29·71	48	45	—	—
		20	W	29·82	29·55	54	42	—	26
		21	NW	30·15	29·82	56	48	—	
		22		30·20	30·15	57	39	—	
	Full M.	23		30·15	30·00	51	40	27	
		24	W	30·00	29·97	52	47	—	
		25	SW	29·97	29·92	53	46	—	
		26	NW	30·40	29·92	54	29	—	5
		27		30·47	30·44	48	25	—	
		28	SE	30·44	30·15	45	31	—	
3d Mo.		March 1	SE	30·19	30·15	52	33	20	1
				30·47	29·28	57	25	80	1·18

NOTES.—Second Mo. 1, 2. Misty dull weather. 3. Cloudy: wind and rain in the night. 4. Temperature 49° at the time of observation: fine: *Cumulus* beneath hazy *Cirrus.* 5. *Cirrostratus* and haze at sunrise, and much dew, which did not evaporate: a very fine day: temperature 56° in the sun: *Cumulostratus* p. m. 6. The same, followed by rain. 7. Overcast a. m.: dripping: Wind NW p. m.: groups of *Cumulostratus, Cumulus,* and *Cirrostratus,* occupying the whole S from one to two, coloured light indigo, with red haze above, and attracting the smoke. 8. The wind rose at S p. m. with the usual hollow sound: a beating rain from that quarter ensued. 9. Misty, from a diffused *Cirrostratus*, and cloudy above. Water from a well, which might be 10° warmer than the air, emitted a visible steam. 10. Cloudy: dripping at intervals. 11. The same: dark *Nimbi* passed in the S, the wind being W p. m. 12. Misty a. m.: cloudy: wind and showers. 13. A wet windy morning, succeeded by a very fine day: *Cumulus*, with *Cirrus*. 14. A very moist *Cirrostratus* a. m.: then fine, with various clouds: a little misty rain at night. 15. Much dew: the rain kept off till evening. 16. Windy: wet: a lunar corona. 17. Morning clear, with dew: then *Cumulus*, which becoming dense inosculated with the clouds above and with the smoke: a little rain p. m.: lunar corona. 18. Slight hoar frost: haze, passing to *Cirrus* and *Cirrostratus*. 19. Heavy *Cumulostrati* through the day: much wind. 20. Windy: wet: lunar halo at night. 21, 22. *Cumulostratus*: windy. 23. Fine. 24. Ramified *Cirrostratus*, indicating wind, which followed. 25. Blustering wind, with *Cumulostratus*, and strong evaporation: a little rain notwithstanding. 26. Showers, which laid the dust: a sudden clearing in the W at sunset. 27. Very white frost: a fine day out of the smoke. 28. Hoar frost, and a frozen mist: a serene day.

Third Mo. 1. Hoar frost: *Cirrus* in a fine elevated veil, passing to *Cirrostratus*, and exhibiting between ten and eleven a solar halo: a very fine day, the pollution of the smoke excepted.

RESULTS.

Winds Variable, but for the most part Southerly.

Barometer:	Greatest height	30·47 in.
	Least	29·28 in.
	Mean of the period	29·785 in.
Thermometer:	Greatest height	57°
	Least	25°
	Mean of the period	44·43°
	Evaporation	0·80 in.
	Rain	1·18 in.

VOLCANIC ERUPTION.

An American paper contains an account of an eruption of the volcano of *Albay*, on the 1st of February, 1815. This volcanic mountain is situated in the province of Camarines, on the southern part of the island of Lucon, or Luconia, one of the Phillippine Isles in the Indian Ocean. Five populous towns were entirely destroyed by the eruption; more than 1200 of the inhabitants perished amidst the ruins; and the 20,000 who survived the awful catastrophe were stripped of their possessions, and reduced to beggary.

(PUB, LEDGER).

TABLE CIV.

1815.			Wind.	Pressure. Max.	Min.	Temp. Max.	Min.	Evap.	Rain, &c.
3d Mo.	L. Q.	March 2	Var.	30·22	30·17	48°	35°	—	8
		3	NW	30·20	30·17	49	37	—	
		4	SW	30·17	30·11	54	41	—	
		5	NW	30·17	30·06	55	40	—	
		6	W	30·17	29·90			—	
		7	SW	29·90	29·35	52	38	—	—
		8	W	29·50	29·35	52	32	—	45
		9	NW	29·46	29·16	49	34	30	15
		10	W	29·32	29·16	48	29	—	6
	New M.	11	SW	29·48	29·32	47	30	—	—
		12	S	29·32	28·86	49	35	—	39
		13	NW	29·52	28·86	52	39	—	10
		14	SW	29·96	29·52	49	32	—	5
		15	Var.	29·96	29·68	55	37	—	19
		16	SW	29·83	29·68	62	43	—	
		17	NW	29·97	29·83	62	42	35	
	1st Q.	18	NW	29·99	29·88	55	46	—	—
		19	NW	29·91	29·88	57	45	—	13
		20	W	29·91	29·75	60	41	—	
		21	SW	29·75	29·54	59	45	—	6
		22	SW	29·54	29·31	61	45	—	6
		23	W	29·42	29·14	59	40	—	15
		24	SW	29·42	29·25	57	40	35	8
		25	SW	29·70	29·25	56	31	—	
	Full M.	26	SW	29·70	29·42	53	46	—	—
		27	SW	29·46	29·34	60	49	—	12
		28	SW	29·95	29·46	60	44	—	26
		29	S	29·93	29·77	62	43	—	
		30	SW	29·93	29·88	59	43	—	
		31	Var.	29·88	29·65	73	46	43	
				30·22	28·86	73	29	1·43	2·33

NOTES.—Third Mo. 2. Small rain at intervals. 3. Misty morning: fine day. 4, 5. *Cumulostratus*. 6. Fine day: *Cirri* appeared, much elevated, and coloured at sunset. 7. Fine morning: cloudy and windy, with some rain p. m.: night very stormy. 8. Windy, wet a. m.: showers by inosculation in the evening. 9. Hoar frost a. m.: turbid sky: rain: fair p. m. 10. Snow early, after which various modifications of cloud, ending in showers of rain and snow p. m. 11. Hoar frost: *Cirrostratus* and *Cumulostratus: Nimbi*, with large hail p. m. 12. Dull misty day: at night very stormy, with rain. 13. Cloudy a. m., with a gale at SW, and rain at intervals: several dense *Nimbi*, thunder,

1

hail, and hard rain p. m.: much wind with distant lightning, at night. 14. *Cirrostratus* and haze: then *Cirri*, passing to dense *Nimbi:* gusts of wind, hail, and rain. 15. The barometer has risen, with an almost uniform motion, about an inch and a quarter in 36 hours ; yet the air has not become clear: it should be observed, that there had been much previous depression : a wet forenoon, with a breeze at E: *Cirrostratus* p. m.: at night much wind. 16. High wind at SW, with *Cumulostratus* a. m.: fair and pleasant. 17. Much dew a. m.: *Cirrostratus* with *Cirrocumulus:* the light clouds after sunset beautifully tinted with lake and purple. 18. After a few drops, the *Cumulostratus* prevailed, followed by rain in the night. 19. Some rain a. m.: then *Cumulostratus:* and at evening *Cirrostratus*, with a lunar corona. 20. Dew a. m.: a light veil of *Cirrostratus:* at evening, the clouds passed to the N. 21. *Cumulus*, beneath *Nimbiform Cirrus*, both elevated: about 5 p. m. during the approach of a squall, the wind was very noisy among the branches (now covered with opening buds), producing an almost vocal modulation of sound: as soon as the trees became wet, this was exchanged for the usual hoarse noise, resembling that of the sea shore. The night was boisterous. 22. Much wind: showers : two strata of cloud : borne very high, as for some days past. 23. Heavy squalls, with some hail in the showers : a singular combination of clouds in the E, p. m.: it was a *Nimbus*, with *Cumuli* adhering and entering at the flanks, while a very lofty columnar *Cumulus* shot up through the midst of the crown, and this again was capped with a small *Cirrostratus*. 24. Various clouds: squally p. m. 25. The same : a brisk evaporation : at sunset *Cumulus* at a considerable height inosculated with *Cirrus* above, after which two distinct *Nimbi* in the S, which went away eastward. 26. Driving showers: at evening a lunar corona, followed by much wind and rain at intervals. 27. Stormy: showers. 28. Fair. 29. Large *Cirri*, which passed chiefly to the *Cirrocumulus* p. m. 30. Misty a. m.: overcast p. m.: little wind. 31. A very fine day: large *Cirri* formed alone at a considerable elevation, and passed in the evening to the NW: much dew followed.

RESULTS.

Winds Westerly.

Barometer : Greatest height .... 30·22 in.
Least...................... 28·86 in.
Mean of the period .... 29·672 in.
Thermometer: Greatest height............... 73°
Least.. .... 29°
Mean of the period .... 47·44°
Evaporation .... 1·43 in.
Rain .... 2·33 in.

TABLE CV.

1815.			Wind.	Pressure. Max.	Pressure. Min.	Temp. Max.	Temp. Min.	Evap.	Rain, &c.
4th Mo.	L. Q. April	1	SE	29·72	29·55	70°	45°	—	3
		2	W	29·82	29·81	63	42	—	
		3	Var.	29·91	29·81	57	34	—	3
		4	W	30·18	29·91	60	35	—	
		5	W	30·20	30·12	59	41	—	
		6	NW	30·12	30·06	66	43	—	
		7	E	30·06	29·91	69	45	—	
		8	E	29·91	29·81	61	45	68	
	New M.	9	NE	29·84	29·81	64	43	—	—
		10	NE	29·90	29·84	54	50	—	50
		11	Var.	29·90	29·89	63	47	—	—
		12	SW	29·89	29·75	68	40	—	4
		13	Var.	29·59	29·56	69	36	—	23
		14	NW	30·00	29·59	46	34	—	—
		15	N	30·13	30·00	50	28	30	
	1st Q.	16	NE	30·13	30·11	53	30	—	
		17	NE	30·25	30·11	54	31	—	
		18	N	30·25	30·18	54	30	—	—
		19	Var.	30·18	29·85	59	38	—	
		20	SW	29·85	29·17	59	36	—	—
		21	Var.	29·17	28·74	49	40	—	72
		22	N	29·06	28·74	50	40	38	20
	Full M.	23	NW	29·24	29·06	52	35	—	16
		24	NE	29·66	29·24	54	36	—	2
		25	Var.	29·91	29·66	53	41	—	8
		26	SE	30·06	30·04	58	33	—	—
		27	NE	30·04	29·92	61	43	—	1
		28	NE	29·92	29·64	64	45	—	
		29	NE	29·64	29·49	54	39	—	—
		30	SE	29·65	29·47	56	40	45	7
				30·25	28·74	70	28	1·81	2·09

NOTES.—Fourth Mo. 1. A summer-like day, with a breeze of wind: *Cirrus* formed about noon, and continued after sunset, passing to *Cirrostratus:* some rain followed. 2. Overcast a. m., apparently with *Cirrostratus* from *Cirrus:* during the middle of the day this gave place to *Cumulostratus.* 3. Overcast, calm, with dew on the grass a. m.: showers followed: after which *Cumulus* and *Cirrus.* 4. Much dew: *Cirrus,* followed by *Cirrostratus:* a few large drops about sunset. 5. Much dew: *Cirrus.* 6. The same: low *Cumuli* prevailed afterwards, some of them capped with *Cirrostrati:* then a shallow bed of *Cumulostratus,* ending at sunset in *Cirrostratus:* the evening twilight opake, dewy, and suffused with red. 8. *Cumulus* beneath

Cirrocumulus: a strong breeze: a mixture of *Cirrostratus* and *Cirrus* gave the clouds an appearance of active electricity; but the whole ended in *Cirrostratus.* 9. A confused mixture of the modifications, as yesterday, with the addition of haze: some drops by 10 a.m., and a shower p.m. 10. Heavy showers a.m.: cloudy p.m. 11. Wind SW a.m : the clouds mingled with the smoke of the city, came back from the E at sunset. 12. a.m. The clouds gathered pêle mêle, the *Cumulus* being capped with *Cirrostratus:* p.m. the *Nimbus* appeared, aad after some distant thunder to SW we had a shower. 13. *Cirrus, Cirrocumulus* a.m.: much dew, the large drops of which sparkled in the sun with the prismatic colours: various modifications of cloud followed: about 1 p.m. it thundered N: then nearer, with lightning, S: the wind shifting to that point, we had a heavy shower from W, with hail: wind and rain at night. 14. Cloudy, windy, a.m.: *Cumulostratus:* a little snow: swallows skimming the meadows. 15. *Cumulus,* succeeded by *Cumulostratus:* the wind moderate. 16. The same: there were traces of hoar frost this morning. 17. Hoar frost: *Cumulostratus:* at night *Cirrostratus,* and a lunar corona. 18. a.m. A few drops during the union of some clouds, which became *Cumulostratus.* 19. Loose *Cumulus* a.m.: at night *Cirrus,* and afterwards a very large *white* lunar halo. 20. Overcast, dripping: a hail shower p.m.: at night a lunar halo, very small, the prismatic ring touching a corona within. 21. Much cloud and wind : wet forenoon and night: the greatest depression of the barometer that has occurred in the present year. 22, 23. Much cloud, windy: rain at intervals. 24. Windy: cloudy: rain a.m., with a little hail: in the evening *Cumuli* dispersed rapidly, with *Nimbi* and *Cirrostrati* in the horizon. 25. Various clouds, ending in rain, of which, however, very little till night. 26, 27. Windy, variable: some dripping. 28. A little rain early: then *Cumulostratus.* 29. *Cumulostratus:* strong breeze: dripping: the *Cirrocumulus* appeared, transiently, as for several days past: a *Stratus* at night. 30. *Cirrocumulus,* which passed to *Cirrostratus:* p.m. (upon the wind becoming SE) rain.

RESULTS.

Winds Variable, but for the most part Northerly.

Barometer: Greatest height .... 30·25 in.

Least...................... 28·74 in.

Mean of the period .... 29·783 in.

Thermometer: Greatest height .... 70°

Least......... .... 28°

Mean of the period .... 48·56°

Evaporation .... 1·81 in.

Rain .... 2·09 in.

TABLE CVI.

1815.			Wind.	Pressure. Max.	Min.	Temp. Max.	Min.	Evap.	Rain, &c.
5th Mo.	L. Q.	May 1	E	29·67	29·65	67°	50°	—	—
		2	SE	29·72	29·70	71	47	—	13
		3	NE	29·70	29·68	73	47	—	
		4	NE	29·67	29·65	67	44	—	3
		5		29·69	29·65	67	48	—	4
		6	S	29·70	29·66	68	49	—	2
		7	S	29·70	29·67	69	49	—	5
		8	SW	29·81	29·67	70	47	61	
	New M.	9	W	29·80	29·79	69	49	—	5
		10	SW	29·85	29·79	70	51	—	
		11	SW	29·79	29·51	73	51	—	5
		12	SW	29·66	29·64	67	47	—	4
		13	SW	29·65	29·55	67	43	—	19
		14	SW	29·81	29·77	66	45	—	6
		15	S	29·90	29·72	69	41	—	13
	1st Q.	16	Var.	30·23	29·90	70	40	52	
		17	W	30·23	30·19	71	56	—	
		18	NW	30·19	30·10	76	50	—	
		19	W	30·10	29·78	77	48	—	
		20	NW	29·78	29·51	76	43	—	
		21	NW	29·77	29·51	65	42	—	2
		22	W	29·88	29·85	61	34	45	
		23	SW	29·85	29·67	61	44	—	20
	Full M.	24	SW	29·97	29·72	66	48	—	—
		25	NW	30·12	29·97	71	51	—	
		26	NW	30·12	30·10	80	47	—	
		27	NE	30·10	29·89	73	52	—	
		28	NE	29·89	29·80	78	58	—	—
		29	Var.	29·85	29·80	69	50	—	9
		30	Var.	29·90	29·85	68	49	50	
				30·23	29·51	80	34	2·08	1·10

NOTES.—Fifth Mo. 1. Dew this morning: a very fine day: light-ning at night far in the S. 2. Dew, with rudiments of thunder clouds, which in increasing became very beautiful: a storm was within hearing the whole midday, to S and SW: p.m. a heavy shower mixed with large hail, followed by lightning in the SE. 3. A strong breeze: thunder clouds, which dispersed in the evening. 4. *Cumulostratus:* some showers, followed by *Cirrostratus.* 5. Much dew: clear morn-ing, succeeded by *Cumulus,* &c.: thunder to the S: a shower in the evening. 6. Dew in large drops: somewhat misty and overcast: a shower in the night. 7. Windy: dripping at eve. 9. *Cumulostratus:*

a few drops by inosculation at sunset : rain in the night. 10. Windy: driving clouds. 11. Dew : windy, a. m. at SE, with large *Cirri,* and below them *Cirrocumulus,* variable and beautiful : p. m. the *Cumulus* was added, with *Cirrostratus* in the region of its base : at sunset a storm in SW, which about nine passed us to NE : the lightning in violet coloured sheets, with delicate white branched streaks on them : the thunder moderate, rolling out to a great length. 12. Much wind: slight showers a. m.: clouds and haze at sunset. 13. wind and rain: at half-past 6 p. m., during a heavy shower which passed to the E, there was a perfect double rainbow, on which I repeated an observation already recorded in this register, under Fifth Month, 4, 1813. The contrast between the space which separated the two bows, and the remainder of the cloud, was on the present occasion very striking. 14. Fair and warm: a shower p. m., with a bow : *Cirrostratus.* 15. Hollow wind at S, with an overcast sky: wet evening. 16. *Cumulostratus:* wind p. m. NW, a milky luminous twilight: much dew. 17. Somewhat misty morning : p. m. *Cirrostrati* advancing from the N overspread the sky, without any other cloud. 18. Windy at N, and overcast with *Cumulostratus,* a. m.: clear and calm p. m.: red sunset. 19. A very fine day: the twilight luminous, and somewhat ruby-coloured, followed by *Cirrostratus.* 20. Windy: various clouds: the sky purplish round the moon at night. 21. Brisk NW wind a. m.: showers. 23. Showers. 28. A brisk wind at SE: *Cirrus,* followed by *Cirrostratus: Nimbi* in the S at sunset: rain by ten at night. 29. Some thunder: rainy afternoon.

RESULTS.

Winds Westerly, with a small portion of Easterly at the beginning and end of the period.

Barometer: Greatest height . . . . 30·23 in.
Least . . . . 29·51 in.
Mean of the period . . . . 29·812 in.
Thermometer: Greatest height . . . . 80°
Least . . . . 34°
Mean of the period . . . . 58·58°
Evaporation . . . . 2·08 in.
Rain . . . . 1·10 in.

The thunder storm of the 2d, it appears, was about Croydon. At Addington, three miles from that place, " the water poured in torrents, and rushing into the valley formed a stream of 50 feet wide. It took its way through the village, forced open the doors, and carried away the furniture of the habitations."—(Pub. Ledger.)

TABLE CVII.

1815.			Wind.	Pressure. Max.	Pressure. Min.	Temp. Max.	Temp. Min	Evap.	Rain, &c.
5th Mo.	L. Q.	May 31	SW	29·90	29·75	72°	54°	—	—
6th Mo.		June 1	NE	30·03	29·80	70	38	—	
		2	NW	30·03	29·93	72	55	—	22
		3	W	29·93	29·83	69	56	—	2
		4	W	29·83	29·55	73	54	—	
		5	SW	29·55	29·41	71	47	—	
		6	SW	29·46	29·33	68	45	45	28
	New M.	7		29·62	29·33	72	42	—	
		8	SE	29·77	29·62	72	46	—	—
		9	SE	29·78	29·77	73	42	—	
		10	NW	29·78	29·74	72	45	—	
		11	SW	29·73	29·67	76	41	—	1
		12	S	29·67	29·43	68	43	—	—
		13	Var.	29·43	29·21	67	50	50	80
	1st Q.	14	W	29·42	29·21	68	45	—	9
		15	SW	29·79	29·42	70	45	—	
		16	E	29·79	29·50	80	55	—	23
		17	SW	29·53	29·46	76	54	—	8
		18	SW	29·63	29·55	74	51	—	
		19	S	29·63	29·59	73	55	—	
		20	N	29·68	29·57	74	46	—	
	Full M.	21	SW	29·81	29·68	73	50	48	
		22	NE	29·86	29·81	74	44	—	4
		23	NE	29·96	29·86	71	52	—	
		24	N	29·96	29·86	74	55	—	8
		25	NE	30·01	29·86	67	40	—	
		26							
		27	W	30·08	29·98	75	44	—	
		28	W	30·17	30·08	79	49	30	
				30·17	29·21	80	38	1·73	1·85

NOTES.—Sixth Mo. 4. Cloudy: windy. 5, 6. Windy, with *Cumulostratus* and *Cirrocumulus:* showers. 7. Heavy clouds: and at noon a sound like distant thunder in the NE: towards evening the dense clouds dispersed, leaving *Cirri* at a great elevation, and a most beautiful *Cirrostratus*, grained like wood, in the NW: hygrometer (De Luc's) 30° to 35°. 8. A grey sky, with *Cirrocumulus*, &c. which formed heavy *Cumulostratus*, threatening thunder: but a few drops of rain falling, the whole dispersed, save some *Cirrostratus*. 9. *Cirrocumulus*, with *Cirrostratus:* a fine day: much dew at night. 10. A very fine day: in the course of which the hygrometer went to 24°.

11. a. m. Clouds and wind, followed by a shower p. m. 12. Overcast: hygrometer, at 8 a. m. 62°; at nine 55°: clouds from SE, the wind being NW; a little rain: the evening obscure, with *Cirrus* and *Cirrostratus* hanging very low. 13. Cloudy morning: showers: after which large *Cumuli* capped and followed by *Nimbi:* from one to 2 p. m. heavy rain, the wind going from SE by S to NE, then back to SW: some thunder followed, and a fine afternoon, but the *Cirrostratus* remained at night. 14. Wet morning: hygrometer at eight 72°; at nine 50°: after some showers, a fine afternoon. 15. Hygrometer at nine 55°: showers and wind: fair p. m. 16. Hygrometer at nine 39°: a fine day: *Cirrus:* a corona round the moon. 17. Hygrometer 71°: wet morning, and rain in the night: a slight shower p. m. 18. Hygrometer 52°: rain in the night: rather cloudy. 19. Hygrometer 43°: a pretty fine day. 20. Hygrometer 40°: rather a dull day. 21. Hygrometer 55°: pleasant: not very sunny: about 1 p. m. a clap of thunder, and a few large drops of rain. 22. Hygrometer 49°: a dull cloudy morning: a little rain p. m. 23. Hygrometer 44°: morning cloudy: pretty high wind. 24. Hygrometer 38°: morning very dull: about twelve cleared up, and the sun shone very hot: p. m. cloudy again. 25. Rain in the night: p. m. sunshine at intervals. 26—28. Very fine days: hygrometer 46° to 52°.

RESULTS.

Winds Variable.

Barometer: Greatest height................30·17 in.
Least........................29·21 in.
Mean of the period...........29·708 in.
Thermometer: Greatest height............... 80°
Least..... .... 38°
Mean of the period ....60·10°
Evaporation........................ 1·73 in.
Rain............................. 1·85 in.

The observations from the 16th inclusive to the end were made by my son, during my absence from home.

FIGURES ON THE SUN'S DISK.

There is now on the Sun's disk the most extraordinary configuration of maculæ, or spots, that ever was seen: they present, when viewed through an astronomical, or inverting telescope, the exact resemblance of the figures 28. If viewed through an erect telescope, they will of course appear inverted, but equally distinct—the 2 in particular is perfectly formed.—(PAPERS, *June* 13.)

TABLE CVIII.

		1815.		Wind.	Pressure. Max.	Pressure. Min.	Temp. Max.	Temp. Min.	Evap.	Rain, &c.
6th Mo.	L. Q.	June	29	SW	30·17	30·11	77°	46°		
			30	NW	30·11	30·02	77	53		
7th Mo.		July	1	NE	30·03	30·02	75	49		
			2	NE	30·03	29·92	71	49		
			3	W	29·92	29·85	67	46		
			4	N	29·96	29·85	72	42		—
			5	Var.	29·98	29·92	70	52		
	New M.		6	Var.	29·88	29·86	70	50	—	10
			7	NE	30·00	29·88	64	42	—	—
			8	NW	30·00	29·96	69	54	—	1
			9	NW	30·04	30·00	70	52	—	
			10	NW	30·05	30·04	75	51	—	
			11	E	30·05	29·99	75	48	—	
			12	SW	29·99	29·93	79	49	62	
	1st Q.		13	SW	29·97	29·87	77	57	—	—
			14	SW	29·97	29·92	79	61	—	—
			15	SW	29·88	29·85	77	57	—	—
			16	W	29·95	29·84	75	55	—	—
			17	W	29·84	29·70	80	54	—	—
			18	SW	29·70	29·62	75	49	—	—
			19	NW	29·65	29·47	68	52	—	69
			20	NW	29·82	29·65	65	49	70	31
	Full M.		21	NW	30·00	29·97	68	47	—	—
			22	NE	30·00	29·98	68	52	—	
			23	NW	30·04	29·98	68	53	—	6
			24	NW	30·13	30·04	69	56	—	—
			25	NW	30·13	30·13	71	52	—	21
			26	NW	30·18	30·13				—
			27	NW	30·19	30·18	66	42	—	
			28	NW	30·18	30·17	74	47	68	
					30·19	29·47	80	42	2·00	1·38

NOTES.—Sixth Mo. 29. A very fine day: the western sky in the twilight bright orange near the horizon, with a purple glow above. 30. Cloudy morning: after which sunshine at intervals.

Seventh Mo. 1. Heavy *Cumulostrati* p. m. 2. Windy, cloudy, a. m.: *Cirrus*, passing to *Cirrocumulus*, &c. p. m.: a luminous twilight, the clouds much coloured. 3. Windy a. m., with *Cumulostratus*. 4. Cloudy: a few drops p. m. 5. *Cumulostratus*, formed by *Cirrocumulus*. 6. The wind passed this morning by SE to SW, but settled at NW, with various clouds: rain fell in the night. 7. Wet this morning early, and windy at NE: p. m. fair, with *Cumulostratus*. 8. *Cumulus* a. m., with

3

Cirrostratus: cloudy evening: some rain by night. 9. *Cumulostratus:* orange twilight. 11. A very fine day: pink-coloured *Cirri* at sunset. 12. Sultry: a. m. *Cumulostratus* by inosculation. About noon, an appearance of distant rain in the NE, which continued till evening: the whole of our own clouds gradually disappeared, with a steady SW breeze. At sunset it was clear, and somewhat orange-coloured to NW, but obscure, with *Cirrostratus,* to NE. 13. Large ill-defined *Cirri,* with nascent *Cumuli,* and afterwards *Cirrocumulus,* at a great height, passed to the NE with a fresh breeze: a little rain fell in the evening. Hygrometer about 40° these three mornings past. 14. Various clouds, threatening rain at intervals, which followed p. m. in quantity scarcely sufficient to lay the dust: windy. 15. *Cumulus* beneath *Cirrostratus* a.m.: windy: some light showers, and a trace of the rainbow at sunset. 16. A slight shower a. m. 17. Various clouds a. m.: a few drops p.m.: at evening a tendency to the rapid formation of *Cirrostratus,* the denser clouds at the same time exhibiting a beautiful gradation of colours: twilight orange. 18. In the morning an extensive sheet of flimsy *Cirrocumulus,* which soon moved away. Hygrometer at 9 a. m. 68°. About 10 p. m. the same kind of cloud: a low murky sky. 19. A steady rain a. m. Hygrometer 70° at 9 p. m. 20. Overcast, with *Cumulostratus:* windy. 21—26. Mostly cloudy: occasional showers. 27, 28. Fine.

RESULTS.

Prevailing Winds Westerly, and these for the most part NW.
Barometer: Greatest height................30·19 in.
　　　　　Least.......................29·97 in.
　　　　　Mean of the period............29·961 in.
Thermometer: Greatest height...............　80°
　　　　　Least................... ....　42°
　　　　　Mean of the period............61·36°
Evaporation (in 23 days, from the 6th inclusive) .... 2 in.
Rain............................. 1·38 in.

Hail Storms in France.

The department of Dordogne has been most dreadfully ravaged by two hail storms. In some parts angular hailstones, some of which were the size of a hen's egg, in less than ten minutes, covered the ground with a solid mass of ice, to the depth of a foot. Every thing was torn up, cut to pieces, and destroyed. Torrents of water, in the sequel, covered these miserable wrecks with sand, pebbles, and rubbish. The whole country presents a scene of desolation not to be described, and the wretched inhabitants are wandering about in despair, without food, and without shelter.—(Pub. Ledger, *July* 24.)

TABLE CIX.

1815.			Wind.	Pressure. Max.	Min.	Temp. Max.	Min.	Hygr. at 9 a.m.	Rain, &c.
7th Mo.	L. Q.	July 29	W	30·17	30·08	75°	55°		
		30	NE	30·15	30·08	72	54		—
		31	NE	30·24	30·15	63	45		2
8th Mo.		Aug. 1	NW	30·25	30·23	71	50		
		2	NW	30·23	30·19	70	59		
		3	NE	30·19	30·10	75	55		
		4	SW	30·10	29·87	79	55		
	New M.	5	W	29·76	29·75	78	52		—
		6	NW	29·96	29·76	65	46		25
		7	NW	29·97	29·96	69	45		
		8	E	29·99	29·97	71	49		
		9	NE	29·99	29·97	70	50		
		10	SW	29·97	29·66	71	58		
	1st Q.	11	NW	29·44	29·35	69	44		—
		12	NW	29·68	29·44	67	50		22
		13	SW	29·93	29·68	71	49	50	
		14	SW	29·94	29·93	78	59	50	7
		15	SW	29·94	29·64	75	59	50	2
		16	NW	29·77	29·51	78	50	47	35
		17		29·89	29·87	70	50	48	
		18	Var.	29·87	29·60	72	54	49	28
	Full M.	19	W	29·77	29·58	67	45	42	
		20	NW	29·86	29·77	68	47	65	
		21	Var.	29·86	29·70	69	49	49	
		22	Var.	29·70	29·58	75	59	43	
		23	NW	29·91	29·58	72	58	45	86
		24	SW	30·02	29·99	79	55	52	
		25	SW	29·99	29·89	76	63	53	—
		26	SW	30·00	29·89	75	50	46	16
				30·25	29·35	79	44		2·23

NOTES.—Seventh Mo. 29—31. Fine.
Eighth Mo. 1—3. Fine. 4. Very bright starlight. 5. Rainy evening. 6. Showery day: a thunder storm, with hail, about 2 p. m. 11. Windy: *Cumulostrati:* and in the evening *Nimbi*, with a little rain. 12. Much wind, with *Cumulostratus:* thunder and rain from N twice p. m., after which more calm. 13. Fine: much wind, with *Cumulus:* coloured *Cirri*, evening. 14. Cloudy morning: temperature 71° at nine: hygrometer at eight 60°: windy: a smart shower by night. 15. Windy: *Cumulus* capped, and *Cumulostratus:* lunar corona at night, followed by rain. 16. Fair and windy a. m., with clouds.

About 4 p. m. at the precise time of the barometer's turning to rise, came a very heavy shower, with two claps of thunder. 17. Fair: somewhat windy: large *Cumulostrati*. 18. Rain till 9 a. m., after which fair: brilliant sunset and moonlight. 20. *Cumulostratus,* low and stationary. 21. *Cumulus,* with *Cirrus* above, having little motion: p. m. the wind went to NE, and the clouds descended, showing a corona round the moon. 22. Overcast a. m., with thunder clouds, the wind SE: very heavy sudden shower before one: wet p. m. 23. Rain and wind early this morning, with thunder, the wind SE: after which sweeping showers from NW, and much wind by night. 24. Fair, with NW wind, and *Cumulus:* then SW, with *Cirrocumulus.* 25. Fine day: *Cumulus,* with *Cirrus:* strong breeze. 26. A little rain early: heavy showers, evening.

RESULTS.

Winds for the most part Westerly.

Barometer: Greatest height . . . . 30·25 in.

Least. . . . . 29·35 in.

Mean of the period . . . . 29·88 in.

Thermometer: Greatest height . . . . 79°

Least . . . . 44°

Mean of the period . . . . 62·13°

Evaporation (at the Laboratory). . . . . . . 4 in.

Rain . . . . 2·23 in.

WATER SPOUT.

On Tuesday the very singular phenomenon of a water spout was observed at Marsden, near Huddersfield. It appeared to be formed of a dense black cloud, and resembled a very long inverted cone, the lowest part of which seemed nearly to touch the ground; above it the clouds were white and fleecy, forming a striking contrast with it, but they appeared strongly agitated. Our correspondent concludes his description with stating, " that the spout, when completely formed, appeared to be round and smooth, and hollow within, and there appeared a stream of water running down the inside, part of which in its descent passed to the outside, and was carried up again very rapidly, by a spiral motion. The whole duration of this phenomenon was about twenty minutes.—(PUB. LEDGER, *August* 18.)

TABLE CX.

1815.			Wind.	Pressure. Max.	Min.	Temp. Max.	Min.	Hygr. at 9 a. m.	Rain, &c.
8th Mo.	L. Q.	Aug. 27	S	30·00	29·75	76°	54°		
		28	NW	29·81	29·72	71	50	65	5
		29	SW	29·97	29·81	68	43	55	
		30	SW	29·97	29·97	70	51		
		31	SW	30·05	29·97	72	49	47	
9th Mo.		Sept. 1		30·02	29·98	73	50	56	
		2		29·96	29·86	76	54	65	—
	New M.	3	W	29·96	29·88	72	56	52	—
		4	NW	29·95	29·86	73	40	54	6
		5	NW	29·97	29·95	63	40	51	
		6		30·06	29·97	62	31	48	
		7	NE	30·11	30·06	61	32		
		8	SW	30·11	30·08	65	38		
		9	SW	30·08	30·04	68	36	60	
	1st Q.	10	NW	30·05	30·00	72	47		
		11	NW	30·07	30·00	74	46	59	
		12	SE	30·07	29·92	70	42	55	
		13	S	29·92	29·80	78	39	67	
		14	SE	29·80	29·75	79	45		
		15	SE	29·75	29·67	77	54		
		16	Var.	29·80	29·67	75	47	62	9
		17	SW	30·01	29·80	70	50	65	
	Full M.	18	S	30·05	30·04	74	51	58	
		19	Var.	30·04	29·94	68	43	56	
		20	SE	29·94	29·87	60	34		
		21	S	29·87	29·69				
		22	S	29·69	29·57	71	50	49	11
		23	NW	29·57	29·50	59	38	69	
		24	SW	29·77	29·46	58	37	68	26
		25	SW	29·80	29·75	61	43	72	
				30·11	29·46	79	31	58	57

NOTES.—Eighth Mo. 27. *Cirrus,* passing to *Cirrocumulus* and *Cirrostratus.* 28. Shower early, and again p. m. 29. Lightning in clouds to the E, between three and 4 a. m. with moonlight westward: a fair day, with *Cumulus.* Hygrometer at 7 a. m. 70°. Near 8 p. m. a large meteor was seen to pass from the NE to the N. 30. Much dew : *Cumulostratus* during the day. 31. Gray morning: then heavy *Cumulostratus:* very clear night.

Ninth Mo. 1. Misty morning: *Cumulostratus,* which gave place at night to the *Stratus.* 3. A little fine rain early : various clouds followed, and some drops p. m.: a *Cirrostratus* exhibited the prismatic colours at sunset, and some elevated *Cirri* remained long red after it. 4. *Cumulostratus*, after large *Cirri*: showers at evening: rainbow: brilliant twilight. 5. Strong breeze: in the evening the new moon appeared with a well defined disk, and a pale phosphoric light, becoming afterwards gold coloured. 7. Hoar frost: hygrometer 78° at 7 a. m. 8. After a fine day, nearly calm and cloudless, the smoke settled over the opposite valley, which was soon afterwards filled with a *Stratus*. 10. A veil of light clouds, a. m.: somewhat hazy air, with a smell of electricity. 12. A *Stratus*. 13. Much dew: the evening twilight of late has been generally coloured, and at times streaked with converging shadows, the origin of which could not be traced to clouds intercepting the light. 14. *Cirrus* only, which increased during the day, and mostly disappeared in the night: the extremes of temperature near the ground were 83° and 45°: the hygrometer receded nearly to 22°. 15. Clear a. m.: in the evening *Cirri*, and obscurity to the W: after which *Cirrostratus*, and a very distant flash of lightning in the SW. 16. A little rain a. m.: much more cloud than of late has been usual: a *Nimbus* forming in the SW: in the evening steady rain. 17. Large *Cirrus*, passing to *Cirrocumulus*: at sunset a sheet of compound *Cirrostratus*, while increasing by rapid propagation from NW towards SE, was most beautifully kindled up, for a short time, with flame colour and orange on a purple ground. 18. Fair, with the lighter modifications. 19. Much wind at ENE this morning: hygrometer 40° at half-past 10 a. m. 20. Hoar frost: strong breeze: hygrometer 30° at 3 p. m. 21. *Cirrus*, followed by the intermediate modifications. 22. The sky filled gradually with clouds, both above and below: in the evening they grew black, but the rain came on without any explosion of electricity here. A *fire-balloon*, which I discovered near the SW horizon this evening, appeared to be impelled by different currents in rising, but passed the zenith going at a great rate and elevation towards the E. 23. Cloudy: wind NW: then N: small rain. 24. Early overcast, with *Cirrostratus*: the swallows went off, as it appears, this morning: after a murmuring sound in the wind, usual before southerly showers, we had a drizzling day till evening: the hygrometer did not recede past 68°. 25. A fine day: hygrometer went to 87° in the night.

RESULTS.

Winds light and Variable.

Barometer: Greatest height .... 30·11 in.

Least...................... 29·46 in.

Mean of the period .... 29·892 in.

Thermometer: Greatest height .... 79°

Least...................... 31°

Mean of the period .... 57·00°

Mean of the hygrometer .... 58°

Evaporation (at the Laboratory)........ 3·65 in.

Rain............................. 0·57 in.

EXTREMES OF WET AND DRY.

The great drought which for so long a time prevailed was most severely felt throughout the country; in many parts the water in the ponds used for cattle was entirely exhausted, and none could be procured except from the rivers, in many instances at a considerable distance from the farmer's dwellings. Grass had become so universally scarce, that the farmer was obliged to feed his cattle with hay as in the severest winter.—(PUB. LEDGER, *September* 25.)

For these three weeks past it has rained in Wales almost every day. A gentleman just returned from thence states, that there was no shooting in the beginning of the month, on account of the constant and heavy rains.

(PUB. LEDGER, *September* 28.)

It appears from the papers, that much damage was done in Scotland by violent thunder storms, attended with inundations from excessive falls of rain on the 24th, 25th, and 26th of the ninth month.

Seventh Month, 28, 1815.

Being on a journey, I lodged this night at the hospitable mansion of a friend in Yorkshire. The scite commands an extensive and pretty deep valley, in which lies the town of Bradford with a considerable scattered population, and, in the bottom of the valley over against the house, some iron-works. When it grew dark, the large coke fires and flaming chimneys of the latter made a conspicuous appearance, representing a mass of buildings in a state of conflagration. Having enjoyed this spectacle over-night, and happening to wake at an early hour next morning, I looked out, to see what appearance the works would then put on. To my surprise for the moment, the various objects which I had before contemplated were not now to be found! An immense *Stratus* had risen, and filled the valley: its level surface, on which the light of the morning began to be spread, lay stretched out like a lake, bounded by the opposite hills. But turning my view a little to the left, I recognized the situation of the iron-works by the smoke and flames of the principal furnace, still rising from under this sea of vapour, in a manner which forcibly recalled the description of a submarine volcano. The smoke spread itself horizontally upon the surface of the cloud, probably by the effect of an opposite electricity; but the sun's rays presently began to disturb this arrangement, the *Stratus* was dissipated by six o'clock, and a fine day, as to this neighbourhood, was the result.

Proceeding, however, the same morning on my journey, and passing the mountainous ridge called Blackstone Edge into Lancashire, I remarked that, precisely at the summit of the ridge, we left the fair weather behind us, and encountered the first of a series of showers (at intervals indeed heavy rain), which continued the whole way to Liverpool.

TABLE CXI.

1815.			Wind.	Pressure.		Temp.		Hygr. at 9 a. m.	Rain, &c.
				Max.	Min.	Max.	Min.		
9th Mo.	L. Q.	Sept. 26	S	29·79	29·63	66°	42°		23
		27	SW	30·06	29·79	61	34	72	13
		28	SE	30·06	29·46	59	42	80	—
		29	W	29·51	29·28	65	44	81	34
		30	SW	29·49	29·45	58	45	60	17
10th Mo.		Oct. 1	SW	29·72	29·49	62	39	71	13
		2	NW	30·11	29·72	60	33	77	—
	New M.	3	S	30·11	29·93	61	43	83	
		4	S	29·97	29·95	62	37	65	
		5	S	29·95	29·81	66	50	80	—
		6	N	30·08	29·81	63	34	90	50
		7	NW	30·20	30·08	59	37	79	
		8	NE	30·22	30·19	56	39	85	
		9	E	30·19	30·02	57	39	80	
	1st Q.	10	E	30·02	29·72	55	39	63	
		11	E	29·72	29·62	55	45	56	54
		12	NW	29·73	29·62	57	37	75	
		13	Var.	29·73	29·65	62	51	70	
		14	SW	29·76	29·65	60	42	60	
		15	SW	29·82	29·79	60	47	73	5
		16	S	29·73	29·71	62	46	75	
		17	SW	29·85	29·73	59	35	73	—
	Full M.	18	SE	29·85	29·52	56	46	77	18
		19	SW	29·52	29·25	63	49	83	16
		20	SW	29·45	29·25	61	45	60	—
		21	SW	29·82	29·45	59	33	70	
		22	S	29·85	29·57	57	42	77	
		23	S	29·57	29·47	59	50	65	39
		24	SW	29·47	29·42	59	42	89	10
				30·22	29·25	66	33	74	2·92

NOTES.—Ninth Mo. 26. A rainy sound in the trees this morning, from a gale at S: this was followed by rain, during which the wind veered westward. 27. Damp a. m., somewhat misty air: *Cirrus*, with *Cirrostratus:* much wind and a heavy shower by noon, with a suspicious sound, like thunder, at a distance: p. m. a second shower, after which a fine bow in the E, and some distinct *Nimbi*, the elevated crowns of which continued to reflect the light for 30 minutes after sunset. 28. A wet mist a. m., very little wind, the vane, which stood to N, turning to SE: sunshine, with *Cumulus* at noon: large *Cirri* p. m., which were permanent. 29. Rainy appearances in the sky a. m., soon followed by a shower, which came over from SW: much *Cirrostratus* followed, with more rain. 30. Clear a. m.: wind NW: p. m. a veil of *Cirrostratus* advancing from W, completely obscured the sky: in the night a beating rain from the southward.

Tenth Mo. 1. As yesterday, a. m.: showers, with rainbow, p. m.: rain by night. 2. Misty morning: much dew: *Cumulostratus*, and a few drops: luminous twilight. 3. Hoar frost: misty air: *Cumulus*, capped with a delicate double sheet of *Cirrostratus: Cirrocumulus* and inosculation followed. 4. *Cirrostratus* in a close veil most of the day. 5. Misty morning: then large *Cirri*, arranged from SE to NW, and passing to *Cirrocumulus*, &c.: rain at night. 6. Wet morning: fair p. m. 7. Hoar frost: slight *Stratus:* a serene day: much dewy haze at sunset, coloured red, first in the E, then in the W above an orange tint. 8. Cloudy a. m. 14. A lunar halo of moderate diameter, which, disappearing, gave place to a portion of a very large one. 15. Rain: the wind fresh at night from SSW. 16. *Cumulus* a. m.: fine day: a number of swallows, which re-appeared at the end of last month, have kept about our neighbourhood to the present time. 17. Large *Cirri*, passing to *Cirrostratus:* a little rain p. m.: *Nimbi.* 18. *Cirrostrati* a. m., with obscurity gradually increasing: wet p. m.: much wind, evening. 19. Coloured sunrise: calm, overcast a. m.: then windy, with driving showers, the sky filled with cloud: a tempestuous night. 20. Coloured sunrise, and much wind: a few drops of rain: cloudy night. 21. Clear a. m.: then *Cumulus* in a very blue sky, passing to *Cumulostratus*, which, with some beds of *Cirrus* above, was finely coloured at sunset: I suspected thunder and rain far to the S this afternoon. 22. Misty: much dew: *Cirrostratus, Cumulostratus, Cirrus.* 23. Maximum of temperature at nine this morning: little dew: cloudy: windy: rain. 24. Misty sunrise a. m., with *radii* through broken clouds: a *Nimbus* in SW: rain: about sunset *radii* again, followed by many distinct *Nimbi.*

RESULTS.

Winds chiefly from the S and W.

Barometer: Greatest height	....	30·22 in.
Least	....	29·25 in.
Mean of the period	....	29·747 in
Thermometer: Greatest height	....	66°
Least	....	33°
Mean of the period	....	50·79°
Mean of the hygrometer	....	74°
Evaporation (at the Laboratory)	....	2·12 in.
Rain	....	2·92 in.

TABLE CXII.

	1815.		Wind.	Pressure. Max.	Min.	Temp. Max.	Min.	Hygr. at 9 a. m.	Rain, &c.
10th Mo.	L. Q.	Oct. 25	SW	29·39	29·33	57°	39°	80	
		26	E	29·45	29·43	53	32	71	
		27	NE	29·65	29·45	50	43	80	26
		28	NE	29·98	29·65	51	43	81	35
		29	NE	29·98	29·86	56	47	75	
		30	NE	29·92	29·86	50	44	57	—
		31	NE	29·94	29·92	51	36	60	
11th Mo.	New M.	Nov. 1	NE	29·93	29·92	50	41	65	
		2	N	30·23	29·93	51	33	60	
		3	N	30·35	30·23	47	25	61	
		4	S	30·35	30·25	44	30	87	—
		5	SW	30·25	30·17	51	34	70	—
		6	W	30·17	30·06	52	40	80	15
		7	W	30·06	29·99	53	32	83	—
		8	W	29·99	29·78	53	41	85	—
	1st Q.	9	SW	30·06	29·78	57	41	75	35
		10	SW	30·05	30·02	56	44		
		11	SW	30·02	29·85	56	48	78	—
		12	S	29·85	28·99	57	48	78	75
		13	W	29·07	28·95	51	35	71	3
		14	NW	29·10	29·01	44	26	65	
		15	Var.	29·28	29·10	41	27	77	—
	Full M.	16	NW	29·50	29·28	35	25	75	8
		17	NW	29·90	29·50	36	21	90	—
		18	NW	30·06	29·90	35	18	85	
		19	Var.	30·06	29·72	35	25	80	
		20	NE	29·62	29·60	42	33	80	
		21	NE	29·80	29·60	42			
		22	NE	30·09	29·80		25		
				30·35	28·95	57	18	76·5	1·97

NOTES.—Tenth Mo. 25. Lightning to the N and W last evening. 27. The wind changed to NE this morning: *Cumulostratus*, with inosculation, a. m.: wet p. m. 28. wet a. m.: fair p. m.: rain again at night: windy. 22. A brisk gale at NE, continued through the day and night: a bank of *Cumulostratus* was visible in the morning in the S, from which quarter *Cumuli* were propagated northwards, changing the state of the superior clouds as they advanced: some *Cirri* in long lines above the whole were not affected: a solar halo appeared from one to 2 p. m., and the sky was turbid beneath the sun. 30. Completely overcast a. m.: windy, drizzling at intervals. 31. *Cirrostratus* tending to *Cirrocumulus*, a. m., beneath large *Cirri* pointing NE and SW: fair.

Eleventh Mo. 1. Low *Cumulostratus* beneath *Cirrostratus:* a breeze at NE, changing at night to NW, without affecting the state of the clouds, which were dense p. m. 2. Breeze at NW a. m.: barometer very steady : *Cumulostratus:* much redness in the twilight. 3. Clear a. m., with *Cirrostratus:* slight hoar frost: coloured sunset. 4. Clear a. m., with a little *Cirrostratus:* very white hoar frost, with ice : a fine day : after sunset, a dull purple in the E, with a little orange in the W: the moon conspicuous, the crescent indifferently defined, and pale. 5. Rain by 7 a. m.: after which low *Cumulostratus.* 6. A few drops a. m., with the wind S: then fine. 7. Cloudy: rain: lunar corona. 8. Some drops a. m.: then much *Cirrostratus:* sunset, with streaks of brown and purple on a yellow ground : moon visible, but its light peculiarly dim : wind and rain in the night. 9. Wet morning: dripping day : lunar corona: wind. 10. Fair, with *Cirrostratus.* 11. A little rain at night: *Cirrostratus.* 12. A fair warm day : various clouds passed over with a moderate wind : at evening the moon showed a lucid corona: to which succeeded (the wind having risen and veered to S) a continued exhibition of coloured halos varying in diameter, formed on low, rapidly passing, curling clouds, with an occasional corona, of pale green or yellow, between : a most tempestuous night followed, with rain. 13.Windy: a shower p.m.: the moon gold-coloured. 14. Clear: wind moderate. 15. Cloudy a. m.: windy: a sensible odour of electricity in the air at 1 p. m. 16. A snowy morning: fair p.m. 17. White frost a. m.: little of yesterday's snow remaining: the wind SW: a breeze: a little rain : p.m. a waggon from the north came thickly covered with snow: wind brisk at N at night. 18. Hoar frost: the moon looks like a map, so great is the transparency of the higher atmosphere. 19. Hoar frost and rime on the trees: bodies of thin mist, probably *Cirrostratus,* moved quickly over us this morning from the SW, rendering the tree tops invisible : a fine day : *Cirrus* and *Cirrostratus* at night. 20. *Cumulostratus* a. m.: maximum temperature at nine. 21. *Cirrus* a m.: *Cirrostratus:* minimum temperature at nine. 22. Fair, with hoar frost.

RESULTS.

Prevailing Winds Northerly, interrupted by a Southerly current, which greatly depressed the barometer, soon after the middle of the period, and was followed by a sharp frost.

Barometer : Greatest height .... 30·35 in.

Least..................... 28·95 in.

Mean of the period .... 29·783 in.

Thermometer : Greatest height................ 57°

Least..................... 18°

Mean of the period .... 41·75°

Mean of the hygrometer... .... 76·5°

Rain .... 1·97 in.

TABLE CXIII.

	1815.		Wind.	Pressure. Max.	Min.	Temp. Max.	Min.	Hygr. at 9 a. m.	Rain, &c.
11th Mo.	L. Q.	Nov. 23	N	30·28	30·09	37°	29°	80	
		24	NW	30·46	30·28	40	29	86	—
		25	NE	30·58	30·46	43	29	95	
		26	NE	30·58	30·29	40	31	90	
		27	NE	30·29	30·04	39	29	87	—
		28	Var.	30·02	29·92	35	23	74	
		29	SW	30·02	29·84	36	25	86	—
	New M.	30	S	29·75	29·70	49	36	90	18
12th Mo.		Dec. 1	SW	30·04	29·75	53	41	93	—
		2	SE	30·04	29·96	53	42	95	14
		3	W	29·96	29·71	50	37	96	22
		4	NW	29·82	29·70	46	39	92	—
		5	NW	29·35	29·31	48	37	66	38
		6	N	29·75	29·30	43	36	63	8
		7	NE	29·98	29·75	36	24	72	
	1st Q.	8	NE	30·05	29·98	30	23	59	
		9	N	30·35	30·05	32	25	61	
		10	N	30·37	30·33	43	30	75	
		11	NW	30·37	30·25	38	29	66	
		12	W	30·36	30·20	40	27	80	
		13	SW	30·27	30·15	44	27	90	
		14	SW	30·15	29·75	44	30	90	14
		15	SW	29·75	29·00	49	44	90	13
	Full M.	16	NW	28·90	28·85	44	31	65	—
		17	W	29·33	28·90	36	24	67	—
		18	SW	29·55	29·33	34	23	78	
		19	S	29·55	29·10	45	24	80	54
		20	SW	29·29	28·98	47	35	96	31
		21	NW	29·56	29·29	39	26	66	
		22	SW	29·70	29·56	33	27	90	—
				30·58	28·85	53	23	80	2·12

NOTES.—Eleventh Mo. 23. Serene, with hoar frost. 24. Hoar frost: light rain for a few minutes, p. m. 25. Hoar frost, with *Cirrostratus* in the horizon: steady breeze. 26. Hoar frost: the clouds coloured at sunrise: clear p. m. 27. Overcast a. m.: some light rain a. m.: *Cumulus* capped, and inosculating with *Cirrostratus*, p. m. 28. Fine: the ground lightly covered with granular snow. 29. Hoar frost: about 1 p. m. a slight snow, granular, and in stars: in the evening a mist over the marshes: and at about 8ʰ 30′ p. m. a brilliant meteor. It resembled a sky-rocket, and fell almost directly down with an uniform motion, blazing out larger before it became extinct. This meteor, with two others which I lately saw in the same quarter (SW) passing in the same track at about a minute's interval, had very much the appearance of a simple electrical discharge between two horizontal beds of cloud at different elevations. 30. Wet a. m.: cloudy p. m., the wind rising at S and SE.

Twelfth Mo. 1. Much wind and early, with rain. 2. Fine a. m., with *Cirrostratus:* then *Cumulus*, with *Cirrus.* 3. Very dark a. m., with clouds: wind SE: p. m. *Cumulostratus*, after which *Nimbus* in the horizon: the new moon conspicuous in an opaque twilight. 4. A wet morning: windy at SW: in the fore part of the night much wind. 5. Notwithstanding the dryness of the air, which was also clear below, there was this forenoon a continuous cloud above at a great height, with a hollow sound in the wind. We had a steady rain after this, and a gale of wind in the night. 6. Much wind: *Cumulus*, with *Cirrostratus:* wet p. m.: a gale through the night, shifting to N and NE. 7. Cloudy a. m.: the barometer, which the NW wind failed to bring up, now rises, with a continued hard gale from NE: the hygrometer receded to 51°: in the evening the moon's disk appeared small, and its light scanty, though no visible cloud intervened. 8. Clear, dry, windy morning. 9. Steady breeze: clear: hygrometer receded to 48°. 10. Sleet a. m.: lunar halo, evening. 12. For these three days past we have had a pleasant, clear air, with a fragrant smell, like that which exhales from the dry turf after showers. 13. Cloudy a. m.: drizzling: the windows of a room without a fire, for the first time this season, collect moisture on the *outside*, remaining dry within: sounds come louder than usual from the NE. 14. Hoar frost: a fine day: after dark, a lunar corona, occasioned by bars of *Cirrus* pointing N and S, and appearing to converge in the horizon. These soon passed to *Cirrostratus*, and were followed by wind and rain from the southward. 15. Much wind: cloudy: some rain: a very stormy night, with showers. 16. *Cumulus*, mixed with *Cirrostratus:* early in the afternoon the lofty summits of the former, rising from a fore ground of the latter on the E horizon, presented the resemblance of an Alpine landscape. In the evening, and on 17, a. m. the wind NW, with *Nimbi*, bringing some snow, followed by much cloud, and a gale at evening. 18. Fine day. 19. Hoar frost: clear: then overcast from the south, and some snow in large loose grains. In the evening more snow, followed by rain from S. 20. Cloudy a. m : much wind at S, with a hollow sound: rain p. m., and a gale through the night. 21. Fine morning: the ground slightly frozen. 22. Very white frost: *Cirrus* above, and *Cirrostratus* to the SE: a little granular snow on the ice. Snow in the night.

RESULTS.

Winds Variable, but with a larger proportion of Northerly than usual at this season.

Barometer: Greatest height................30·58 in.

Least........................28·85 in.

Mean of the period...........29·839 in.

Thermometer: Greatest height................ 53°

Least.,...................... 23°

Mean of the period...........35·96°

Mean of the hygrometer .... 80°

Rain .... 2·12 in.

St. Petersburgh, *November 22.*—The weather is still mild, and the navigation quite open; merchantmen arrive and depart from Cronstadt daily.

Hamburgh, *December 8.*—By the severe frost which we have had for these two days, the Elbe is covered with ice almost to Blankenese, and the navigation entirely interrupted.

The Lark packet arrived at Harwich on Friday from Cuxhaven. A dreadful hurricane was experienced on the opposite coast, which continued from the 7th to the 9th instant.—(Pub. Ledger.)

It appears by the papers, that this gale was exceedingly destructive to the shipping on our own eastern coast.

Meteor in Connexion with an Earthquake.

Extract of a letter dated *Lisbon*, February 2, 1816.—" At about five minutes before one (this morning) I felt my bed move up and down for about a minute, or a minute and a half: the shaking increased after this, changed its direction from side to side, and was very severe.—I opened my window, the atmosphere was dense (misty); a thick fog covered the whole city, yet I could see the lamps of the further end. On a sudden every thing was light, and *a meteor* was seen, which approached the earth, and of itself dissipated, and all was again in darkness. My thermometer in the room was at 60° or 62°. Every thing then became quiet until 7 p.m., when another more trifling shock was felt. For two days past we have all been noticing the oddity of the weather. All the morning dull, close, and very cloudy—no sun; wind north; no rain: at about one o'clock wind changed to south, blowing a hurricane, and dreadful rains. Last night the rain cleared off; and although the wind did not change, a thorough calm followed.

" The rain since the shock at seven has been incessant, and particularly eavy—a loud rumbling noise attended the great shock."

(Annual Register.)

Temperatures of Tottenham and Plymouth compared.

In the Results of Table XLII. I have inserted a comparison (from data furnished by a friend) of the respective Temperatures of Tottenham and *Exeter*, during the second month, 1810. My friend, James Fox, jun. has since gone into a comparison of the Temperatures of Tottenham and *Plymouth*, for the years 1814 and 1815, the Results of which may be seen in the Annals of Philosophy, vol. viii. p. 434–5.

The Mean Temperature of Plymouth for these two years, taken together, is 49·95°; of Tottenham, for the same period, 48·31°; difference in favour of the former 1·64°. This is a result which I have deduced from the above communication; and it is what we might have expected, Plymouth being near a degree *south* of Tottenham. But it appears that the greater warmth of Plymouth is chiefly found in the *nights* (which are warmer than ours nearly throughout the parallel), as also in the days of *winter*, when these are inclined to frost. In *summer* the days attain a higher Temperature about London, and the difference is considerable; the heat at Plymouth being then kept down by the same means (the vicinity of a great exposure of sea) which moderates to that part of the island the cold of winter. Thus it is proved by the evidence of the thermometer, that the climate of Devon is entitled (in conformity to general opinion heretofore) to the character of a more *equable*, and consequently, in the case of invalids, a more salubrious temperature than our own.

TABLE CXIV.

1815.			Wind.	Pressure. Max.	Min.	Temp. Max.	Min.	Hygr. at 9 a. m.	Rain, &c.
12th Mo.	L. Q.	Dec. 23	SW	29·70	29·27		32°	83	
		24	SW	29·50	29·25	42°	27	80	
		25	SW	29·78	29·53	36	21	76	
		26	SW	29·53	29·00	43	40	83	—
		27	Var.	29·86	29·00	41	25	80	—
		28	SW	29·88	29·75	46	32	77	39
		29	NW	30·09	29·88	50	41	77	—
	New M.	30	SW	30·52	30·09	43	22	55	
1816.		31	SW	30·52	30·33	35	24	90	
1st Mo.		Jan. 1		30·33	30·20	39	22	94	
		2	NW	30·20	30·01	33	21	80	
		3	SW	30·32	30·00	38	24	90	
		4	SW	30·32	30·14	37	26	95	
		5	SW	30·14	29·79	45	34	77	
		6	W	29·78	29·60	49	35	91	6
	1st Q.	7	NW	29·78	29·63	40	32	61	13
		8	S	29·40	29·30	49	41	94	13
		9	SW	29·34	29·31	48	41	67	13
		10	W	29·31	28·90	50	41	70	24
		11	NW	29·42	28·90	47	36	61	
		12	SW	29·42	28·96	43	32	55	17
		13	SE	29·15	28·87	39	29	79	2
	Full M.	14	S	29·32	29·26	43	31	86	5
		15	W	29·56	29·08	42	33	91	27
		16	SW	29·35	29·20	47	34	67	44
		17	SW	29·63	29·35	39	30	75	
		18	SW	29·64	29·60	42	29	80	—
		19	SW	29·64	29·45	40	29	98	10
		20	W	29·45	29·16	38	29	85	5
				30·52	28·87	50	21	79	2·18

Notes.—Twelfth Mo. 23. A thaw p. m., with a little rain: windy night. 24. Dew and *Cirrostratus*, a. m.: cloudy at intervals: windy at SW, yet it froze in the evening. 26. Maximum of temperature at 9 a. m., and beginning to rain: much wind, especially about sunset. 27. It is said to have lightened much, early this morning: a stormy day, with rain and snow. 28. Temperature at the minimum at nine: snow: sleet: rain. 29. Temperature at maximum at nine: a little rain: a gale through the night. 30. Fine morning, though with a pale sky: *Cirrostratus*, coloured at sunset: the river Lea rose higher, apparently by the tide, than at any time since 1809. 31. Hoar frost: a frozen mist, with *Cirrostratus* above, followed by a fine day.

1816. First Mo. 1. Misty air: *Cirrostratus*. 2. Hoar frost: a frozen mist, depositing much rime: the middle of the day fine. 3. Fine morning: the roads icy, it having thawed some part of the night. 4. Hoar frost: *Cirrostratus* in flocks: a breeze. 5. Coloured sunrise: fine, with *Cirrostratus*. 6. Maximum temperature at nine: cloudy: the wind rising: very heavy *Cumulostrati*, after some rain: clear windy night. 7. Minimum of temperature at nine: elevated *Cirrostratus* in bars, just visible: wind and clouds: a lunar halo. 8. Maximum temperature at nine: wet morning. 10. Much cloud, with *Nimbi* forming, p. m.: stormy night. 11. A gale through the day and night: much evaporation evident in consequence: lunar halo. 12. *Cirrostratus* descending from above: a gale, with rain after. 13. Fair day. 14. After frost in the night, a shower early: drizzling p.m. 15. Overcast with *Cirrostratus:* rain: clear at night. 16. A slight ground frost: large *Cumuli*, mixed with other modifications, p.m. which going off, showed elevated in the N and NE: to these succeeded linear *Cirri*, filling the sky, and crossing each other almost at right angles: these appearances were followed by a most violent storm of wind and rain in the night. 17. Fair: wind at night. 18. Minimum temperature at nine: hoar frost: misty horizon: *Cirrocumulus*, followed by denser clouds, and rain at evening. 19. Maximum temperature at nine: very misty: at noon a bank of dense clouds of various modifications in the S: windy at evening: rain in the night. 20. Fair day. 21. Wet morning: the wind SE.

RESULTS.

Prevailing Wind SW.

Barometer: Greatest height................30·52 in.
Least.........................28·87 in.
Mean of the period............29·615 in.
Thermometer: Greatest height................ 50°
Least........................ 21°
Mean of the period............36·52°
Mean of the hygrometer at 9 a. m....... 79°
Rain .... 2·18 in.

THUNDER STORM.

A few days since, was experienced such a storm of thunder and lightning, at *Gainsborough*, as scarcely ever was known at this season of the year.

(Pub. Ledger, *December* 30.)

TABLE CXV.

1816.			Wind.	Pressure. Max.	Min.	Temp. Max.	Min.	Hygr. at 9 a. m.	Rain, &c.
1st Mo.	L. Q.	Jan. 21	SE	29·30	29·15	43°	37°	92	16
		22	SE	29·42	29·30	44	36	90	
		23	SE	29·42	29·16	43	37	93	—
		24	E	29·16	29·01	43	34	81	17
		25	NE	29·17	29·01	39	34	86	
		26	NE	29·52	29·17	40	34	96	3
		27	N	30·00	29·52	41	29	76	—
		28	N	30·30	30·00	36	29	78	—
	New M.	29	NE	30·38	30·30	33	23	84	
		30	SE	30·38	30·27	32	21.	85	
		31	SW	30·27	29·93	34	21	74	
2d Mo.		Feb. 1	SE	29·93	29·64	35	19	75	
		2	SW	29·64	29·42	40	26	85	25
		3	S	29·47	29·43	48	37	100	
		4	SW	29·40	29·37	45	33	98	—
		5	S	29·37	29·09	39	35	100	15
	1st Q.	6	SE	29·09	28·90	38	31	97	—
		7	NE	29·31	28·90	31	15	77	1·38
		8	N	29·62	29·31	24	7	76	
		9	E	29·68	29·62	20	—5	80	
		10	SW	29·77	29·65	30	19	75	
		11	N	30·25	29·75	37	18	60	—
		12	N	30·35	30·25	32	11	56	
	Full M.	13	Var.	30·31	30·24	36	22	72	
		14	W	30·35	30·31	39	25	88	
		15	SW	30·35	29·96	44	32	99	—
		16	NE	29·82	29·73	47	33	75	
		17	NW	30·04	29·82	38	26	56	—
		18	SW	29·88	29·77	41	27	54	—
		19	SW	29·96	29·88	45	37	80	7
				30·38	28·90	48	—5	81	2·21

NOTES.—First Mo. 21. A dripping day. 22. *Nimbi* grouped with other clouds: fine at midday. 23. Overcast. 24. Drizzling : rain in the night. 25. Overcast. 26. The same : a fog on the Thames appeared from hence to be a dense bank of cloud in the horizon: a little rain by night. 27. Fresh breeze : cloudy: p. m. a shower, with hail : night frosty. 28. Fresh breeze: drizzling rain : snow : fair p. m. 29. Hoar frost : cloudy : fair. 30. Very white frost : misty horizon : sunshine after. 31. As yesterday, a. m.: at noon hygrometer 50°: the dust flies : wind SW p. m., with the usual sound for rain.

Second Mo. 1. Hoar frost: fair. 2. Ice now about two inches thick: after hoar frost, a misty thaw: wet and windy evening. 3. Fair: moisture on the outside only of the windows. 4. Strong breeze at S: misty: rain, followed by *Cumulus*, with *Cirrostratus*. 5. Very misty: the trees dripping. 6. Small and heavier rain by intervals: sleet at evening. 7. A gale from NE, which came on last night, has brought a deep snow: snowy at intervals through the day. 8. A smart breeze, with clear sunshine: the roads sloppy at midday: some distant clouds in the horizon at sunset. 9. A continued sunshine produced not the least effect on the ice to-day: hygrometer at 3 p. m. 47°: there was a mistiness perceptible, to a certain height, round the horizon: the wind a gentle breeze. 10. For remarks on this night, see the sequel. 11. Hygrometer as yesterday nearly: sleet, snow, and rudiments of hail, in minute quantity. 15. After three days of clear sky (a little *Cirrostratus* excepted), an extremely misty air: different clouds followed, and a few drops by inosculation. 16. *Cirrostratus* in flocks a. m.: wind changed to SW, then to NW, and blew strong at night: hygrometer receded to 46°. 17. Frost on the ground, from evaporation merely: the air by two thermometers not below 33°: the snow mostly gone; but a very thick ice remains on the water: *Cumuli* rose this morning, and passed to large spreading *Cumulostrati*. 18. Obscurity to the NE: snow, p. m. which melted in the night.

RESULTS.

Winds Northerly and Easterly.

Barometer: Greatest height...............30·38 in.
Least......................28·90 in.
Mean of the period..... ......29·696 in.
Thermometer: Greatest height.......... .....   48°
Least........ ....   −5°
Mean of the period...........   32°
Mean of the hygrometer at 9 a. m.......   81°
Rain and melted snow.................   2·21 in.

A night on which Fahrenheit's thermometer remains for some hours below *Zero*, is, in this climate, a rare occurrence: probably not above five of them fall within a century; the last appearing to have been 19 years ago. It is observable that this extreme low temperature is not, as might be expected, peculiar to long continued frosts, but happens rather at an interval of one winter after such a season. Such was the frost of 1794-5, which lasted 44 days, one whole day's remission excepted, immediately before which the thermometer had descended to

— 2; but in 1796-7, I find a temperature recorded of — 6, 5, with circumstances that indicate its having continued below *Zero* for some hours. Again, the character of the winter before last will be fresh in remembrance : the minimum of that season appears to have been not lower than 5°; and we have now a depression reaching to — 5. I do not, however, lay much stress on this analogy, which is pointed out rather for the use of future observers.

I was prepared to expect the intense cold of the night of the 9th to 10th of Second month, by the circumstance of a temperature of 7°, (or probably 5°) on the night before, being followed by a clear sky, with the wind at E, and a maximum for the day of only 20°. Early in the evening, on trying the experiment of placing a wet finger on the iron railing without, it was found to adhere immediately and strongly to the iron. I exposed several thermometers in different situations. At 8 p.m., a quicksilver thermometer, with the bulb supported a little above the snow, stood at *Zero:* at 11 p.m., a spirit thermometer in the same position indicated — 4; the former, which had a pretty large bulb, had not sunk below — 3. At half-past 7 a.m., the 10th, a quicksilver and a spirit thermometer, hung over-night about eight feet above the ground, indicated respectively — 3, and were evidently rising. The thermometer near the surface of the snow had fallen to — 5, and probably lower; but at the usual height from the ground of my standard thermometer, the temperature was at no time *below* — 5. The exposure is north, and very open.

From 8 a.m. the thermometer continued to rise steadily: at noon a temperature of 25° was pleasant, by contrast, to the feeling, and it was easy to keep warm in walking without an upper coat. Even at *Zero*, however, the first impression of the air on the skin was not disagreeable, the dryness and stillness greatly tending to prevent that sudden abstraction of heat which is felt in moist and quickly flowing air. Early in the afternoon the wind changed all at once to SW; some large *Cirri*, which had appeared all the day, passed to *Cirrocumulus* and *Cirrostratus*, with obscurity to the south. I now confidently expected *rain* (as had happened in former instances), but was deceived; and the thaw took place with a dry air for the most part, and with several interruptions by night.

During these two days the barometer, which had risen rapidly, fluctuated between 29·6 and 29·7 inches, and immediately afterward resumed its course, and rose at the same rate as before.

The mean temperature of this period is precisely 32°; and it is remarkable, that the mean temperature of each of three similar periods of frost, comprehended in the long winter of 1813-4, does not vary a degree from 32°; though preceded and followed by periods which respectively exhibit a mean of about 44°. On examination, I perceive that this analogy might be extended further.

The gale at NE, with which this frost came in, brought with it abundance of snow, which loaded the trees to their tops, and weighed down the smaller shrubs to the ground. The peculiar clinging quality of some snows merits inquiry. It is in part the result of the needly crystallized texture, aided by a degree of moisture attending, which afterwards freezes in the mass; but as light volcanic ashes have been found likewise to possess this quality, and in a still higher degree, perhaps we ought to attribute something to the electric charge with which each of these light bodies arrives at the earth. The seasonable covering which snow affords to the vegetable kingdom is matter of common remark; but it is not so generally understood in how great a degree the very circumstance of its production abates the first rigour of the cold. Just before this snow the air was extremely moist; the snow cleared it of an inch and a half, nearly, of water, and it has since indicated considerable dryness. Now it is quite probable that the *vapour* which afforded this water was found, by the supervening NE current, diffused in our local atmosphere, and by it decomposed. In this case the *latent or constituent heat* given out by the vapour in passing to the solid state, must have gone in great part to raise the temperature of that current. Hence a considerable interval, of gradually increasing cold, before we experienced its extreme effects; during which, too, the earth got provided with its accustomed covering.

After a copious fall of snow, an observer may find, in the scenery which it forms, some things on which to exercise his powers of reflection. The pensile drifts, which in a mountainous country are objects of just alarm, may be contemplated, here, to discover the principles of their construction, and the manner in which they rest on so narrow a base. When the sun shines clear, and the temperature is at the same time too low for it to produce any moisture, the level surface may be found sprinkled with small polished *plates of ice,* which refract the light in colours as varied and as brilliant as those of the drops of dew. At such times, there are also to be found on the borders of frozen pools, and on small bodies which happen to be fixed in the ice and project from the surface, groups of feathery crystals, of considerable size, and of an extremely curious and delicate structure. From the moment almost that snow alights on the ground, it begins to undergo certain changes, which commonly end in a more solid crystallization than that which it had originally. A notable proportion evaporates again, and this at temperatures far below the freezing point. On the night of the 10th of the second month I exposed 1000 grains of light snow, spread on a dish (which had previously the temperature of the air), of about six inches diameter. In the first hour after dark it lost five grains; in the second, four grains; in the third it acquired a grain, the wind having changed, and the temperature, which had been falling from 25°, inclining to rise again. The hygrometer was at 50°, with a gentle breeze at east. In the course of the night the loss was about 60 grains. This evaporation from snow may very well supply the water for forming those thin mists, which appear in intense frost: and the slight increase during a part of the time, in this experiment, may throw light on the formation of the secondary icy crystallizations above-mentioned. It appears that the air in a still frosty night becomes partially loaded, either with spiculæ of ice, or with particles of water, at a temperature below freezing, and ready to become solid the moment they find a support. Hence the rime on trees, which is found to accumulate chiefly on the windward side of the twigs and branches.

As to those more copious mists, of the modification *Stratus*, which accompany the setting in of long frosts, I conceive them to originate in part from the yet unfrozen rivers, and other waters, near which they are most abundant; in part from the moisture of the earth itself: for it is contrary to experience to suppose, that the frozen state of the surface can prevent the ascent of vapour from the porous soil below: which will continue to emit it, until its temperature becomes, by the gradual penetration of the frost, nearly on a level with that of the cold air then constantly flowing over it, though too gently to disperse the cloud formed.

The snow on the Grampian Hills, in Scotland, is at this time of a greater depth than has been known for the last 20 years. Vast flocks of grouse have come down from them for shelter.—(Pub. Ledger, *February* 23.)

EARTHQUAKES FELT AT SEA.

Extract of a letter from *Madeira*, dated February 5.—" On the 1st instant, about 12 o'clock at night, a severe shock of an earthquake was felt all over the island; and the following morning, at four o'clock, another shock was felt."

The shock of an earthquake, which we lately stated, was experienced at *Lisbon*, about one in the morning of the 2d February, was repeated rather more slightly at seven. The first is supposed to have lasted from two and a half to three minutes, and was generally and regularly felt in the whole city.

February 27.—A few days since the Ann transport, A. Clarke, master, arrived at Portsmouth, from Antigua. On the 2d February at 45 minutes past 11 a.m. being a little to the southward of the latitude of Lisbon, and about 150 miles to the eastward of *St. Mary's* (one of the Western Islands), she experienced the shock of an earthquake. The spot precisely was long. 19° 30′ W, lat. 37° 30′ N. The sensation produced was what would have been felt had the ship touched the ground, or her motion been impeded by a strong counter-undulating current; the masts trembled, as they would preceding their fall over the ship's side. The ship was thrown aback, with the sails; and the lead being thrown, with 150 fathoms of line, no ground was touched. At twenty after three o'clock, on the same day, she experienced another shock, which produced the same sensations, but in a much less degree; the ship was then in long. 17° 4′ W, lat. 39° N ; the lead was again thrown, and no bottom could be found with 200 fathoms of line. After this period, until three o'clock on the following morning, the 3d instant, several other similar convulsions were felt—but every successive one producing a less effect: the whole number of shocks was 12.

Extract of a letter from Captain Welsh, of the Claudine, arrived in the Downs from Batavia.—" On the 9th of February, off *St. Michael's,* we experienced very tempestuous weather, with a tremendous confused sea. The wind shifting from SW to SE and NE, with constant lightning and heavy rain. On the 10th, at half-past 8 p. m. the ship then under reefed fore-sail and main-stay-sail, we were much alarmed by a severe shock of an earthquake, which lasted four or five seconds.—(PUB. LEDGER.)

TABLE CXVI.

	1816.		Wind.	Pressure. Max.	Pressure. Min.	Temp. Max.	Temp. Min.	Hygr. at 9 a. m.	Rain, &c.
2d Mo.	L. Q.	Feb. 20	NE	30·00	29·88	47°	34°	70	
		21	W	30·07	30·00	46	36	80	4
		22	SW	30·19	30·07	49	34	75	—
		23	S	30·19	30·11	50	27	80	
		24	SW	30·11	29·99	50	35	98	
		25	SW	30·04	29·85	53	32	70	12
		26							
		27	W	29·75	29·45	51	29		8
	New M.	28	NW	29·88	29·75	40	22		
		29	NW	29·86	29·82	35	25	53	
3d Mo.		March 1	S	29·82	29·43	41	28	59	
		2		29·43	29·15	42	36	75	25
		3	SW	29·25	29·06	45	29	80	10
		4	SW	29·19	29·13	46	26	76	21
		5	S	28·97	28·90	44	32	74	—
		6	SW	29·10	28·92	46	33	67	34
	1st Q.	7	S	29·13	29·02	50	32	76	27
		8	NE	29·32	28·99	43	33	80	18
		9	N	29·67	29·32	38	30	75	—
		10	NW	29·87	29·86	41	26	62	
		11	SW	29·65	29·62	50	41	77	39
		12	SW	29·64	29·50	52	40	90	8
	Full M.	13	W	29·87	29·82	52	38	70	—
		14	SW	29·82	29·32	52	43	83	—
		15	S	29·76	29·32	52	26	58	23
		16	W	29·74	29·67	49	30	73	
		17	SW	29·74	29·49	47	34	60	—
		18	NW	29·56	29·43	52	35	78	17
		19	NW	29·91	29·56	47	34	51	3
				30·19	28·90	53	22	72	2·49

NOTES.—Second Mo. 20. Light clouds. 21. Several birds sing: *Cirrostratus* beneath large *Cirrus*. 22. Cloudy: drizzling: fair: windy. 23. White frost, which speedily went off: there appears to have been a dripping mist in the night: *Cirrocumulus:* fair. 24. *Cirrus: Cirrostratus:* cloudy: hollow wind. 25. A gale from SW, with showers: changed to NW in the night. 27. A snow shower early, which was followed by sleet and rain: much wind in the night. 28. Light *Cirrostratus*, a. m.: windy: *Cumulus* and *Cumulostratus* succeeded. 29. Slight hoar frost: fair, with light clouds: hygrometer went to 42°.

Third Mo. 2, 3. Rain at intervals. 4. *Cirri*, a. m. consisting of streamers rising from a horizontal base, with *Cirrostratus* below: heavy clouds: wind: p. m. hail, sleet, rain: lastly, upon the wind getting somewhat northerly, a heavy short storm of snow. 5. Clear a. m.: the ground crusted with yesterday's snow. 6. Various modifications of cloud, a. m.: heavy showers, p. m. 7. Cloudy: some rain, a. m.: *Nimbi*. 8. Rainy. 9. Snow storm: much evaporation: fair night. 10. Fair. 11. Stormy: very wet, p. m. 12. Temperature 50° at 9 a. m.: wind and rain. 13. Much wind: rain at intervals: *Nimbi*. 14. Wet morning: the wind SE: stormy day and night. 15. Much wind: rain, p. m.: calm at night. 16. Hoar frost: fair: calm: hygrometer went to 45°; and although it was overcast through the day, with the usual indications of rain in the sky, yet none fell. 17. *Cirrus*, with other light clouds, a. m.: wet, p. m. 18. Wet morning: hollow southerly wind, which changed to NW, with *Nimbi* at night, and blew strong. 19. A raw blustering day, with much evaporation evident.

RESULTS.

Prevailing Winds Westerly.

Barometer: Greatest height . . . . 30·19 in.
Least . . . . 28·90 in.
Mean of the period . . . . 29·606 in.
Thermometer: Greatest height . . . .   53°
Least . . . .   22°
Mean of the period . . . . 39·46°
Mean of De Luc's hygrometer . . . .   72°
Rain . . . . 2·49 in.

Character of the period: cloudy, wet, and windy: vegetation has made but little progress.

EARTHQUAKE.

On Sunday, March 17, about half-past twelve o'clock, a violent concussion of the earth was sensibly felt at Doncaster, and at Bawtry, Blyth, Carlton, Worksop, Sheffield, Chesterfield, Mansfield, Nottingham, Lincoln, Gainsborough, &c.—(DONCASTER PAPER.)

On Sunday a smart shock of an earthquake was perceptibly felt in Lincoln, at about ten minutes before one o'clock in the day. The undulation appeared to be from west to east, and lasted from about a minute and a half to two minutes. The wind was at the time SE, cold, and with every appearance of rain. Pictures and other articles hanging on the walls were set in a swinging motion. At Newark, also, and the neighbouring villages, the shock was distinctly felt, as well as at Leicester and Loughborough.—(STAMFORD PAPER.)

TABLE CXVII.

1816.			Wind.	Pressure. Max.	Pressure. Min.	Temp. Max.	Temp. Min.	Hygr. at 9 a. m.	Rain, &c.
3d Mo.	L. Q.	March 20	NW	29·96	29·91	47°	32°		
		21	E	30·01	29·96	48	28		
		22	SE	30·20	30·01	53	35	70	
		23	E	30·27	30·20	48	34	55	
		24	NE	30·27	30·08	39	34	55	
		25	NE	30·08	30·05	41	34	58	
		26	NE	30·14	30·08	42	35	60	
		27	NE	30·14	30·10	42	33	62	
	New M.	28	E	30·12	30·10	42	33	53	
		29	E	30·16	30·12	43	32	47	
		30	E	30·15	30·12	45	25	51	
		31	SE	30·15	30·07	47	26		
4th Mo.		April 1	SE	30·07	29·80	50	29	61	
		2	SE	29·79	29·75	47	29	53	
		3	SE	29·91	29·79	43	27	49	
		4	E	29·97	29·95	51	26	60	
		5	Var.	29·95	29·58	57	29	75	
		6	SW	29·58	29·11	55	32	55	16
	1st Q.	7	W	29·03	28·95	49	33	60	31
		8	NE	29·16	29·12	48	31	62	
		9	N	29·15	29·09	46	39	65	21
		10	SE	29·44	29·34	55	35	69	—
		11	NE	29·61	29·44	50	38	82	52
		12	NW	29·70	29·61	49	33	86	32
	Full M.	13	N	29·70	29·49	40	24	60	—
		14	NW	29·61	29·49	40	28		3
		15	NW	29·62	29·55	45	32	55	
		16	NW	29·55	29·38	51	40	50	1
		17	NW	29·52	29·38	56	30	59	
		18	SE	29·64	29·48	59	36	56	
				30·27	28·95	59	24	60	1·56

NOTES.—Third Mo. 21. Breeze: sunshine. 22. The same.
23. About sunset, a body of shallow *Cumulostratus*, with an abrupt
boundary forward, advanced from the E. 24. Cloudy: breeze.
25. The same. 26. The same. 27. The same. 28. Breeze stronger,
unsteady: *Cumulus*. 29. Breeze: *Cumulus* passing to *Cumulostratus*,
which cleared off at night, leaving a little *Cirrus* above. 30. Close
Cumulostratus, a. m. resembling drapery, as frequent in cold spring
weather: p. m. more open sky. 31. Large *Cumuli*, a. m.: wind SE,
gentle: the temperature was 45° at 10 a. m.: the roads are now dusty
to an extreme: *Cirrus* passing to *Cirrostratus* at evening.

Fourth Mo. 1. Hoar frost: sunshine: *Cirri*, with haze above.

2. *Cirrostratus,* with *Cirrus:* breeze much stronger. 3. Windy: hoar frost: *Cirrus.* 4. Hoar frost: sunshine: *Cirrus,* with *Cumulus:* drains emit an offensive gas. (This is a very common circumstance after long settled weather, before a change, and depends unquestionably in great measure on renewed electrical action on the general surface.) 5. White frost: misty from the N: the wind NE: sunshine: at night a lunar halo of the largest diameter: *Cirrostratus.* 6. The higher atmosphere filling, a. m.: *Cirrus, Cirrocumulus,* &c.: wind N: a smart breeze: then SW: wind and rain in the night. 7. Dripping a. m.: sleet: cloudy: windy: *Cumulostrati,* succeeded by numerous *Nimbi,* letting falling showers of large opaque hail, followed by rain: three distinct peals of thunder, p. m.: one N, another S, and a third near at hand, with lightning. 8. Cloudy: windy. 9. Windy at N, and more so in the night, seemingly from the westward: rain. 10. *Cumulostratus:* some dripping: rain by night. 11. Obscurity early a. m., with *Cirrostratus* beneath to S: rain and wind chiefly from the NE: p. m. moderate weather. 12. Sky as yesterday, but the *Cirrostratus* to NE: rain at mid-day: in the night a gale from NW, with snow for two hours. 13. The high ground to the W and NW is white with snow, a. m.: with us none remains. 14. White frost (8 a. m.), yet cloudy overhead, and a group far to the N, in which were *Nimbi:* in an hour's time this group reached us, and we had showers of heavy granular snow by intervals. 15. Clear morning: dew: fair, though with *Nimbi* in sight: very high tides, and much water out in the marshes. 16. A moderate gale at S and SW: some rain by night. 17. Cloudy a. m.: calm: mild. 18. *Cumulus, Cirrus:* sunshine, with cool breeze.

RESULTS.

Winds for the most part Easterly, non-electric, keen, and drying.

Barometer: Greatest height............30·27 in.

Least.....................28·95 in.

Mean of the period...........29·762 in.

Thermometer: Greatest height.............. 59°

Least..................... 24°

Mean of the period...........39·66°

Mean of De Luc's hygrometer at 9 a. m. 60°

Rain............................... 1·56 in.

The mean temperature of this period is full 8° lower than that of the corresponding portion of 1815. It has accordingly presented a striking contrast to the latter in its effects on the vegetable kingdom; not a single day having occurred in it of that which cultivators emphatically denominate " growing weather," when a moist air co-operates with a rising temperature to stimulate vegetable life, and make way for the unfolding of its products.

TABLE CXVIII.

	1816.	Wind.	Pressure. Max.	Min.	Temp. Max.	Min.	Hygr. at 9 a. m.	Rain, &c.
4th Mo. L. Q.	April 19	SW	30·03	29·64	56°	28°	52	
	20	SE	30·03	29·68	58	36	50	
	21	NE	29·70	29·68	52	44	50	
	22	SE	29·69	29·65	59	51		32
	23	E	29·75	29·69	70	43	65	
	24	NE	29·78	29·69	70	42	58	
	25	E	29·99	29·78	68	39	46	
	26	NE	29·99	29·94	68	39	61	
New M.	27	NE	29·94	29·81	68	37	56	
	28	SE	29·81	29·50	68	44	56	
	29	W	29·50	29·44	67	47	46	—
	30	SE	29·54	29·42	61	36	57	5
5th Mo.	May 1	SE	29·66	29·54	62	42		
	2	NW	29·87	29·66	63	34	41	—
	3	NW	29·93	29·83	59	48	61	18
1st Q.	4	SW	29·96	29·78	54	45	60	—
	5	NW	29·80	29·62	61	45	75	22
	6	SW	29·86	29·80	58	40	60	
	7				58	45	46	—
	8	SW	29·86	29·44	60	39		50
	9	SW	29·52	29·40	56	39	60	
	10				54	39	63	—
Full M.	11	NW	29·40	29·17	54	32		55
	12	NW	29·49	29·17	49	29	53	1
	13	W	29·75	29·49	57	30	53	1
	14	SE	29·77	29·75	62	37	55	5
	15	SW	29·78	29·77	64	44	74	1
	16	NE	29·78	29·64	72	49	54	1
	17	W	29·71	29·64	68	38	48	
	18	NW	29·76	29·71	52	41		
			30·03	29·17	72	28	54	1·91

NOTES.—Fourth Mo. 19. Cloudy a. m.: cool dry wind. 20. Warm forenoon: about noon, a murmuring S wind, with traces of a solar halo. 21. a. m. Obscurity above, with rudiments of the *Cumulus* beneath it: after this, thunder clouds in the S horizon: rain followed these appearances, and continued during most of the forenoon: swallows appeared to-day: the hygrometer went to 70°. 22. Fine: *Cirrus, Cirrocumulus,* &c. 23. Very fine day: blue sky, with large *Cumuli,* and the lighter modifications above. 24. Warm forenoon: a smart easterly breeze p. m.: the hygrometer went to 35°: *Cirrus* predomi-

nated. 25. Brisk wind at NE and SE: the sky clear and pale.
26. Fine day: steady breeze. 27. Much dew: clear morning: then
Cumulostratus, with a breeze. 28. Dew: clear morning: *Cirrostratus*
appeared, passing afterwards to *Cumulostratus:* at sunset, *Cirrus* appeared above. 29. Little or no dew: the sky full of a confused
mixture of *Cirrus, Cirrocumulus,* &c.: some drops of rain, followed
by more in the night. 30. Overcast: dripping.

Fifth Mo. 1. Fair. 2. Cloudy at intervals, with a few drops: much
Cirrostratus to the westward. 3. Rain at intervals, chiefly in the
night. 4. Completely overcast a. m., with *Cirrostratus:* a wet day.
8. Very rainy p. m., after a little hail about noon. 9. A little rain
a. m.: some sunshine p. m. 10. Rainy the whole day. 11. Fair in
the evening. 13. A little rain p. m. 17. Very fine day: cool evening. 18. Fair, but cold.

RESULTS.

Prevailing Winds Easterly in the fore part, and Westerly, with rain,
in the latter part, of the period.

Barometer: Greatest height................30·03 in.
Least........................29·17 in.
Mean of the period...........29·686 in.
Thermometer: Greatest height............... 72°
Least.. .... 28°
Mean of the period...........50·83°
Mean of the hygrometer .... 54°
Rain...................... .... 1·91 in.

During this period the leafing of the more forward trees has proceeded, for the most part, under the retarding influence of cold
breezes. Twice, the temperature having risen for a few days, the accumulation appears to have gone off in local thunder storms. In
travelling on the 17th of fifth month from Bristol to Southampton, I
had the opportunity of observing, from a convenient distance, the
gradual formation and discharge of a prodigious *Nimbus*, forming part
of a series of clouds, which for several hours continued to pour a
flood of rain, accompanied by large hail, thunder, and lightning, on
the country about Andover and Winchester. As the sun, which was
declining, strongly illuminated these clouds, they reflected a lively
copper tint above the indigo ground which marked the heavy rain:
the electrical light which fills the striking cloud at each discharge was,
therefore, with the stroke itself, imperceptible: but I assured myself
of the above-mentioned effects from subsequent information, as we
passed over the tract thus plentifully irrigated.

2 Q 2

TABLE CXIX.

		1816.		Wind.	Pressure.		Temp.		Hygr. at 9 a. m.	Rain, &c.
					Max.	Min.	Max.	Min.		
5th Mo.	L. Q.	May	19	NE	29·82	29·76	55°	35°	49	
			20	E	29·84	29·82	65	36	51	
			21	NE	29·84	29·77	62	41	48	
			22	NE	29·82	29·77	66	40		
			23	N	29·82	29·77	65	48	60	
			24	N	29·77	29·75	56	50	75	—
			25	SW	30·03	29·77	62	38	70	20
			26	SW	30·03	29·90	66	43	45	28
	New M.		27	NE	30·05	29·90	62	46	73	—
			28	SE	30·12	30·10	66	51	45	
			29	SE	30·10	29·87	69	54		
			30	NW	29·87	29·80	67	49		
			31	NE	30·00	29·80	64	46		
6th Mo.		June	1	NW	30·00	30·00	70	55		
			2	NW	30·00	29·90	72	49		
	1st Q.		3	NW	30·08	29·97	65	46		
			4	SW	30·05	29·90	65	50		
			5	NW	29·96	29·85	67	41		—
			6	NW	29·90	29·80	64	44	38	—
			7	Var.	29·76	29·48	61	46	45	20
			8	W	29·40	29·15	62	42	58	27
			9	NW	29·52	29·35	57	37	70	1
	Full M.		10	NE	29·86	29·52	58	41	47	22
			11	SW	30·00	29·86	65	39	47	
			12	SE	30·00	29·97	70	37	45	
			13	SW	29·97	29·91	75	54	39	
			14	N	29·91	29·87	53	48	53	8
			15	N	29·92	29·86	59	44	57	
			16	NW	29·92	29·92	67	36	47	
					30·12	29·15	75	35	52	1·26

Notes.—Fifth Mo. 19. Hoar frost: a fine day. 20. Clear morning. 23. Cloudy a. m.: much wind at N. 24. Misty: small rain at intervals. The hygrometer, these two days, noted at 7 a. m. 25. Overcast a. m.: wind at SW: rain, evening and night. 26. *Cumulus* cloud by day: *Cirrostratus* at evening. The hygrometer noted at half-past 6 a. m. 27. A wet morning, succeeded by close *Cumulostratus* through the day. 28—31, inclusive. Fair days.

Sixth Mo. 2. A fine breeze: large *Cirri* and *Cumuli*. 3. *Cirrostratus* prevails, with a cooler atmosphere. 5, 6. Showery. 7—10. Rain. 12. The hygrometer receded to 30°: *Cumulus* prevailed, and was succeeded by *Cirrus* in the evening. 13. This afternoon, there was a fine, but transient display of *Cirrocumulus*: in the N and NW there was an obscurity, mixed with rudiments of *Nimbi*. 14. After a warm, still night, a cold blowing day, with small rain at intervals. 15. Overcast, with *Cumulostratus*: cool breeze: in the evening *Cirrostratus*. 16. A fine day: the air becomes calmer.

RESULTS.

Winds rather Variable, but for the most part Northerly.

Barometer: Greatest height ....30·12 in.

Least....................... 29·15 in.

Mean of the period........... 29·850 in.

Thermometer: Greatest height .... 75°

Least..................... 35°

Mean of the period .... 54·15°

Mean of the hygrometer (for 20 days) .. 52°

Rain .... 1·26 in.

The character of this period has been, on the whole, ungenial: though not one frosty night has occurred, yet cloudy weather, with blighting winds, mostly predominated; and the mean temperature turns out nearly 5° lower than that of the corresponding portion of 1815.

COLD IN NORTH AMERICA.

A letter from Quebec, dated June 10, says—"We had a fall of snow here on the 8th instant, several inches in depth. The weather is still cold, and it snows at intervals; the trees are not yet in bloom, and the oldest inhabitant does not remember such a season."

There has not been, for upwards of 40 years, so backward a season known in Nova Scotia as the present. Although now in the middle of June, but little vegetation has taken place, and there is scarcely any seed sown in the ground. Ice was seen on the morning of the 11th June, in the harbour, and a few days since snow was falling in different parts of the country.

From New York it is stated, under date of the 15th of June, that the cold weather, and even frosts, continued: in the upper part of the States, large icicles were pending, and the foliage of the forests was blasted by the frost.

(PUB. LEDGER.)

FROM A FRIEND.

Wilmington, State of Delaware, Fifth Month, 1817.

" Our winter has been extremely severe. It is supposed to have equalled in cold and duration the winter of 1779—80. Our thermometer during the first and second months was often near cypher, and many times several degrees below it. I think it was twice ten degrees below 0, in this town. In New Hampshire, the mercury froze in the bulb—the thermometer was consequently *thirty-eight* degrees below 0."

TABLE CXX.

		1816.		Wind.	Pressure.		Temp.		Evap.	Rain, &c.
					Max.	Min.	Max.	Min.		
6th Mo.	L. Q.	June	17	SW	29·92	29·82	67°	48°	—	
			18	SW	29·88	29·79	71	53	36	
			19	W	29·96	29·88	69	47	—	
			20	N	29·96	29·95	74	55	25	
			21	NE	29·95	29·95	71	51	—	
			22	NE	29·95	29·94	78	53	21	—
			23	SW	29·85	29·82	69	50	—	1·06
			24	NW	29·97	29·85	63	48	—	
	New M.		25	NE	29·97	29·86	73	56	37	
			26	Var.	29·60	29·54	70	54	—	2·05
			27	NE	30·00	29·60	63	50	—	
			28	NW	30·07	30·00	69	47	—	
			29	Var.	30·08	29·90	78	58	39	
			30	Var.	29·90	29·76	76	53	16	19
7th Mo.		July	1	NW	29·80	29·75	63	51	—	
			2	SW	29·84	29·80	73	53	—	—
	1st Q.		3	NW	29·86	29·84	64	50	—	13
			4	Var.	29·75	29·72	65	50	44	18
			5	Var.	29·88	29·75	66	46	—	
			6	Var.	29·89	29·75	69	56	—	—
			7	S	29·75	29·70	69	52	—	—
			8	S	29·70	29·69	70	52	45	26
	Full M.		9	SW	29·69	29·66	70	51	—	—
			10	SW	29·66	29·63	73	51	—	05
			11	NW	29·72	29·66	66	54	—	—
			12	NW	29·90	29·72	65	48	63	—
			13	NW	29·94	29·90	67	49	—	—
			14	SW	29·90	29·69	65	58	—	32
			15	SW	29·69	29·68	71	52	—	12
			16	Var.	29·69	29·66	63	52	36	77
					30·07	29·54	78	46	3·62	5·13

NOTES.—Sixth Mo. 23. Cloudy morning: showery day: evening cold. 24. Cloudy morning: a strong cold wind from the NW. 26. The early part of the morning was fine: wind NE: changed to SW between 10 and 11, and began to rain, which continued without intermission all day: in the evening and night, was extremely heavy. 27. Morning very much overcast: the rain fallen from nine yesterday morning to nine this morning amounts to 2·95 inches, a very unusual quantity for the neighbourhood of London. 29, 30. Foggy mornings: overcast.

Seventh Mo. 1. A *Stratus* on the marshes at night. 2. A little rain about 10 p. m. 4. Showery day: some hail about 3 p.m.: wind variable, chiefly SW. 5. Showery morning: the day fine. 6. Morning overcast: heavy dew. 8. Showery morning: fine day: a heavy shower of rain between 9 and 10 p.m. 10. Showery morning: fine afternoon. 12. Cloudy morning: squally afternoon. 13. A heavy shower of rain about 10 p.m. 14. A gentle rain nearly all the day. 15. Rainy morning: showery day. 16. Very rainy day.

RESULTS.

Winds Variable: for the most part Westerly.

Barometer: Greatest height . . . . 30·07 in.
Least. . . . . 29·54 in.
Mean of the period . . . . 29·816 in.
Thermometer: Greatest height . . . .   78°
Least . . . .   46°
Mean of the period . . . . 60·3°
Evaporation . . . . 3·62 in.
Rain . . . . 5·13 in.

WHIRLWIND.

June 25.—At two o'clock, being a still sultry day, a whirlwind passed over the nursery ground of Mr. Henderson, in the Edgeware-road, which lifted seven lights from the green-houses and carried them to the height of the highest elm-trees; each of the lights weighs 50 or 60 pounds at least. At the same time two garden mats were carried to an immense height, so that the eye could not distinguish them.

SNOW REMAINING ON THE MOUNTAINS.

The *Kendal Chronicle* of July 4, says—" A traveller, who has visited the top of Helvellyn this day, brought to the office a lump of snow from that summit. The gentleman informs us, that he saw three or four patches of snow, varying in extent in different directions."

On a hill, the property of Sir A. Ramsey, in the parish of Fettercairn, at the distance of little more than twelve miles from the German Ocean, there was a remnant of a wreath of snow, which measured on the 5th July five feet deep and eighty yards in circumference.—(PUB. LEDGER.)

I may add to the above proofs of the coldness of the higher atmosphere during this summer, that in passing through Switzerland, I saw the snows of the preceding winter lying in very large masses, in hollows on the chain of the *Jura*, and on the *Mole* near Geneva, from whence they usually vanish in summer; and this at a time when the new snows had already begun to fall on the same summits.

TABLE CXXI.

	1816.		Wind.	Pressure. Max.	Min.	Temp. Max.	Min.	Evap.	Rain, &c.
7th Mo.	L. Q.	July 17	Var.	29·66	29·48	67°	50°	—	11
		18	SW	29·56	29·48	66	51	—	—
		19	SE	29·75	29·55	70	58	—	21
		20	SW	29·76	29·57	81	65	66	—
		21	SW	29·78	29·59	70	54	—	—
		22	SW	29·79	29·65	70	58	—	32
		23	SW	29·65	29·65	73	55	—	—
	New M.	24	Var.	29·66	29·65	64	52	59	45
		25	NW	29·86	29·66	65	54	—	05
		26	NW	29·95	29·86	68	53	—	—
		27	NW	29·96	29·80	64	53	26	—
		28	NW	29·80	29·74	64	46	—	
		29	Var.	29·74	29·63	63	45	—	
		30	NW	29·63	29·55	64	41	25	—
	1st Q.	31	NE	29·65	29·54	65	48	10	03
8th Mo.		Aug. 1	NW	29·80	29·65	63	49	—	—
		2	W	29·85	29·80	67	51	—	27
		3	SW	29·88	29·85	68	49	41	
		4	SW	29·88	29·85	69	47	—	
		5	NW	29·95	29·80	70	51	—	51
		6	SW	29·97	29·88	68	57	28	03
		7	SW	29·88	29·79	70	57	—	—
	Full M.	8	SW	29·79	29·77	74	55	—	—
		9	SW	29·98	29·79	67	53	42	—
		10	NW	30·06	29·98	65	57	—	—
		11	SW	30·06	30·06	70	57	—	—
		12	NW	30·06	30·00	67	56	35	26
		13	SW	30·00	29·80	66	52	—	
		14	SE	29·80	29·58	68	58	—	08
		15	SE	29·59	29·53	71	56	25	09
				30·06	29·48	81	41	3·57	2·41

NOTES.—Seventh Mo. 18. Squally day. 19. Rainy morning: very boisterous wind all day, with showers. 20. Fine morning. 21. Showery day: a strong breeze from the SW. 24. Wind variable: very rainy day: some thunder in the afternoon: a *Stratus* on the marshes at night. 25. Foggy morning: a *Stratus* on the marshes at night. 30. A heavy shower of rain between 1 and 2 p. m.: some hail. 31. Very foggy morning: a thunder storm in the evening.

Eighth Mo. 2. Showery day. 5. Foggy morning: trees dripping: some thunder in the afternoon: very rainy night.

RESULTS.

Winds Variable: chiefly SW and NW.

Barometer: Greatest height . . . . 30·06 in.
Least. . . . . 29·48 in.
Mean of the period . . . . 29·771 in.

Thermometer: Greatest height. . . . .   81°
Least. . . . .   41°
Mean of the period . . . . 60·4°

Evaporation . . . . 3.57 in.
Rain . . . . 2·41 in.

STORMS, &c.

Our naval column this day bears the aspect of winter—strong gales, ships on shore, and loss of anchors, are rather unusual in the month of July.
(PUB. LEDGER, *July* 24.)

From all parts of the country we hear of damage done by the late storms, and floods occasioned by the heavy rains.—(PUB. LEDGER, *July* 26.)

An earthquake was felt at Martinique on the 15th of August, which lasted a considerable time, and being unusually severe, excited great alarm among the inhabitants.—(PUB. LEDGER.)

EARTHQUAKE IN SCOTLAND.

The letters and papers received from the North bring accounts of this phenomenon, so uncommon in our country. The shock appears to have extended over the counties of Ross, Inverness, Moray, Banff, Aberdeen, Kincardine, Forfar, Perth, and Fife; and was indistinctly felt in Edinburgh and Glasgow. By all accounts, there has been no loss of lives, although considerable damage to property. The following is an extract of a letter from Inverness, dated August 14:—" Last night, exactly a quarter before 11 o'clock, the town of Inverness and the surrounding country was fearfully shook by an earthquake. We fled to the street, where we found almost every inhabitant; women and children screaming, and a very considerable portion of them naked. Many fled to the fields, and there remained for the greater part of the night. Chimney tops were thrown down or damaged in every quarter of the town. The Mason Lodge, occupied as an hotel, was rent from top to bottom, the north stalk of the chimney partly thrown down—one of the coping stones, weighing, I should think, from 50lb. to 60lb. was thrown to the other side of the street, a distance not less than 60 feet. The spire of the steeple, which I think one of the handsomest in Scotland, has been seriously injured, and must in part be taken down. The spire is an octagon; and within five or six feet of the top, the angles of the octagon are turned nearly to the middle of the square or flat side of the octagon, immediately under it. Notwithstanding the vast quantities of stones and bricks that have been thrown from such a height, not one person has received any hurt. I have only further to remark, that it was not attended with any of those phenomena that have been said to accompany earthquakes. The day had been beautiful and serene, and still continues so; no agitation or rising was observable in the river. It has been frequently observed, that in countries subject to these awful visitations, the mercury suddenly falls in the barometer: this I instantly attended to, but no alteration whatever took place."

See also a most circumstantial account of this earthquake, certainly one of the most violent on record in this island, communicated to Dr. Thomson, by Tho. Lauder Dick, Esq., in the Annals of Philosophy, vol. viii. p. 364, &c.

TABLE CXXII.

1816.		Wind.	Pressure.		Temp.		Hygr. at 9 a. m.	Rain, &c.
			Max.	Min.	Max.	Min.		
8th Mo. L. Q.	Aug. 16	Var.	29·84	29·59	65°	52°		—
	17	NW	29·98	29·84	61	49		21
	18	NW	30·15	29·98	62	51		—
	19	NW	30·16	30·10	67	55		13
	20	N	30·18	30·10	64	42		—
	21	N	30·18	30·14	66	50		—
	22	NW	30·14	30·10	68	53		
New M.	23	NW	30·10	30·10	66	56		—
	24	NE	30·18	30·10	70	45		—
	25	NE	30·20	30·18	69	44		
	26	NE	30·18	30·15	67	47		
	27	E	30·19	30·15	66	44		
	28	E	30·19	30·17	70	46		
1st Q.	29	NW	30·17	29·95	64	52		
	30	NW	29·95	29·36	61	53		23
	31	SE	29·46	29·30	59	46		1·09
9th Mo.	Sept. 1		29·57	29·22	49	40		92
	2	NW	29·67	29·57	55	30!		
	3	SW	29·67	29·33	56	43		—
	4	NW	29·49	29·31	57	37	66	—
	5	W	29·86	29·49	60	40	73	18
Full M.	6	SW	29·86	29·79	61	50	75	
	7	W	29·79	29·77	67	52	62	
	8	SW	29·77	29·65	65	47	52	
	9	SW	29·48	29·38	60	54	77	55
	10	SW	29·80	29·77	64	47	60	18
	11	W	29·93	29·80	65	42	49	
	12	SW	30·13	29·93	65	41	70	
	13		30·13	29·99	65	55	70	—
			30·20	29·22	70	30		3·49

NOTES.—Eighth Mo. 19. Clear morning. 23. A little rain, evening. 24. Cloudy morning: smart shower at noon: a considerable appearance for thunder, evening. 25. *Stratus:* fine sunset. 26. Much dew, a. m : cloudy. 28. Foggy morning: fair: *Stratus.* 29. Very foggy: fair. 30. Cloudy morning. 31. Wet morning: stormy night, with heavy rain.

Ninth Mo. (1. This was a very wet day at *Paris.*) 2. Hoar frost this morning: there is said to have been thick ice formed in several exposed situations. 3. Rain, with much wind in the night. 4. A hard shower of hail, followed by rain about noon. 5. Wet morning:

fair day after. 6. Cloudy a. m.: misty to the S: overcast day. 7. Maximum of temperature at 9 a. m.: fair day. 8. Fair: wind rising at SW in the evening, and *Nimbi* about. 9. It began to rain about 7 a. m. and continued till 3 p. m.: after which a very stormy night, the lower clouds moving much faster than the higher. 10. Fair day. 11. Much dew: fair: *Cirrocumulus* for two days past. 12. Much dew: fair. 13. Dew: misty a. m.: a few drops at midday, after much cloud and wind: the sky at sunset appeared to be clearing gradually.

RESULTS.

Winds Variable in the fore part, Westerly in the latter.

Barometer: Greatest height................	30·20 in.
Least........................	29·22 in.
Mean of the period............	29·872 in.
Thermometer: Greatest height................	70°
Least.......................	30°
Mean of the period............	55·29°
Rain	3·49 in.

The two preceding periods, with so much of the present as is comprehended in the Eighth Month, apply in point of local circumstance, to the scite of the Laboratory, where the Observations were conducted by my friend and partner, John Gibson.

During a tour of nine weeks, in this interval, extending from Amsterdam to Geneva, I had ample occasion to witness the fact, that the excessive rains of this summer were not confined to our own islands, but took place over a great part of the continent of Europe. From the sources of the Rhine among the Alps, to its embouchure in the German ocean, and through a space twice or thrice as broad from east to west, the whole season presented a series of storms and inundations. Not meadows and villages alone, but portions of cities and large towns, lay long under water: dikes were broken, bridges blown up, the crops spoiled or carried off by torrents, and the vintage ruined by the want of sun to bring out and ripen the fruit.

While the middle of Europe was thus suffering from wet, the North for a time, and to a certain extent, was parched with drought, and public prayers appear to have been ordered, about the same time, at Dantzic and Riga for rain, and at Paris for sunshine! The probable natural causes of this unequal distribution may form a subject for discussion in another part of the work: it would in this place be premature. I have deduced the principal part of the evidence respecting it from numerous circumstantial accounts of the weather given in the public papers during the summer months.

TABLE CXXIII.

	1816.		Wind.	Pressure. Max.	Min.	Temp. Max.	Min.	Hygr. at 9 a. m.	Rain, &c.
9th Mo.	L. Q.	Sept. 14	SW	29·99	29·95	67°	48°	53	
		15	S	29·96	29·93	72	46	56	
		16	S	29·95	29·87	74	44	61	
		17	E	29·85	29·78	72	48	70	
		18	NE	30·02	29·78	70	47	72	—
		19	NE	30·02	29·98	60	43	59	
		20	E	29·98	29·64	62	40	53	
	New M.	21	SE	29·60	29·57	60	50	62	
		22	NE	29·74	29·60	66	48	63	
		23	N	29·73	29·64	63	51	66	
		24	S	29·97	29·73	61	51	72	0·13
		25	SE	30·09	29·97	67	42	59	
		26	N	30·12	30·06	64	48	72	—
		27	Var.	30·06	29·96	63	47	56	—
	Full M.	28	S	29·96	29·70	63	43	60	—
		29	SW	29·65	29·32	60	39	80	0·43
		30	SW	29·65	29·42	60	47	65	—
10th Mo.		Oct. 1	SW	29·58	29·55	63	46	89	66
		2	W	29·84	29·41	62	45	78	16
		3	Var.	29·84	29·80	58	49	65	16
		4	SW	29·82	29·80	66	53	85	—
		5	W	29·82	29·78	66	52	77	
	1st Q.	6	SE	29·78	29·74	68	56	71	0·54
		7	E	29·87	29·74	59	55	85	0·27
		8	E	29·95	29·87	66	54	80	
		9	E	29·98	29·95	63	55	65	—
		10	NW	29·95	29·92	65	51	64	
		11	NE	30·06	29·92	60	44	69	—
		12		30·06	30·04	57	47	73	—
		13		30·04	30·03	55	39	65	
				30·12	29·32	74	39	68	2·35

Notes.—Ninth Mo. 14. The sky overcast, with numerous beds of *Cirrocumulus*, which at sunset changed to *Cirrostratus*, and became red. 15. An electrical smell a. m.: much dew: there appears to have been a *Stratus* in the night: serene day. 16. Much dew: large plumose *Cirri*: very fine. (A dreadful hurricane at *Guadaloupe*. Houses were levelled, plantations destroyed, and the soil driven about like dust in a whirlwind.) 17. Much dew: misty: fine day: in the evening a solitary *Cumulus* cloud in the W, spired up to inosculate with a *Cirrostratus* above it, a *Stratus* at the same time appearing in the meadows nearer to us: several discharges of lightning in the NW followed these appearances: the barometer, which had fallen a little, now rising. 18. Overcast day: a little rain perceptible in the evening. 19. Fair, with a grey sky, and a few distinct *Cirrostrati* beneath. 20. Grey sky a. m.: then sunshine, with a breeze. 21. Cloudy a. m.: some dripping at midday: fair evening. 22. Cloudy, in different modifications. 23. (At *Paris*, abundance of rain, with a SE wind.) 24. Showers: breeze at SE, evening: misty air. 25. Misty to S: fair, with *Cumulus* beneath *Cirrocumulus:* a *Stratus* at night. 26. Various modifications of cloud: some rain in the night. 27. *Cumulus* beneath *Cirrocumulus.* 28. Cloudy morning, the wind increasing from the westward: rain, midday: fair evening. 29. Wet morning: much wind at S till evening: stormy night. 30. Fair, with *Cirrostratus:* much wind in the night.

Tenth Mo. 1. Wet morning: much wind: lunar halo at night. 2. A plentiful dew: cloudy afterwards, with much wind: drizzling rain at intervals. 3. Cloudy a. m.: wind N: drizzling: in the night easterly, with misty air. 4. Overcast: small rain. 5. Fair. 6. Misty morning: much dew: *Cirrocumulus* in the superior stratum, as for some days past at intervals: rain p. m. and night. 7. At twenty minutes before one this morning a loud explosion of electricity, which kept the ground in a sensible tremor for several seconds: it was followed by thunder in long peals, and vivid lightning to the south and east for above an hour: also by much wind and rain: the day was cloudy and drizzling after. 8. Fair: mostly cloudy: *Stratus.* 9. Cloudy: breeze at E: large *Cirri:* a few drops of rain: a well-formed mushroom was brought me, which measured twelve inches over the crown, and weighed twenty ounces. 10. *Cirrocumulus* above *Cumulus:* calm. 11. Fair, save a few drops. 12. Cloudy. 13. Cloudy: breeze.

RESULTS.

Winds Variable.

Barometer:	Greatest height	30·12 in.
	Least.........	29·32 in.
	Mean of the period	29·840 in.
Thermometer:	Greatest height....	74°
	Least............	39°
	Mean of the period	55·9°
Mean of the hygrometer at 9 a. m.......		68°
Rain		2·35 in.

Extract of a letter from Lloyd's Agent at Gottenburgh, dated the 2d Oct.— " On the 30th ult. at daylight, it came on to blow heavy at WSW; at nine it veered to the W, and blew a tremendous hurricane, which continued till almost 4 p. m. yesterday, when it moderated a little, but still blew a gale from the WNW. At five this morning it abated, and now blows only a fresh breeze."

TABLE CXXIV.

	1816.		Wind.	Pressure.		Temp.		Hygr. at 9 a. m.	Rain, &c.
				Max.	Min.	Max.	Min.		
10th Mo.	L. Q.	Oct. 14	SE	30·13	30·03	60°	38°	88	
		15	SE	30·13	29·99	57	41	82	
		16	SE	29·99	29·71	58	34	90	—
		17	Var.	29·72	29·69	52	35	93	12
		18	NW	29·85	29·72	53	30	77	
		19	NW	29·85	29·56	56	41	78	17
	New M.	20	W	29·58	29·52	50	39	60	
		21	Var.	29·63	29·61	49	40	55	—
		22	NW	29·89	29·63	54	30	86	5
		23	Var.	29·92	29·74	46	30	80	
		24	S	29·74	29·41	52	34	65	10
		25	SE	29·53	29·30	54	29	95	44
		26	SE	29·62	29·59	53	37	95	
	1st Q.	27	E	29·59	29·57	58	46	67	
		28	E	29·55	29·51	58	38	72	
		29	SE	29·51	29·15	56	42	88	—
		30	SE	29·16	29·09	57	42	90	37
		31	S	29·25	29·16	55	39	83	19
11th Mo.		Nov. 1	Var.	29·32	29·25	53	32	95	—
		2	S	29·22	29·17	47	35	94	81
		3	SW	29·63	29·22	48	36	85	11
		4	NE	29·63	29·62	51	44	86	19
	Full M.	5	SE	29·62	29·36	53	40	85	11
		6	Var.	29·33	29·29	50	36	90	—
		7	N	29·45	29·29	41	20		—
		8	SW	29·45	28·87	43	24	90	34
		9	SW	29·30	28·72	47	34	83	—
		10	NW	29·70	29·30	34	20	66	—
		11	Var.	29·23	29·03	44	26	73	54
				30·13	28·72	60	20	81	3·54

NOTES.—Tenth Mo. 14. Much dew these two mornings past: gossamer. 15. Dew: somewhat misty air, with an electric odour: sunshine. 16. Misty: the trees dripping: calm: then sunshine, with a breeze at SE, and *Cumuli*, &c. 17. Morning overcast: rain by nine: rainbow: drips at intervals: fair. 18. Fair: a fine breeze through the day: twilight pale orange, with *Cirrostratus.* 19. Wind and rain in the night. 20. Cloudy day. 21. Obscurity to the S, indicating rain there: after dark, a flash of lightning to the SE, with a small meteor. 22. Rain a. m., succeeded by a fair day and night. 23. Hoar frost: clear a. m.: fine day. 24. Cloudy: windy at S: wet evening:

clear night. 25. Cloudy: windy at SE: very wet, p. m. 26. Misty morning: fair day. 27. Various clouds: fair. 28. Overcast, with *Cirrostratus:* clear night. 29. Very heavy dew: misty at night, with a *Nimbus* to the SW. 30. Rainy. 31. Misty: close, electric air: rain, p. m. and evening.

Eleventh Mo. 1. Misty a. m.: *Cumulostratus, Nimbus:* a little rain. 2. Misty morning: much rain this day and night: hail. 3. Misty a. m.: fair day: wet evening. 5. *Cumulostratus:* small rain. 6. *Nimbi,* with other clouds. 7. Snow this morning in the high lands: shower, with hail: snow again at night. 8. Very white frost: much smoke and cloud accumulated over the city: cloudy evening: the *Cirrostrati* appeared convergent to the rising moon: in the night the temperature, which had not passed 33° in the day, advanced to 43°, with much wind and rain from the southward. 9. Fair morning: squally, p. m.: after dark a small meteor and lightning to the S, in which direction *Nimbi* were visible at sunset. 10. Snowy morning: clear day: the hygrometer receded to 45°. 11. Clear, a. m.: wind NW: the ground crusted over with some snow, which fell last night: a bank of clouds in the SW, and some attenuated *Cirrostrati* aloft: at night rain: the wind violent at SW.

RESULTS.

Prevailing Winds SE and NW.

Barometer: Greatest height................30·13 in.
Least.........................28·72 in.
Mean of the period............29·512 in.
Thermometer: Greatest height................ 60°
Least....................... 20°
Mean of the period............43·12°
Mean of the hygrometer at 9 a. m....... 81°
Rain.............................. 3·54 in.

The barometer has been throughout unsteady, and its oscillations towards the end considerable, chiefly in depression. Eight days only were without rain.

The city of *Chester* on Wednesday (November 6), at noon, nearly enveloped in darkness; candles or lamps were obliged to be lighted in all the houses; this was succeeded by a slight shower. On Thursday successive falls of hail and rain took place; on Friday the frost was uncommonly severe, and on Sunday the snow which fell was above two feet deep in the streets.

An article from *Rochelle,* of the 14th November, states, that a tempest occurred on the French coast, in the night of the 11th and 12th instant, which occasioned many shipwrecks. Wind first N, then SSE.—(PAPERS.)

TABLE CXXV.

	1816.		Wind.	Pressure. Max.	Min.	Temp. Max.	Min.	Hygr. at 9 a.m.	Rain, &c.
11th Mo.	L. Q.	Nov. 12	NW	29·70	29·23	52°	38°	69	1
		13	W	29·68	29·64	56	39	75	
		14	W	29·64	29·44	45	27	63	
		15	NW	29·73	29·44	37	28	67	
		16	NW	30·04	29·73	38	28	65	
		17	SW	30·04	29·68	43	28	70	
		18	SW	29·75	29·68	47	32	90	7
	New M.	19	SW	29·93	29·75	46	39	93	
		20	S	30·00	29·97	51	37	95	
		21	SE	29·97	29·78	44	29	76	
		22	E	29·78	29·75	38	26	68	
		23	NE	29·89	29·74	33	17	67	
		24	NW	29·93	29·89	30	18	82	
		25	SE	29·93	29·88	40	25	90	—
	1st Q.	26	N	30·22	29·93	42	30	98	41
		27	W	30·30	30·22	45	32	96	
		28	NW	30·43	30·30	44	32	80	
		29	N	30·56	30·43	44	30	98	
		30	N	30·62	30·56	38	30	77	
12th Mo.		Dec. 1	NW	30·62	30·40	36	25	78	
		2	W	30·35	30·33	38	28	73	
		3	E	30·35	30·32	40	32	92	
	Full M.	4	E	30·32	30·08	42	36	91	
		5	SE	29·52	29·45	43	36	80	11
		6	SW	29·45	29·35	40	29	68	21
		7	SW	29·52	29·47	39	27	75	18
		8	SW	29·69	29·52	37	25	95	—
		9	W	29·55	29·46	42	27	95	17
		10	Var.	29·25	29·20	46	35	88	13
		11	NW	29·36	29·25	40	27	77	
				30·62	29·20	56	17	81	1·29

NOTES.—Eleventh Mo. 12. Windy. 13. Strong breeze: sunshine: much water out in the marshes. 14, 15. Breezy: sun and clouds. 16. Slight hoar frost. 17. *Cirrostrati* at a great elevation, in which a solar halo appeared for an hour or two, a.m. 18. Maximum temperature at 9 a.m.: windy: overcast: dripping forenoon. 19. The diurnal temperature disturbed by the solar eclipse (of which see the particulars in the Note). 20. Fair at 2 p.m.: *Cumulostrati* formed rapidly, and passed off at a great elevation, the wind veering S. 21. Fair: rather windy. 22. Cloudy a.m.: steady breeze: p.m. *Cumuli*. 23. Cloudy: steady breeze. 24. A serene sky. 25. Hoar frost, the third morning: overcast, with *Cirrostratus* and haze: rain at night. 26. Very misty a. m.: *Cirrostratus* sweeps the ground: rain. 27. Much *Cirrostratus*, especially to the N, of delicate texture: fair day. 28. The sky was so completely shrouded in a *Cirrostratus*, without the smallest opening to admit the sun's rays, that from nine to three the temperature did not ascend 2°: at sunset the sky cleared pretty suddenly, showing red *Cirri* above for a considerable time: the lower air, which had been transparent, now filled with mist. 29. Misty a. m.: calm: very bright sun at midday: lunar halo. 30. Hoar frost: breeze: very fine day.

Twelfth Mo. 1. Cloudy morning. 2. Lightly clouded. 3. Misty by *Cirrostratus*, soon after sunrise, during which a hoar frost formed: grey lofty sky. 4. Grey: little wind. 5. Idem: the lunar eclipse not visible for clouds: much wind, with rain after. 6. Fine. 7. Very fine day: but stormy at night. 8. Rain, the middle of the day: clear night. 9. Very white frost a. m., and rime on the shrubs: *Cirrostratus* floating at an elevation of two or three yards: the temperature rose quickly, and it rained, p. m. and night. 10. Fine day, with *Cirrus* and *Cirrostratus:* wet and stormy fore part of night. 11. Fair: at sunset a lofty and wide-spread *Nimbus* in the NW at 10 p. m.: a bright shooting star to the W: some large flakes of moist snow in the night.

RESULTS.

Winds Variable and moderate till after the Full Moon.

Barometer: Greatest height .... 30·62 in.

             Least..................... 29·20 in.

             Mean of the period .... 29·866 in.

Thermometer: Greatest height .... 56°

             Least..................... 17°

             Mean of the period .... 35·80°

Mean of the hygrometer .... 81°

Rain .... 1·29 in.

The rain fell at three distinct intervals, and chiefly by night, increasing greatly each time in quantity and continuance.

The atmosphere was so darkened yesterday morning in the vicinity of the metropolis by the thick fog, combined with smoke, that in some parts it appeared like a cloudy night. In the neighbourhood of Walworth and Camberwell it was so completely dark, that some of the coachmen driving stages were obliged to get down and lead their horses with a lantern.

(EVENING MAIL, *November* 27)

The Inverness Journal, of November 17, says—" The winter has commenced with a severity almost beyond example: frost, rain, and snow have been almost incessant during the last week; and the greater proportion of corn, still uncut or in stooks, has suffered material injury. We regret to say, that several lives have been lost. A postboy of Bennet's, coming from the south, was obliged to leave his chaise on the road, and would have been lost but for the lights shown from the windows of Moyhall, which he reached nearly in a frozen state. *The obvious advantage of keeping lights, in stormy nights, in the windows of houses in the country, has thus been illustrated.*"

The winter, it appears, has set in with extreme severity in the interior of the Continent. At Augsburg, on the 19th of November, the eclipse of the sun was entirely obscured by a fall of snow, which commenced at seven o'clock in the morning, and lasted till noon. The ground was covered with snow a foot in depth. There was a great fall of snow at Frankfort on the same day, and Reaumur's thermometer showed at from 9 to 10 degrees below the freezing point.

METEOR.

The meteor lately seen at Glasgow, was also visible at Perth. It made its appearance in the SW, in the form of a small star, and gradually increased in magnitude till it reached the zenith, when it subtended an angle nearly equal to that of the full moon. In shape it resembled a paper kite. After passing the zenith, it again seemed to diminish in size, owing, no doubt, to its gradually receding from the observer, till its altitude was equal to about 30°, when it exploded like a sky-rocket. The report of the explosion, which is described as more tremendous than the noise of the loudest thunder, reached the ear about three minutes after the meteor vanished, so that it could not be less than 40 miles distant, and probably about 27 miles above the surface of the earth. From the various particulars collected concerning it, its diameter must have been about 240 yards. The light which it yielded was very considerable, being sufficient to render the smallest objects visible.—(PUB. LEDGER, *December* 12.)

Influence of a Solar Eclipse on the Diurnal Temperature.

The radiation from the sun is so manifestly the cause of a diurnal elevation of temperature on the earth, that in a considerable eclipse of that luminary we ought to expect some diminution, for the time, as well of its heating, as of its illuminating effect; but I do not know that any one has yet submitted this consequence to the test of observation by the thermometer. In determining to do this, I thought it right to have a proximate standard wherewith to compare the results I might obtain; and, therefore, although the day was by no means so favourable for the purpose as some preceding clear ones, I caused a number of observations to be taken of the progress of the diurnal temperature on the 18th of the eleventh month; and devoted the forenoon of the 19th, on which the eclipse of the sun took place, to a similar investigation by myself.

It will not be necessary very minutely to detail the observations of the 18th, which, as well as those during the eclipse, were made with the Six's thermometer which I constantly use. At 6 a. m. the thermometer stood at 40°, the sky being overcast with *Cirrostratus* clouds pretty close and dense, and a steady breeze blowing from SSW. At $6^h 45^m$, and for half an hour after, the temperature did not exceed 41°. At $7^h 45^m$, the sun being up, it was 42°; and from this time to 11^h it advanced (with some interruption, but no depression intervening) to 47°. In the interval before noon occurred a depression of *half a degree*, which being over, the temperature attained its maximum for the day, of 47·5°, at half-past twelve. During the time there fell about 0·07 inch of rain. The afternoon was fair, the temperature descended more rapidly than it had risen, and in the fore part of the night (as it appears) touched upon 32°.

A. M.   *Eleventh Month, 19. Day of the Eclipse.*

h. m.   *Ther.*

6 30 — 35° Temperature going down, having risen some degrees in the night: dawn of day perceptible: light *Cirrostratus* clouds, with a gentle breeze at SW. Barometer 29·68 in.

7 0 — 34·5 Day Breaks. Barometer 29·69 in.

7 20 — 34 Breeze increasing: a veil of clouds above, passing off with a definite edge to NE, but leaving detached streaks below.

7 42 — 33 Barometer 29·72 in. Sun not yet visible, being hid by a low mist.

8 0 — 34 The sun has been some time shining; but is now among thin streaks of cloud.

8 20 — 35 The sun among streaks of cloud, the disk scarcely visible; so that the commencement of the eclipse was not observed.

8 45 — 36 The sun still behind a light skreen of *Cirrostratus*, through the interstices of which the disk is at times distinctly seen eclipsed. The thermometer now ceases to rise.

8 55 — The temperature now begins to decline, although there is less cloud than heretofore.

9 0 — 35·5 Barometer 29·75 in.: but hesitating, as if it would fall.

9 15 — 35 A somewhat thicker bed of cloud now coming over the sun increases the obscurity of the eclipse, which is yet not very considerable. There is perhaps as much *light* as when the sun was 20 min. high.

A. M.

h. m.   *Ther.*

9 20 —   The thermometer now tends to rise again.

9 30 — 35·5° Barometer 29·77 in. Cloudiness, by *Cirrostratus*, becoming general.

9 45 — 37 Eclipse visibly going off.

10 0 — 38

10 15 — 39 The clouds again lighter.

10 30 — 40 Near the termination of the eclipse.

10 50 — 41 Barometer stationary: wind more to the W: the clouds thicken again.

11 5 — 42

11 40 — 43 The clouds tend to form *Cumulostratus*.

P. M.

12 20 — 45 Barometer rising.

1 30 — 45 In this interval the temperature has fallen half a degree, and risen again: as it did yesterday an hour earlier.

2 0 — 46 Temperature the same at this hour yesterday: cloudy.

3 0 — 46 Idem.

4 0 — 44 Barometer 29·80. No rain or strong wind has occurred during the observations.

The foregoing observations, I presume, will be found satisfactory. The temperature on this day was falling (as is commonly the case) before sunrise; presently after which it began to rise. This effect continued until a considerable portion of the sun's rays became intercepted, when it fell again, to near the middle of the eclipse (by my watch, which had not been adjusted), and in proportion as the latter went off, resumed its former movement, rose steadily, and attained its maximum at nearly the same degree as the day before, though later in the afternoon. Had the elevation proceeded from 8^h 45^m to 9^h 15^m, at the rate which it had assumed previous to this interval, the temperature at 9^h 15^m, instead of 35°, would have been 38°, and the progress of the diurnal elevation would have been still more uniform than the day before; which was to be expected from the greater uniformity of the sky. Now as the depression coincides sufficiently with the time of the sun's being under eclipse, and as no other disturbing cause is apparent, we may conclude that there resulted from this cause an interruption to the diurnal accumulation of heat at the place of observation, the amount of which, at six or eight feet from the surface, was equal to 3° of Fahrenheit's thermometer.

Since these Observations were published (in the Annals of Philosophy), I have met with the following passage, in an " Observation of the Solar Eclipse, April 1, 1764, by Dr. John Bevis." Philosophical Transactions, abridged, vol. xii. p. 113. " A full digit of the sun, or more, remained uneclipsed. The day-light was but inconsiderably diminished.—Fahrenheit's thermometer, placed without door to the north, stood at 50° when the eclipse began, *and fell but one division while it lasted.*"

Thus it appears that the thermometer has been heretofore resorted to on this occasion, and with a result which agrees very well with the present; the *depression* in our own case being *but one degree,* and the digits eclipsed nine.

In an *annular* eclipse of the sun observed at Edinburgh, by Colin Maclaurin, the middle of the time being about half-past 3 p. m., February 18, 1736–7, the observer remarks, " It was *very cold* at this time; a little thin *snow fell,* and some small pools of water in the college area, where there was no ice at two o'clock, *were frozen* at four.—Some curious gentlemen found that a common burning-glass, which kindled tinder at $3^h 59^m$, and burned cloth at $4^h 8^m$, had no effect during the annular appearance, and for some time before and after it." Idem. viii. 171.

The Moon on this occasion was scarcely to be discerned on the middle of the sun's disc, without the help of a dark glass: and with glasses it appeared much smaller than it should have done, their respective apparent diameters considered. The moon therefore stops fewer of the sun's rays than previous theory would suggest: or it collects and sends forward into the shadow a portion of those which would otherwise be only tangents to its orb. And in proportion as the illumination thus passes by the obstacle, so must the heat likewise. We want still the observations on a delicate thermometer during a *total* eclipse, to enable us to ascertain fully the power of the moon to intercept our heat.

It is somewhat curious, that in the observations of the Lunar period to which these notes are annexed, we have occasion to see, on two nearly successive days, the still greater effect of a complete skreen of cloud in stopping heat (Eleventh month, 28, *Note*), and of the smoke of the city in bringing on darkness.—See the note under the Results.

TABLE CXXVI.

1816.			Wind.	Pressure.		Temp.		Hygr. at 9 a. m.	Rain, &c.
				Max.	Min.	Max.	Min.		
12th Mo.	L. Q.	Dec. 12	SW	29·00	28·65	47°	32°	95	40
		13	NW	29·35	29·00	43	28	70	—
		14	SW	28·82	28·53	47	33	76	52
		15	W	29·38	28·82	41	29	65	
		16	Var.	29·49	29·31	38	32	81	—
		17	Var.	29·28	29·22	49	35	97	20
		18	N	30·09	29·28	44	33	92	—
	New M.	19	N	30·47	30·09	36	25	70	
		20	NE	30·47	30·35	32	22	74	
		21	N	30·35	30·07	28	14	85	
		22	Var.	30·15	30·00	31	17	75	
		23	SW	30·00	29·66	46	32	93	10
		24		29·72	29·53	48	33	83	
		25		29·72	29·38	48	35		
	1st Q.	26		29·42	29·27	50	33		37
		27	NW	29·78	29·42	43	27	90	—
		28	SW	29·40	29·30	49	32	90	18
		29	W	29·91	29·40	42	34	64	—
		30	SE	29·67	29·62	44	37	83	1·08
1817.		31	SW	29·67	29·51	48	39	99	9
1st Mo.		Jan. 1	S	29·35	29·30	48	36	88	35
		2	SW	29·49	29·45	44	32	90	72
	Full M.	3	S	29·63	29·20	48	31	94	—
		4	W	29·73	29·12	52	36	87	76
		5	W	29·73	29·52	44	32	72	70
		6	SW	30·25	29·52	45	33	80	14
		7	NW	30·42	30·25	38	22	65	—
		8	N	30·53	30·43	30	26	90	
		9	E	30·58	30·53	30	21	82	
				30·58	28·53	52	14	83	5·61

NOTES.—Twelfth Mo. 12. A wet day after a frosty night: the fore part of this night a violent storm of wind from the westward, the barometer rising fast. 13. Calm a. m., with a turbid sky: about noon a clap of thunder followed by some heavy sweeping hail. 14. The day fine, with *Cirrus*: after dark, the sky being suddenly overcast, the wind rose to an excessive degree of violence, with rain: the barometer had fallen since noon rapidly, the minimum (which is also the lowest point for the year) occurred very early in the morning of 15. During the storm in the night I was twice sensible of a tremor of the earth, distinct from the effects of the wind, and lasting perhaps a quarter of a minute. This I found reason to attribute to the shock of electrical discharges, as I found it had thundered twice about the time. 15. A gale a. m., with clouds: the day fine, and windy afterwards. 16. Hoar frost: fair, with *Cirrostratus:* at night a small meteor moving eastward. 17. Wet a. m.: the wind SE. 18. The wind passed by W to N, and gradually rose to a moderate gale: a few drops about noon. 19. Wind inclining to NE, a stiff breeze: snow p. m., part of which lay on the ground. 20. A brilliant evening twilight, which was reflected by a haze in the eastern sky. 22. Clear, save a little *Cirrostratus:* wind gentle and variable. 23. Wind rising a. m., the air turbid: sleet and rain followed, with a windy night. 25. Very fine day, the barometer nearly quiescent at 29·72 till evening: at night the wind rose, and was boisterous till the morning. 26. Much rain, in squalls, p. m.: a lunar corona at night. 27. *Nimbi:* the sun set fiery red, and much enlarged: windy. 28. Hoar frost: fair day: night very tempestuous, with rain from the southward, which began, with the rise of the barometer, at 10 p. m. 29. Wind, followed by *Cirrocumulus,* and a calm night. 30. A very wet day and night. 31. Misty: little wind.

1817. First Mo. 1. Windy: wet p. m. 2. Fair: at 5 p. m. hygrometer 65°, and the moon yellow: notwithstanding these indications, there fell much rain and snow after it in the night. 3. Fair day, save a slight shower: the night (after bright moonlight) very stormy, with rain. 4. Small driving rain: at night another gale of wind. 5—7. Much the same alternations as for several preceding days. 8. Very white rime: misty about noon: *Cirrostratus.* 9. The wind has now gained the E, having gradually shifted round by N: hoar frost: misty air.

RESULTS.

Barometer: Greatest height..............30·58 in.
          Least........ ....28·53 in.
          Mean of the period ....29·649 in.
Thermometer: Greatest height.............. 52°
          Least...................... 14°
          Mean of the period ....36·10°
Mean of the hygrometer. . ... 83°
Rain.......... .... 5·61 in.

The wind, though chiefly westerly, has been very variable in direction, and equally so in force; presenting a succession of heavy gales, with intervals of frost and rain. This enormous quantity of rain, being *twice* as much as usually constitutes a *wet moon* in this part of the island, had the usual effect of inundating the low lands to a great extent, especially when met by the spring tide after the full. By a mark preserved at the Laboratory, however, I find that in the inundation of 1809, the river Lea rose 15 inches higher than on the present occasion.

HAMBURGH, *December* 13.—It has blown a hurricane here the whole of this day.

December 15.—A dreadful storm of wind and rain raged during the whole of last night in *Paris*, and did considerable damage.

January 2, 1817.—The accounts from Holland, of the effects of the weather, are deplorable: the winter season seems to have set in with heavy rains, and in consequence almost all the rivers and canals have overflowed their banks.

(PAPERS.)

SOOT ON SNOW.

I have observed, that the flakes of soot which are deposited on the surface of snow, and remain there exposed to the sun's rays, disappear after some hours, leaving a cavity, the bottom of which is visible and clean. There is therefore probably a real oxidation of the carbon, after which it is dissolved in the water, in the way in which the colouring matter of cloth is destroyed in bleaching.

LEECHES UNHURT BY FROST.

Among the cold-blooded animals which resist the effects of a low temperature, we may reckon the common leech; which is otherwise interesting to the meteorologist, on account of its peculiar habits and movements under different states of the atmosphere. A group of these animals, which I left accidentally in a closet without a fire, during the frost of 1816, not only survived, but appeared to suffer no injury from being locked up in a mass of ice for many days.

END OF THE FIRST VOLUME.

C. Baldwin, Printer,
New Bridge-street, London.

MONTHLY MEAN TEMPERATURE in LONDON for Ten Years, from 1797 to 1806.

Year	First Mo. Jan.	Sec. Mo. Feb.	Third Mo. Mar.	Four. Mo. April	Fifth Mo. May	Sixth Mo. June	Sev. Mo. July	Eight Mo. Aug.	Nin. Mo. Sept.	Ten. Mo. Oct.	Elev. Mo. Nov.	Twel. Mo. Dec.
1797	37·32	37·33	39·85	47·41	53·96	57·56	65·48	61·80	56·95	48·95	43·39	42·66
1798	30·62	30·94	42·96	51·60	56·31	64·00	63·86	65·62	58·89	52·17	41·61	35·19
1799	35·09	38·21	39·33	44·06	52·41	58·04	62·32	60·49	56·45	49·67	44·68	34·30
1800	38·67	35·99	39·41	50·99	57·02	57·98	65·58	66·41	60·08	50·04	44·06	40·03
1801	41·05	40·39	46·07	47·64	55·30	60·85	63·01	65·36	61·11	52·72	41·96	37·49
1802	34·62	40·83	43·15	50·98	52·15	59·58	59·14	67·56	60·23	52·48	42·38	39·30
1803	35·27	38·27	44·38	50·41	53·01	59·05	66·28	61·57	55·14	51·07	43·70	42·78
1804	44·98	38·94	43·23	46·29	59·59	63·46	62·80	63·19	61·75	53·46	45·93	37·14
1805	36·17	40·67	41·01	47·58	52·43	57·70	62·09	64·99	61·71	49·59	41·76	40·75
1806	42·45	43·14	42·73	45·70	57·77	62·50	63·96	64·51	59·49	53·19	49·13	48·75
Greatest variation of the Mean	10·36	7·45	6·74	7·64	7·44	6·44	7·11	7·07	6·61	4·51	7·52	14·45

MONTHLY MEAN TEMPERATURE in the COUNTRY for Ten Years, from 1807 to 1816.

Year	First Mo. Jan.	Sec. Mo. Feb.	Third Mo. Mar.	Four. Mo. April	Fifth Mo. May	Sixth Mo. June	Sev. Mo. July	Eight Mo. Aug.	Nin. Mo. Sept.	Ten. Mo. Oct.	Elev. Mo. Nov.	Twel. Mo. Dec.
1807	34·14	38·37	36·14	46·00	50·78	58·91	64·72	63·27	53·08	53·06	37·54	36·39
1808	35·99	35·91	37·19	43·05	59·91	59·08	67·19	63·51	56·41	47·27	42·13	34·96
1809	36·42	44·92	43·64	43·21	57·01	58·75	61·14	61·49	57·46	50·17	39·63	40·41
1810	35·06	39·42	43·19	48·09	50·98	60·21	61·25	61·62	59·06	51·01	44·34	39·85
1811	32·64	42·08	45·99	51·69	61·10	61·58	61·84	59·33	57·83	56·04	45·40	38·75
1812	36·88	42·37	40·75	43·85	54·75	55·78	58·79	57·83	55·43	49·41	41·53	35·51
1813	34·84	43·67	43·96	48·36	56·72	58·64	63·30	61·33	57·69	48·67	41·33	38·43
1814	26·71	33·17	37·82	50·64	50·56	55·99	64·75	62·17	55·68	46·86	39·85	40·20
1815	32·77	41·48	47·22	48·36	58·72	60·11	61·00	61·94	55·38	49·70	38·34	36·25
1816	36·13	53·39	39·24	45·21	51·30	57·54	59·74	59·00	54·21	49·95	37·26	35·89
Greatest variation of the Mean	10·17	11·75	11·08	8·64	10·54	5·80	8·40	7·41	5·98	9·18	8·14	5·45

EXTREMES of TEMPERATURE in each Month for Ten Years, with the attendant Winds.   [TABLE B. Temp.]

Year	First Mo. Jan.	Sec. Mo. Feb.	Third Mo. March	Fourth Mo. April	Fifth Mo. May	Sixth Mo. June	Seventh Mo. July	Eighth Mo. Aug.	Ninth Mo. Sept.	Tenth Mo. Oct.	Elev. Mo. Nov.	Twelfth Mo. Dec.
1807	51° NW. SW	57° W. SW	60° Var.	80° SW a E	85° Var. a E	79° NW a NE	*87° Var.	82° E	72° W. NW	66° SW	55° SW a N	54° SW
	+13 W a NW	18 N	18 NW a N	22 N	39 NE	42 NW	39 E	46 W	26 N	33 N	22 NW	17 NW
1808	51 SW a S	52 SW	54 E	66 W	87 Var. a SW	76 W. NW	*96 N	80 SW	71 NE	65 SW	66 S	53 NW a SW
	+12 NW	17 N	18 E a NE	22 NW	38 Var. E	44 SW. NW. V	44 NE	43 S	34 NW a N	34 NW	25 Var. a NE	14 N
1809	56 S	57 SW	66 S	59 W a SW	80 SE	79 E a SE	81 NE	*82 E a SE	74 V. Ely	67 NE	53 N	54 SW
	+18 E	29 W	31 E. S	24 NW	33 W a NW	42 SW	41 NW	45 SW	35 NW	27 NW	22 NW	28 Var. SW
1810	51 SW	66 SW	60 SW	75 NE	74 E	83 NE a SE	81 Var. Wly.	83 NE a NW	*85 E a SE	71 E	58 SW	52 Wly.
	+10 NW a N	11 NW a N	24 E a NE	30 NE	32 N	37 NE	41 SW	40 NW	38 N a E	27 NW	29 W	25 SW. NW
1811	51 NW	54 S. SE	62 Var. NE	77 SE	84 E.	*88 V. a S	80 V. a SE	76 NE. S. SW	80 V. E	73 S	62 SW	54 SW. W
	+14 NW	25 NW	26 V. NE	26 N	39 NE a E	43 N	43 NW	42 NW	39 W	38 S	25 E. NW	21 SW. N
1812	30 S	54 S	59 SW	58 V. NW. NE	76 SE	75 N W	75 SE. W	*78 SE	73 W a SW	69 NW a SW	55 SW	52 S
	26 NW	26 E a N	24 NE	25 NE	32 Var. a NE	39 N	41 NW	43 N	34 NW	32 SW	24 W	+18 NE
1813	50 SW	57 SW a S	67 NW. SW	69 E	78 NW a S	*85 E. NE	82 S	80 SW	75 S	60 W. SW	58 SW	54 SW
	20 Var. a N	30 Var. NW	24 NE	27 SW	30 W	37 N	42 NW	40 N	40 NE	27 Var. a NW	25 E	+19 E. NW
1814	41 SE. SW	50 SW. S	60 SW	74 SE a N	70 SE. NE	85 W a E	*91 SE	80 NW	75 SW a SE	67 SW a SE	54 SW	56 SW
	+ 8 NW. N	18 NE. SE. E	21 E. NE	32 SW. NE	31 NW	36 NE	43 NW	37 N	33 NE a N	24 NW a N	19 NW a N	25 SW
1815	44 NW	57 W	73 Va. SW	70 SE	*80 NW	*80 E a SW	*80 W	79 SW	79 SE	66 S	57 SW. S	53 SW. SE
	+17 Var. N	25 W a NW	29 W a NW	28 N	34 W a NW	38 NE	42 N.NE, NW	43 SW	31 NE	32 E	18 NW	21 SW
1816	50 W a SW	53 SW	53 SE	70 E. NE	72 NE	78 NE. Var.	*81 SW a SE	71 SE	74 S	68 SE	56 W	50 S
	21 NW,SE,SW	+5 E a N	25 E	26 E a SE	29 NW	36 NW a N	41 NW	44 NE. E	30 NW	29 SE	17 NE	14 N

Note : the mark * denotes the greatest elevation of the *year*, and the mark † the greatest depression.

The material originally positioned here is too large for reproduction in this reissue. A PDF can be downloaded from the web address given on page iv of this book, by clicking on 'Resources Available'.

[TABLE C. Barom.]

GREATEST and LEAST HEIGHT of the BAROMETER in each Month for Ten Years, with the attendant Winds.

Year	First Mo. Jan.	Second Mo. Feb.	Third Mo. March	Fourth Mo. April	Fifth Mo. May	Sixth Mo. June	Seventh Mo. July	Eighth Mo. Aug.	Ninth Mo. Sept.	Tenth Mo. Oct.	Elev. Mo. Nov.	Twelfth Mo. Dec.
1807	*30·60 N	30·56 N	30·56 NE	30·23 SW	30·30 NE	30·28 NW	30·19 N	30·15 NW	30·16 NE	30·25 NW	30·4 NW	30·41 NW a SE
	28·82 NEa SW	28·90 S. SW	29·28 NW	29·22 V. a SW	28·90 SW	29·55 SW	29·55 SW	29·50 SE	29·36 SW	29·03 NW	†28·68 SW	29·24 SW
1808	30·51 SW. NE	*30·71 NE	30·46 E a NW	30·29 NW	30·24 SW	30·24 NE	30·15 NE	30·16 NW	30·99 W	30·33 NE	30·30 NE	30·36 W. N
	28·93 S	29·20 SW	29·55 SE	29·09 SW	29·52 SE	29·64 W	29·43 W	29·50 SW	29·98 S	29·15 SW	†28·72 S	29·15 SW
1809	30·12 E a NE	30·17 SW	*30·49 W a N	30·36 N	30·32 SW	30·39 NE	30·16 NW	30·06 Var.	30·13 SW a NE	30·33 NW	30·47 NW a N	30·25 SW
	28·50 S	28·70 SW	29·41 SE	29·06 SW	29·32 NW	29·25 S	29·43 V. Nly.	29·24 S	29·20 SE	29·89 SE	29·10 SW	†28·25 SW a SE
1810	30·48 E a NW	30·50 NW. V	30·17 NE	30·18 E	30·41 E	30·40 E. SE	30·31 W a N	30·21 NW	30·40 E. NE	30·35 N. NE	30·15 NE	*30·51 N
	29·83 S. SE	28·98 SW	28·81 SW	29·30 SE	29·30 Var.	29·90 NE	29·44 S	29·32 NW	29·70 SW	29·30 SW	†28·50 E. SW	29·33 W a SW
1811	30·51 NW	30·20 NW	*30·61 NE	30·23 Var.	30·10 W	30·40 SEa NW	30·19 NW	30·25 SW. NW	30·29 N. NE	30·21 Var. SW	30·41 W	30·20 NW
	29·68 E	29·04 SE	29·30 W	29·22 SE	30·48 SE	29·19 NE	29·75 SE	29·35 SW	28·86 W	†28·65 S. Var.	29·22 SW	28·90 ·W a S
1812	30·25 N	30·06 NW	30·35 SE a E	30·18 s a E	30·27 E. SE	30·40 NE. N	30·39 N	30·15 N. NE	30·28 NW	29·98 W. SE	30·38 N	*30·51 NE
	29·28 SE	29·30 S	29·10 SW a E	29·55 Var.a SW	29·30 S	29·32 SW	29·40 V. a SW	29·76 SE. SW	29·67 SE	†28·53 SW	28·96 NE	28·98 E
1813	*30·50 NW	30·45 NW	*30·50 NW	30·34 NE	30·10 NW	29·43 NWa NE	30·18 W	30·26 NW	30·49 NW	30·12 NE	30·34 W a SW	30·49 NW
	29·30 NW	29·27 SW	29·18 SE	29·18 SW	29·39 Var. SW	30·07 NEaNW	29·40 W a SW	29·42 S	29·25 SW a S	†28·64 SW	29·02 NE. NW	29·09 SE
1814	30·17 N	*30·42 NE	*30·42 NE	30·40 SE.NE.N	*30·42 NE	30·07 NEaNW	30·15 SWa NW	30·24 NW	30·24 E	30·20 NE	30·28 NW	30·18 E a S
	†28·22 SE. SW	29·12 SW a SE	29·07 SW	29·23 S	29·28 Var.a SE	29·38 NW	29·56 SWa NW	29·40 N a SW	29·52 S	29·03 SW a SE	29·12 N a W	29·25 S
1815	30·45 NE	30·47 NW	30·22 Var.	30·25 NE	30·23 Var.	30·17 W	30·24 NE	30·25 NW	30·11 NE	30·22 NE	*30·38 NE	30·52 SWaNW
	29·88 E a SE	29·28 Var. SE	28·86 S	28·74 Var. N	29·51 NW	29·51 Var. a S	29·47 NWaSW	29·35 NW	29·28 W a SE	29·25 SW	28·95 W a S	†28·85 NWaSW
1816	30·38 NE	30·35 N.W.SW	30·27 E	30·07 SE a E	30·12 SE	30·08 NW	29·06 NW	30·20 NE	30·13 SW	30·13 SE	*30·62 N	*30·62 NW a N
	28·87 SE a SW	28·90 SE a S	28·90 S	28·95 W a SW	29·17 W	29·15 W	29·48 Var. SW	29·30 SE	29·22 E a SE	29·09 SE	28·72 SW	†28·53 SW

Note : the mark * denotes the greatest elevation of the year, and the mark † the greatest depression.

TABLE of the WINDS and RAIN, with the number of Days on which Rain fell in each Month, for Ten Years.

The table records, for each year (1807 through 1816, with an average at the foot), the winds and rain for each month (1. Jan., 2. Feb., 3. Mar., 4. April, 5. May, 6. June, 7. July, 8. August, 9. Sept., 10. Oct., 11. Nov., 12. Dec.). For each month the wind directions are given (N–E, E–S, S–W, W–N) together with the total rain in inches and the number of days on which rain fell.

Note: The results marked † were obtained at about 43 feet elevation from the ground, and those to which no mark is annexed, upon or near the ground. The mark * denotes that the result is in part an estimate.

The material originally positioned here is too large for reproduction in this reissue. A PDF can be downloaded from the web address given on page iv of this book, by clicking on 'Resources Available'.

MONTHLY AMOUNTS of EVAPORATION, in Inches and Decimal parts.

Year	First Mo. Jan.	Sec. Mo. Feb.	Third Mo. March	Fourth Mo. April	Fifth Mo. May	Sixth Mo. June	Seventh Mo. July	Eighth Mo. Aug.	Ninth Mo. Sept	Tenth Mo. Oct.	Elev. Mo. Nov.	Twelfth Mo. Dec.	Annual Averages
	With the Gauge at 40 feet elevation.												
1807	0·52 Frost; damp air.	1·64	2·66	3·60 much Electricity.	5·08	4·52	6·03	5·04	4·17	3·98	1·80	0·91 Frosty with Snow.	
1808	1·19	1·65	3·23 Rain very deficient.	3·22 Rain an average	5·39	3·99	5·51	3·82	2·95	2·39	2·02	1·14	37·85 Inches.
1809	1·24	2·40 Mean Temp. 48°92	2·79 Mean Temp. 43°64	2·49 Mean Temp. 43°21	6·07 Mean Temp. 57°01	†4·14		†4·74	3·02	2·63	1·27	1·38	
	With the Gauge variously situated.												
1810	1·04	†1·28	2·71				4·85	4·56	3·45	2·76	0·79 Rain in excess.	1·68	
1811	†1·05	†2·24	2·85	3·14	3·75	4·53	3·66	3·64	3·91	2·65	1·64	1·53	33·37 Inches.
1812	1·25	2·30 Stormy.	1·95	3·19	4·35	4·23	3·65	2·38	2·70		0·61		
	With the Gauge at the ground, the first two months excepted.												
1813	0·45	2·14 Mean T. 43°67° much wind	1·64 Mean T. 43°96° much wind	2·20	2·25	2·93	3·22	3·04	2·31	0·90 much Rain after fair weather	0·63	0·91 Fogs, preceding frost	90·28 Inches.
1814	0·93 Severe Frost	†0·56 Frosty	0·83 Frosty, with Snow	2·16	2·14	1·89	3·42	4·48	2·04	1·15 wet	0·53 wet	0·71 wet and frosty	
1815	0·50	0·78	1·45	1·81	2·14	†1·83	†2·55						30·50 Inches.
Monthly averages	0·832	1·613	2·234	2·726	3·696	3·507	4·111	3·962	3·063	2·208	1·168	1·112	

Note: the amounts marked † are in part estimated, some days in each month having been omitted to be taken.

The material originally positioned here is too large for reproduction in this reissue. A PDF can be downloaded from the web address given on page iv of this book, by clicking on 'Resources Available'.

MEAN RESULTS of LUNAR PERIODS, arranged by the Solar Year.

Year	Period 1 (Solstice)	BRUMAL PERIODS (Solstice)			VERNAL PERIODS (Equinox)			ESTIVAL PERIODS (Solstice)			AUTUMNAL PERIODS (Equinox)			Extra
	1	2	3	4	5	6	7	8	9	10	11	12	13	
1806–7	Period 1 — 29·54 in.													
1807–8		11 — 29·80	15 — 29·82	16 — 30·02	17 — 30·16	18 — 29·86	19 — 29·87	20 — 29·89	21 — 29·97	22 — 29·76	23 — 29·76	24 e — 29·78	13 — 29·47 in.	
1808–9	26 — 29·85	27 — 29·88	28 — 29·41	29 — 30·17	30 — 29·81	31 — 29·83	32 — 29·73	33 — 29·92	34 — 29·75	35 — 29·66	36 — 29·84	37 — 30·08	23 — 29·76	
1809–10		39 — 29·76	40 — 30·07	41 — 29·91	42 — 29·61	43 — 29·89	44 — 29·80	45 — 30·08	46 — 29·94	47 — 29·85	48 — 30·02	49 e — 29·91	50 — 29·39	
1810–11	31 — 29·66	53 — 29·88	53 — 29·80	54 — 29·82	55 — 30·12	56 — 29·70	57 — 29·88	58 — 30·02	59 — 29·83	60 — 30·02	61 — 29·73	62 — 29·61	63 — 29·89	
1811–12		64 — 29·71	65 — 29·90	66 — 29·74	67 — 29·74	68 — 29·60	69 — 29·81	70 — 29·88	71 — 29·97	72 — 29·97	73 — 30·04	74 — 29·47	75 — 29·68	
1812–13		76 — 29·88	77 — 30·02	78 — 29·93	79 — 30·11	80 — 30·00	81 — 29·68	82 — 29·89	83 — 29·87	84 — 29·80	85 — 30·01	86 e — 29·75	87 — 29·62	88 — 29·73
1813–14		89 — 29·76	90 — 29·92	91 — 29·89	92 — 29·84	93 — 29·77	94 — 29·91	95 — 29·90	96 — 29·92	97 — 29·88	98 — 29·94	99 — 29·65	100 — 29·63	
1814–15		101 — 29·66	102 — 29·77	103 — 29·78	104 — 29·67	105 — 29·78	106 — 29·81	107 — 29·71	108 — 29·96	109 — 29·88	110 — 29·89	111 — 29·75	112 — 29·78	113 — 29·84
1815–16		114 — 29·61	115 — 29·69	116 — 29·60	117 — 29·76	118 — 29·68	119 — 29·85	120 — 29·81	121 — 29·77	122 — 29·87	123 — 29·84	124 — 29·51	125 — 29·86	
1816–17		126 — 29·65	127 — 29·84	128 — 29·39	129 — 30·07	130 — 29·03	131 — 29·53	132 — 29·75	133 — 29·74	134 — 29·63	135 — 29·81	136 — 29·88	137 — 29·88	
1817–18		138 — 29·51	139 — 29·78	140 — 29·97	141 — 29·47	142 — 29·64	143 — 29·76	144 — 30·00	145 — 30·04	146 — 30·05	147 — 29·86	148 e — 29·70	149 — 29·83	
1818–19		151 — 30·07	152 — 29·52	153 — 29·77	154 — 29·74	155 — 29·83	166 — 29·90							150 — 29·92
Average of each column on 10 years		29·745	29·788	29·874	29·870	29·814	29·812	29·899	29·879	29·854	29·883	29·736	29·725	General average of 10 years. 29·823

Note.—Period 1 is not included in the average for the Winter Solstice: the mean of 39 is calculated up to the New Moon in the following period: 83 has the Summer Solstice, and the periods marked e the Autumnal Equinox. In 45 and 46, the mean is taken in each case 0·06 in. lower than it stands in my Results, the Barometer employed for them being known to be too high.

MEAN RESULTS of LUNAR PERIODS, arranged by the Solar Year.

		BRUMAL PERIODS.			VERNAL PERIODS.			ESTIVAL PERIODS.			AUTUMNAL PERIODS.			
	Period 1	Solstice			Equinox			Solstice			Equinox			
	1	2	3	4	5	6	7	8	9	10	11	12	13	
1806–7	1 / 41·60	2 / 42·53	3 / 34·75	4 / 38·21	5 / 36·28	6 / 51·12	7 / 55·41	8 / 59·00	9 / 66·08	10 / 64·96	11 / 59·94	12 / 53·20	13 / 37·92	
1807–8	26 / 41·01	14 / 36·26	15 / 36·98	16 / 35·91	17 / 38·11	18 / 41·82	19 / 53·18	20 / 59·41	21 / 65·60	22 / 65·30	23 / 60·34	24 / 48·84	25 / 45·36	
1808–9		27 / 33·68	28 / 40·85	29 / 42·46	30 / 44·01	31 / 48·58	32 / 58·89	33 / 59·37	34 / 61·95	35 / 61·15	36 / 55·20	37 / 47·74	38 / 39·32	
1809–10		39 / 40·82	40 / 36·43	41 / 37·63	42 / 42·08	43 / 47·00	44 / 50·53	45 / 54·20	46 / 62·00	47 / 59·98	48 / 60·65	49 e / 56·00	50 / 45·48	
1810–11	51 / 41·15	52 / 35·86	53 / 38·05	54 / 43·93	55 / 46·75	56 / 57·19	57 / 63·19	58 / 60·00	59 / 61·00	60 / 59·20	61 / 57·85	62 / 54·85	63 / 42·95	
1811–12		64 / 38·05	65 / 38·00	66 / 41·73	67 / 41·50	68 / 43·57	69 / 65·45	70 / 55·87	71 / 58·34	72 / 57·83	73 / 54·93	74 / 51·46	75 / 41·31	
1812–13		76 / 36·68	77 / 36·25	78 / 40·58	79 / 42·50	80 / 49·11	81 / 54·79	82 / 57·93	83 s / 61·69	84 / 63·88	85 / 58·44	86 / 55·28	87 / 43·41	88 / 39·63
1813–14		89 / 32·36	90 / 31·31	91 / 31·93	92 / 44·14	93 / 51·30	94 / 50·50	95 / 60·01	96 / 65·50	97 / 60·20	98 / 53·79	99 / 46·43	100 / 39·05	
1814–15		101 / 40·13	102 / 32·66	103 / 41·43	104 / 47·44	105 / 48·56	106 / 58·58	107 / 60·10	108 / 61·36	109 / 62·13	110 / 57·00	111 / 50·79	112 / 41·75	113 / 35·96
1815–16		114 / 36·52	115 / 32·00	116 / 39·46	117 / 39·66	118 / 50·83	119 / 54·15	120 / 60·30	121 / 60·40	122 / 55·29	123 / 55·90	124 / 43·12	125 / 35·80	
1816–17		126 / 36·10	127 / 40·03	128 / 42·06	129 / 41·50	130 / 43·85	131 / 50·70	132 / 61·38	133 / 59·32	134 / 57·65	135 / 55·76	136 / 43·27	137 / 47·11	
1817–18		138 / 32·66	139 / 28·46	140 / 31·20	141 / 39·70	142 / 45·36	143 / 52·84	144 / 62·36	145 / 67·24	146 / 63·32	147 / 58·68	148 e / 53·31	149 / 48·75	150 / 41·20
1818–19		151 / 35·86	152 / 39·31	153 / 40·78	154 / 49·20	155 / 51·67	156 / 56·90							
Average of each column on 10 years	....	37·92	35·73	39·63	49·25	48·92	55·67	58·62	62·39	60·59	56·70	50·75	40·68	

NOTE.—Period 1 is not included in the average for the Winter Solstice: the mean of 39 is calculated up to the New Moon in 40: period 83 has the Summer Solstice, and the periods marked e have the Autumnal Equinox, about their beginning: the rest include the points under which they are placed in the column.

MONTHLY AMOUNTS of RAIN for 14 Years, corrected for the elevation of the Guage.   [TABLE H. Rain, corrected.]

Rate of addition for the ground.	0·50	0·45	0·40	0·30	0·20	0·10	0·05	0·10	0·20	0·30	0·40	0·45	
Year	1. Jan.	2. Feb.	3. Mar.	4. April	5. May	6. June	7. July	8. Aug.	9. Sept.	10. Oct.	11. Nov.	12. Dec.	Amount for the year.
1797	1·440 in.	0·317 in.	1·087 in.	2·416 in.	1·723 in.	4·645 in.	1·352 in.	3·067 in.	4·873 in.	2·601 in.	2·062 in.	2·335 in.	27·918 in.
1798	1·657	1·004	0·466	0·672	1·945	1·056	3·022	1·677	2·924	4·456	4·278	1·242	24·399
1799	1·423	3·240	0·606	2·172	2·098	0·607	3·058	2·429	3·388	2·848	2·221	0·506	24·596
1800	3·687	0·377	0·427	3·750	1·304	1·096	0·000	1·612	3·250	1·670	5·322	2·422	24·917
1801	1·839	0·788	1·551	0·488	1·810	0·870	3·700	1·725	1·516	1·916	4·611	3·651	24·465
1802	0·219	2·175	0·555	1·282	1·435	2·048	2·956	0·568	0·806	2·133	1·419	1·734	17·330
1803	2·316	1·078	0·628	1·422	2·022	3·694	1·436	0·830	1·102	0·616	3·417	4·486	23·047
1804	2·509	1·938	2·154	2·065	1·497	0·574	3·890	3·081	0·000	2·659	5·590	0·732	26·709
1805	2·269	1·515	1·234	2·057	1·021	3·615	2·279	3·888	1·830	1·797	1·113	2·608	25·226
1806	2·755	0·775	1·859	0·318	1·224	0·535	5·133	2·295	2·304	1·030	3·571	3·939	25·758
1807	0·720	1·370	0·860	0·320	2·830	1·580	0·320	1·650	1·770	1·520	3·960	3·200	20·140
1808	1·620	0·980	0·290	2·020	1·680	0·910	3·370	2·240	3·010	3·950	2·190	0·940	23·240
1809	5·740	1·560	0·370	3·820	0·870	1·030	2·890	1·930	2·530	0·230	1·930	2·200	25·280
1810	0·165	1·260	2·530	1·260	1·360	0·550	3·680	2·910	0·650	3·180	5·320	5·020	28·070

N. B.   The Monthly results at the level of the ground, from 1810 forward, are in Table D.

The material originally positioned here is too large for reproduction in this reissue. A PDF can be downloaded from the web address given on page iv of this book, by clicking on 'Resources Available'.

Printed in the United States
By Bookmasters